THERMAL ENERGY STORAGE AND REGENERATION

SERIES IN THERMAL AND FLUIDS ENGINEERING

JAMES P. HARTNETT and THOMAS F. IRVINE, JR., Editors
JACK P. HOLMAN, Senior Consulting Editor

Cebeci and Bradshaw • **Momentum Transfer in Boundary Layers**
Chang • **Control of Flow Separation: Energy Conservation, Operational Efficiency, and Safety**
Chi • **Heat Pipe Theory and Practice: A Sourcebook**
Delhaye, Giot, and Riethmuller • **Thermohydraulics of Two-Phase Systems for Industrial Design and Nuclear Engineering**
Eckert and Goldstein • **Measurements in Heat Transfer, 2d edition**
Edwards, Denny, and Mills • **Transfer Processes: An Introduction to Diffusion, Convection, and Radiation, 2d edition**
Fitch and Surjaatmadja • **Introduction to Fluid Logic**
Ginoux • **Two-Phase Flows and Heat Transfer with Application to Nuclear Reactor Design Problems**
Hsu and Graham • **Transport Processes in Boiling and Two-Phase Systems, Including Near-Critical Fluids**
Hughes • **An Introduction to Viscous Flow**
Kestin • **A Course in Thermodynamics, revised printing**
Kreith and Kreider • **Principles of Solar Engineering**
Lu • **Introduction to the Mechanics of Viscous Fluids**
Moore and Sieverding • **Two-Phase Steam Flow in Turbines and Separators: Theory, Instrumentation, Engineering**
Nogotov • **Applications of Numerical Heat Transfer**
Richards • **Measurement of Unsteady Fluid Dynamic Phenomena**
Schmidt and Willmott • **Thermal Energy Storage and Regeneration**
Siegel and Howell • **Thermal Radiation Heat Transfer, 2d edition**
Sparrow and Cess • **Radiation Heat Transfer, augmented edition**
Tien and Lienhard • **Statistical Thermodynamics, revised printing**
van Stralen and Cole • **Boiling Phenomena**
Wirz and Smolderen • **Numerical Methods in Fluid Dynamics**

THERMAL ENERGY STORAGE AND REGENERATION

Frank W. Schmidt
The Pennsylvania State University

A. John Willmott
University of York

Hemisphere Publishing Corporation
Washington New York London

McGraw-Hill Book Company
New York St. Louis San Francisco Auckland Bogotá
Hamburg Johannesburg London Madrid Mexico
Montreal New Delhi Panama Paris São Paulo
Singapore Sydney Tokyo Toronto

To
JOAN *and* MARY

This book was set in Press Roman by Communication Crafts Ltd.
The editors were Mary A. Phillips and Lynne Lackenbach.
BookCrafters, Inc. was printer and binder.

THERMAL ENERGY STORAGE AND REGENERATION

1 2 3 4 5 6 7 8 9 0 B C B C 8 9 7 6 5 4 3 2 1 0

Library of Congress Cataloging in Publication Data

Schmidt, Frank W. date
Thermal energy storage and regeneration

Bibliography: p.
Includes index.
1. Heat storage devices. 2. Heat regenerators. I. Willmott, A. J., date joint author. II. Title
TJ260.S315 1981 621.402'8 80-19964
ISBN 0-07-055346-7

CONTENTS

Preface ix
Nomenclature xiii

1 Thermal Energy Storage 1

1.1 Introduction 1
1.1.1 Utilization of Energy Storage Devices 2
1.1.2 Specific Areas of Applications of Energy Storage 2
1.2 Selection of the Type of Energy to Be Stored 4
1.3 Thermal Energy Storage Units 5
References 9

2 Single-Blow Operating Mode: Infinite Fluid Heat Capacity and Simplified Models 10

2.1 Introduction 10
2.2 Infinite Fluid Heat Capacity 12
2.2.1 Negligible Temperature Gradients in the Storage Material 12
2.2.2 Internal Temperature Gradients in Storage Material 13
2.3 Simplified Model 17
2.4 Simplified Model Employing a Modified Heat Transfer Coefficient 27
References 29

3 Single-Blow Operating Mode: Finite Conductivity Model 30

3.1 Introduction 30
3.2 Finite Conductivity Model: Slab Configuration 31
3.3 Finite Conductivity Model: Hollow Cylinder 51
3.4 Comparison of Results for Finite Conductivity Models of Hollow Cylindrical and Slab Configurations 74
3.5 Analysis of the Effects of Finite Thermal Conductivity 75
References 83

4 Prediction of Transient Response of Heat Storage Units with Timewise Variations in Inlet Fluid Temperature and Mass Flow Rate 85

4.1 Introduction 85
4.2 Superposition: Timewise Variations in Inlet Fluid Temperature 86
4.3 Superposition: Arbitrary Initial Temperature Distribution in the Storage Material 93
4.4 Superposition: Arbitrary Variation in Fluid Mass Flow Rate 97
References 104

5 Basic Concepts in Counterflow Regenerators 105

5.1 Introduction 105
5.2 Mathematical Model 110
5.3 Discussion of Design Parameters 113
5.4 Effect of Cycle Time on Regenerator Performance 116
5.5 Imbalance in Regenerator Performance 118
5.6 Solution of Differential Equations 123
References 127

6 Finite Conductivity Models of Counterflow Regenerators 128

6.1 Introduction 128
6.2 Lumped Heat Transfer Coefficients 128
6.3 Further Considerations of the Effect of Checkerwork Conductivity 135
6.4 Longitudinal Conduction in the Regenerator Packing 136
6.5 Note on Numerical Solution of Equations Including Longitudinal Conduction 140
6.6 Latitudinal Conduction in the Regenerator Checkerwork 141
6.7 Relationship between the Two-Dimensional and Three-Dimensional Models 143
6.8 Possible Improvements in the Lumped Heat Transfer Coefficient 149
References 153

7 Nonlinear Models of Counterflow Regenerators 154

7.1 Introduction 154
7.2 Mathematical Representation 155
7.3 Quasi-linearization of Nonlinear Models 158
7.4 Methods for Simulating Nonsteady-State Performance of Regenerators 159
7.5 Variable Gas Flow Problems in Regenerators 159
7.6 Radiative Heat Transfer between Gas and Solid Surface in Regenerators 164
References 167

8 Improved Computational Methodology for Regenerators 168

8.1 Introduction 168
8.2 Open Methods: Numerical Solution of the Regenerator Equations 169
8.3 Control of Truncation Error 174

8.4 Open Methods: Integral Methods for Simulation of the Regenerator 176
8.4.1 Evaluation of the K Function 179
8.4.2 Calculation of Regenerator Effectiveness 180
8.5 Closed Methods 180
8.5.1 Further Consideration of Reversal Conditions for the Symmetric Case 180
8.5.2 Development of the Closed Methods 180
8.5.3 The Method of Nahavandi and Weinstein: Symmetric Case 181
8.5.4 The Method of Nahavandi and Weinstein: Unsymmetric Case 182
8.5.5 Choice of Data Points ξ_i 183
8.5.6 The Closed Method of Iliffe 184
8.6 Numerical Acceleration of Regenerator Simulations 186
References 191

9 Parallel-Flow Regenerators 192

9.1 Introduction 192
9.2 Method of Analysis 194
9.2.1 Symmetric Regenerators: Reversal Conditions 195
9.3 The Heat Pole Method 196
9.4 Refinement of the Heat Pole Method 201
9.5 Application of the Nahavandi and Weinstein Method to Parallel-Flow Regenerators 202
9.6 Parallel-Flow Regenerator Performance 203
9.7 The Approach of Kumar 204
9.8 Unbalanced Regenerators 214
References 218

10 Transient Performance of Counterflow Regenerators 219

10.1 Initial Considerations 219
10.2 Response to a Step Change in Operation Conditions 221
10.3 Step Changes in Inlet Gas Temperature 222
10.3.1 Symmetric Case 222
10.3.2 Unsymmetric-Balanced Regenerators 228
10.3.3 Unbalanced Regenerators 230
10.4 Interpretation of the Relation between the Transient Performance of a Regenerator and Its Dimensionless Parameters 230
10.5 Step Changes in Gas Flow Rate 232
10.6 Further Parameterization of Transient Response 235
10.7 Thermal Inertia of Variable Gas Flow Regenerator Systems 240
10.8 Conclusions 241
References 242

11 Heat Storage Exchangers 243

11.1 Introduction 243
11.2 Two-Fluid Heat Exchangers 244
11.2.1 Single-Blow Operation: Mathematical Model 244
11.2.2 Transient Response Parameters 248

11.2.3 Results: Step Change in One Inlet Fluid Temperature 250
11.2.4 Method of Superposition for Arbitrary Variations in Inlet Fluid Temperature for a Two-Fluid Heat Storage Exchanger 263
11.3 Heat Flux and Single-Fluid Heat Storage Exchanger 266
11.3.1 Mathematical Model 266
11.3.2 Results 269
11.3.3 Methods of Superposition for Arbitrary Timewise Variations in Heat Flux 275
References 277

12 Packed Beds 279

12.1 Introduction 279
12.2 Mathematical Models: Single Blow 280
12.2.1 Negligible Thermal Resistance 280
12.2.2 Simplified Model 281
12.2.3 Intraparticle Conduction and Dispersion Model 290
12.3 Arbitrary Inlet Fluid and Initial Bed Conditions 294
12.3.1 Arbitrary Time Variation in Fluid Inlet Temperature 294
12.3.2 Arbitrary Variation in the Initial Temperature Distribution of the Packed Bed 296
12.3.3 Arbitrary Time Variation in Flow Rate 300
12.3.4 Periodic Operation of Packed Bed Units 300
References 301

13 Design Optimization 302

13.1 Introduction 302
13.2 Complex Optimization Method 303
13.3 Optimization of a Slab Heat Storage Unit for Single-Blow Operating Mode 307
13.4 Results from Optimization Study of Slab Heat Storage Units 310
13.5 Optimization of a Packed Bed Heat Storage Unit for Single-Blow Operating Mode 322
13.6 Results from Optimization Study of Packed Bed Storage Units 325
References 332

14 Heat Transfer and Pressure Drop Correlations 333

14.1 Introduction 333
14.2 Heat Transfer and Pressure Drop Correlations for Channel Flow 333
14.2.1 Pressure Drop Calculation 334
14.2.2 Heat Transfer Coefficient Correlation 338
14.3 Heat Transfer and Pressure Drop Correlations for Flow in Packed Beds 342
14.3.1 Pressure Drop Calculations 342
14.3.2 Heat Transfer Correlations 344
References 348

Index 350

PREFACE

The industrial countries of the world have recently become more conscious of their energy requirements. They have found it economically necessary to develop usage patterns that incorporate waste heat recovery and to explore alternative sources of energy to reduce dependence on other nations for energy-producing fuels. This book is concerned with the management of thermal energy storage systems both in the area of heat recovery and for circumstances where the availability of energy does not coincide chronologically with demand. The significance of this presentation lies in the drawing together of many, varied applications of the principles of heat storage in solids. All share a common theoretical background whose exploitation in one industrial field, perhaps heat recovery in glass manufacturing furnaces or the drying of powders, can be applied possibly immediately with little modification to other processes such as the drying and storage of foodstuffs, the heating/air conditioning of buildings, or the development of solar energy systems.

These are not new problems. In the economic crisis that beset Germany following World War I, the Institute of Heat Recovery, directed by Kurt Rummel, emerged in Dusseldorf. The present energy crisis has again focused international attention upon the efficient use of available supplies of energy.

However, two important factors distinguish our current situation from that faced by Kurt Rummel and his team of research workers. First, there appears to be little or no theoretical work published in the scientific literature before the 1920s on the design and performance of heat recovery devices. Rummel's theory of regenerators, which was the subject of his lecture to the British Institute of Fuel in February 1931 upon the occasion of his being awarded the first Melchet Medal, broke new ground. Today, the reviewer is hard pressed to summarize adequately the extensive literature covering the storage of heat in solids and the performance of thermal regenerators.

Second, the success of the Rummel approach to these problems lay in his minimization of the arithmetic needed to solve the theoretical problems posed in this area. Needless to say, such economics in calculation effort necessarily involved simplifying assumptions about the internal working of heat recovery devices. The advent of the digital computer obviated the need to deal with these matters in the manner proposed by Rummel, and a new more flexible approach to the prediction of the performance of thermal energy storage devices has been developed.

Although one purpose of this book is to suggest that the rigorous and flexible methodologies, which can be readily implemented as computer programs, should be adapted widely in place of the procedures suggested in the 1920s and 1930s by Rummel, Nusselt, Hausen, and Schack among others, the contribution of those workers 40-50 years ago cannot be underestimated. It is not without significance that one of the authors of this book (F. W. S.) teaches in the Mechanical Engineering Department at The Pennsylvania State University, the other (A. J. W.) teaches in the Department of Computer Science at the University of York. The comparatively recent developments in high-speed digital computers and in numerical analysis have facilitated a theoretical approach not open to earlier workers. Indeed Hausen revealed informally, during a visit to the University of York in 1971, that in order to accomplish certain regenerator calculations during the 1930s, he "programmed his family" to perform the necessary arithmetic; these computations occupied much of his family's spare time for 12 months at a time. The same calculations have been repeated in a matter of minutes on a digital computer.

There is a further problem to be solved at present; whereas the theoretical work of the precomputer era was concerned with the prediction of the performance of existing types of thermal energy storage and recovery devices, it is becoming increasingly important to be able to examine the feasibility of new heat storage systems as new methods of energy acquisition are developed. Nevertheless, it is vital to recognize that such new storage systems may well be very similar to existing devices and that present theory can be used immediately.

This book treats the heat storage medium as a solid through whose channels or pores passes the heat-transporting fluid, usually a gas. The presentation is restricted to the storage units that utilize only sensible heat storage in a stationary material. Quite different techniques are used for predicting the performance of units that use sensible heat storage in a fluid that not only serves as the storage medium but also transports the stored energy from one location to another. The use of storage materials that utilize a phase change to improve the stored energy to volume ratio is receiving considerable attention, but the main efforts are associated with the development of low cost materials that can undergo extended periods of operation with little change in their performance. Until the physical chemistry aspect of the problem is solved, little definitive work can be done with regard to the prediction of the thermal performance of units employing phase change materials. This topic will thus be left for a future edition of this book.

The early chapters deal with what is called the "single blow problem," where the gas passes through heat-storing packing continuously in one direction. Solutions to this problem facilitate the control of the drying or cooling of beds of grain, other foodstuff, and industrial granular or powder products. The same solutions can be used to examine, for example, the heat storage capabilities of particular constructions in buildings and thermal stores built into solar collectors. Because of the linear nature of the mathematical models of these systems, they can be used with superposition techniques to predict the performance of systems experiencing timewise variations in inlet fluid temperature and mass flowrate.

It has been long recognized that it is more thermally effective to have the heat transporting fluids flow in opposite directions during the storage and retrieval processes. This feature is embodied in the counterflow regenerator. Traditionally such regenerators have been used very widely as heat exchangers; they are used to preheat the air for the iron making and zinc/lead smelting processes, and they are embodied in various forms of refrigeration equipment; regenerators are used to recover waste heat in power station boilers and in boilers aboard ships. Comparatively recently, they have been employed for energy recovery in ventilating systems. However, the heat storage mechanism embodied in the regenerator can be exploited to enable energy to be retained in the packing and demands for energy to be satisfied even at times when the energy supply is cut off. For example, nighttime "cold" can be stored in regenerative beds and recovered subsequently during the day as part of the air conditioning process for offices and other buildings. Obviously another candidate in cooler climates for this type of application is the storage of solar thermal energy for the warming of domestic premises in the evening.

There are six chapters dealing with thermal regenerators. The implicit force of the argument developed there is that the methodology of the Rummel era, with all its necessary restrictive assumptions about the internal behavior of the regenerator, should be discarded in favor of the methods developed over the last 20 years for digital computers, which enable more realistic modeling of the regenerator systems to be attempted. Nonetheless, the foreword to Neville Shute's *Trustee from the Toolroom* is not forgotten. This may be paraphrased as: an engineer is a person who can do for twenty five pence (cents) what any damned fool can do for five pounds (dollars). Neville Shute spoke in terms of the shillings and pence of a former era! Indeed a chapter is devoted to computational economics that can be introduced in regenerator calculations without simplifying the mathematical models employed.

Heat storage exchangers, or what might be called in Europe thick-walled recuperators, provide an important bridge between regenerator and recuperator theory. These devices, as well as providing a short-term thermal energy storage capability in circumstances where the supply of energy and/or its demand are irregular and spasmodic, have dynamic characteristics similar to those of the equivalent regenerator system and thus provide an insight into the mechanisms governing the performance of both devices.

We have felt it necessary to include a chapter on packed beds; they are employed sometimes as regenerators and heat storage devices, but more particularly, they represent that class of problem including the drying or cooling of granular materials in both the industrial and agricultural sectors.

Little work has been published on the optimization of heat storage systems. The material presented here is not exhaustive but is intended to suggest to the reader, on the one hand, which factors govern the optimality of design and operation of these systems and, on the other hand, the sort of semantic and methodological approach that present studies reveal likely to be most useful.

The last chapter deals with correlations useful for the estimation of the heat transfer coefficient and the pressure drop in storage units. These relationships, when supplemented with more specific correlations found in the technical literature, will assist engineers in obtaining reasonable accuracy in their predictions.

The authors of this book are listed in alphabetical order. The basic division of effort has been that F. W. S. has prepared the material on single blow problems, whereas A. J. W. has concentrated on thermal regenerators.

Many of our co-workers have made significant contributions to the work reported in this book. Although appropriate credit for specific contributions can be determined from the citations of published articles, the authors wish to note that many invaluable discussions with our colleagues have occurred on a continuing basis and have greatly assisted us in our work on thermal energy storage. It is therefore fitting that specific acknowledgments be made.

The excellent works of Dr. James Szego provided the keystone for the study in which the thermal conductivity of the storage material was considered to be finite in all directions. His work in the evaluation of the geometric effects and the transient response of thermal storage exchangers was most helpful. The contributions of Dr. M. Kumar and Dr. R. Somers II to certain aspects of the thermal energy storage studies at The Pennsylvania State University are also gratefully acknowledged.

Significant advances in regenerator theory have been achieved with the collaboration of several research students over the years; in particular the work of Dr. R. J. Thomas on the instability of closed methods must be recalled as well as the far-ranging work of Dr. A. Burns on the transient performance of regenerators. The early work of Mr. David R. Green on the transient response problem cannot be forgotten; neither can the very recent work of Miss Clare Hinchcliffe who developed a time-varying bulk heat transfer coefficient. Dr. B. T. Kulakowski visited the University of York in 1975 and completed a useful piece of work on the acceleration of regenerator simulations.

A patient mentor over some 20 years has been Dr. A. E. Wraith of the University of Newcastle-upon-Tyne. His help and encouragement are much appreciated.

The contents of this book were first brought together for a continuing education service course, "Prediction of Performance of Sensible Heat Storage Units," held at The Pennsylvania State University in September 1977. We are grateful for this opportunity for trans-Atlantic collaboration.

Frank W. Schmidt
A. John Willmott

NOMENCLATURE

A	Heat transfer surface area, m^2
AE	Available energy in fluid stream (Eq. 2.39), W
A^+	Fraction of the available energy stored (Eq. 2.40)
Bi	Biot number, hw/k_m or $h(r_o - r_i)/k_m$
Bi_Δ	Biot number, $h\Delta/k_m$
C^+	Capacity rate ratio of fluids (Eq. 11.4)
c	Specific heat at constant pressure, kJ/kg K
D_h	Hydraulic diameter, m
D_p	Diameter of bed particles, m
d	Semiwidth of flow channels, diameter sphere in packed bed, m
d_e	Diameter of sphere with equivalent volume, m
E	Heat capacity of fluid, $\dot{m}_f c$, J/s
e	Surface roughness, m
f	Fanning friction factor
Fo	Nondimensional time (Eq. 3.3)
G	Internal heat transfer coefficient, $W/m^2\ °C$
G	Mass velocity, $kg/m^2\ s$
G^+	Nondimensional (Eq. 3.3)
g_c	Dimensional constant, $kg\ m/N\ s^2$
H	Heat storage at cyclic equilibrium, W
h	Convective film coefficient, $W/m^2\ °C$
$\bar{h}$	Overall heat transfer coefficient, $W/m^2\ K$
h_e	Effective heat transfer coefficient for packed bed (Eq. 12.23)
h_v	Volumetric heat transfer coefficient, $W/m^3\ °C$
i	Current, A
k	Thermal conductivity, $W/m\ °C$
L	Length of unit, m

M	Total mass of storage material, kg
m_f	Mass of fluid in storage channels, kg
$\dot{m}_f$	Mass rate of flow, kg/s
N	Ratio of time periods, τ_h/τ_c or τ_c/τ_h
NTU	Number of transfer units (Eq. 11.24)
Nu	Nusselt number
P	Heated perimeter of flow channel, m
P	Duration of hot or cold period for regenerator, s
P	Pressure, cm H_2O
Pe	Peclet number (Eq. 12.22) or Re Pr
Pr	Prandtl number, $c_p\mu/k$
Q	Total heat stored, W
Q_{max}	Maximum heat storage (Eq. 2.6), W
Q^+	Nondimensional heat storage (Eqs. 2.8, 11.10, and 11.49)
q	Heat flux, W/m^2
R	Nondimensional radius (Eq. 3.15)
R	Resistance, Ω
R	Effective heat transfer coefficient for heat storage exchanger, °C/W
R^+	Convective resistance ratio (Eq. 11.4)
Re	Reynolds number
r	Radial coordinate, m
r_i	Inner radius, m
r_o	Outer radius, m
S	Cross-sectional area, m^2
St	Stanton number, Nu/Re Pr
T	Nondimensional temperature (Eqs. 2.3, 11.4, and 11.42)
$\bar{T}_m^{ss}$	Nondimensional mean storage material temperature (Eq. 11.36)
t	Temperature, °C, K
$\bar{t}_m$	Mean material temperature, °C
U	Overall heat transfer coefficient, W/m^2 °C
U^+	Nondimensional radius ratio, r_w/r_a (Eq. 3.15)
V	Voltage, V
V	Volume of storage material, m^3
V^+	Nondimensional, w/L (Eq. 3.3) or $(r_o - r_i)/L$ (Eq. 3.15)
V_H	Heat capacity ratio, packed bed (Eq. 12.22)
v	Fluid velocity, m/s
v_a	Interstitial fluid velocity, m/s
Z	Nondimensional (Eq. 6.36)
w	Semithickness of storage material for heat storage units, thickness of storage material for heat storage exchangers, m
X	Nondimensional axial coordinate, X/L
X	Nondimensional distance (Eq. 12.2)

x	Axial coordinate, m
Y	Nondimensional transverse distance, y/w
y	Transverse coordinate, m

Greek

α	Thermal diffusivity, m^2/s
β	Unbalance factor (Eq. 10.28)
β	Thermal capacity value (Eq. 12.22)
γ'	Longitudinal conduction factor (Eq. 6.7)
Δ	Characteristic length (Eq. 2.41)
ϵ	Effectiveness (Eq. 11.10)
ϵ	Porosity of packed bed
ϵf_1	Transient operating parameter (Eq. 10.1)
ϵf_2	Transient operating parameter (Eq. 10.2)
ϵ_s	Emissivity
η	Nondimensional time (Eqs. 2.3 or 3.15)
η_{REG}	Thermal ratio, dimensionless (Eqs. 5.15 and 5.16)
Θ	Nondimensionless time (Eq. 12.2)
Θ	Time needed to reestablish cyclic equilibrium
κ	Eq. (6.48)
Λ	Reduced length $hA/\dot{m}_f c_f$
Λ_H	Harmonic mean reduced length (Eq. 5.22)
λ	Nondimensional length of heat storage unit (Eqs. 2.26 or 3.15)
μ	Absolute viscosity, kg/m s
ξ	Nondimensional axial distance (Eqs. 2.26 or 3.15)
Π	Reduced period, $\bar{h}A(P - m_f/\dot{m}_f)/M_m c_m$, dimensionless
Π_H	Harmonic mean reduced period (Eq. 5.23)
ρ	Density, kg/m^3
σ	Stefan Boltzmann constant
ϕ	Overall heat transfer correction factor (Eqs. 6.5 and 6.6)
τ	Time, s
τ^*	Harmonic mean (Eq. 9.24), s
τ_d	Dwell time $= L/v$
Ω	Reduced time (Eq. 6.31)

Subscripts

a	Adiabatic surface
B	Blast temperature
c	Cold fluid
ci	Cold fluid entering
co	Cold fluid leaving

env	Environment
f	Fluid
fc	Cold fluid entering
fh	Hot fluid entering
fi	Fluid at entrance to unit
fr	Frontal cross-sectional area of packed bed
H	Harmonic mean
h	Hot fluid, heated surface
hi	Hot fluid entering
ho	Hot fluid leaving
iso	Isothermal flow
m	Storage material
o	Initial condition
R	Radiation
s	Surface
ss	Steady state
w	Surface in contact with fluid
wc	Wall temperature, cold fluid side
wh	Wall temperature, hot fluid side

Superscripts

p	Period
fo	Fluid at exit of one unit
ss	Steady-state condition
$'$	Hot period for regenerator
$''$	Cold period for regenerator

CHAPTER
ONE

THERMAL ENERGY STORAGE

1.1 INTRODUCTION

In 1973 the major oil producing countries in the Middle East established an oil embargo. This action dramatically drew the attention of the industrial countries of the world and their inhabitants to the fact that their economic well-being depended on their energy resources. To a very large extent these were controlled by other countries. Immediate action was taken by a large number of these countries to study the ways in which they could reduce their dependence on others and, as an ultimate objective, to become, energywise, completely self-sufficient. In the United States, local, state, and federal committees were appointed to study the "energy crisis" and make recommendations with regard to the ways in which the country might reduce its energy reliance on foreign sources while still continuing to increase its industrial productivity. Similar developments took place in the United Kingdom and Western Europe.

The results of these studies indicated that there was no single solution to the problem. The development of alternative fuel and energy sources to oil and natural gas was proposed. It was suggested that a high priority be given to the increased use of coal and nuclear power, particularly fusion, and to the development of energy systems utilizing solar, wind, and geothermal energy as well as the temperature gradients present in the oceans. Industrial and commercial organizations were requested to establish energy management systems whose prime objectives were to reduce energy demands through the installation of energy conservation measures.

As the results of these studies became available, it became obvious that energy storage devices could greatly assist in improving the overall efficiency of large energy producing units through load leveling, decrease energy demands through the use of waste heat recovery and allow alternative sources of energy to be utilized more effectively.

1.1.1 Utilization of Energy Storage Devices

In 1975 the Energy Research and Development Administration (ERDA) of the U.S. government requested that the National Research Council (NRC) undertake a study on the potential of advanced energy storage systems. The committee conducted an in-depth analysis of the criteria that advanced energy storage systems would have to meet in order to become economically acceptable to the industrial, commercial, and residential communities. The report, published in 1976 [1], indicated that the ultimate decision on the installation of the storage system would be based on the following operating characteristics of the storage unit:

1. Storage capacity
2. Charge/discharge rate
3. Replacement lifetime
4. Weight, volume, and other physical limits
5. Critical safety parameters
6. Environmental standards
7. Acceptable capital and operating costs

The committee also recognized that the areas where these units could possibly be used are so varied that no specific device could be expected to have the proper operating characteristics for all possible applications.

The general areas where energy storage devices will prove economically attractive are load leveling, utilization of alternative energy sources, and waste heat utilization. The load-leveling applications appear to be in the electric utilities segment. The utilization of alternative energy sources can be in the utility, residential/commercial, or industrial areas. Energy storage devices can be used with energy management systems to reduce energy consumption in commercial and industrial establishments by using available waste heat.

To be economically attractive, an energy storage installation must satisfy one or more of the following objectives: (1) reduce energy consumption, (2) reduce energy costs, and/or (3) allow the substitution of a more readily available energy source.

1.1.2 Specific Areas of Applications of Energy Storage

Electric Utilities Electric utilities must have sufficient electricity generating capabilities to meet all demands placed on the system. Energy storage devices could provide a means of meeting peak demand without installing peaking or intermediate generating capacity. In such systems the energy is stored during off-peak hours and released during periods of high demand. Such a system would allow on-line generating units to operate at design conditions where their efficiency is greater. The overall effect of an electric utility using a storage system would have to be a decrease in the cost of the generated electric energy resulting from a reduction in the capital investment needed to meet the peak demands and the operation of the unit at the design conditions. The total energy consumption may indeed be greater for this system, since only about 75% of the energy stored in the energy storage unit can be recovered.

The NRC report [1] indicated that if economically efficient storage units can be developed, as much as 10% of the installed primary generation equipment required by electric utilities in the time period 1985 to 1990 might be displaced by storage systems. It is predicted that the capacity of these units would range from 15 MW to several thousand megawatts. The required storage duration will be from 2 h to 2 days, and the ratio of the charge time to the discharge time will be from 0.2 to 4.0.

Residential/Commercial Approximately 85% of the total residential/commercial requirement is for thermal purposes. Energy storage units can be used in several ways in systems using electrical energy to supply space and water heating requirements. The use of storage devices to assist in satisfying the peak demands (load leveling) has already been discussed. This is particularly important in utility systems that experience peak demands in the winter. This type of storage system, involving the installation of the storage unit in individual residences or business establishments, has been used extensively in Europe. These units usually utilize ceramic storage materials that are electrically heated during off-peak periods at a much cheaper energy charge rate. Descriptions of these units and a discussion of possible savings resulting from their use are presented in Refs. 2-9. The economic feasibility of these units is dependent on the rate structure used by the utility. The total energy consumption of a system using these units will be increased because of the loss of heat incurred during the storage process and the difficulties encountered in retrieving more than 75% of the actual heat stored in the unit.

In heat pump installations, thermal energy storage can be used to improve the ability of the unit to meet the demand when the surrounding air, usually used as the heat sink, falls below 0°C (32°F). The stored energy can be supplied from electrical resistance units using the low cost of off-peak power or from solar collectors. It is very difficult at this time to estimate the full impact on the market of heat pumps with thermal energy storage devices, since it involves a new area of application which has not been thoroughly evaluated.

The use of solar energy to meet the thermal demands of residential and commercial establishments is currently under intensive study, and a significant number of new units utilizing this form of energy are being constructed. A very important component in such systems is the thermal energy storage unit. The storage system must be adequate to supply heat not only during the night when no energy is collected but for long periods of time, 2 to 5 days, when cloudy weather makes it impossible to collect substantial amounts of the energy required. Most applications require that an auxiliary source of energy be available to supplement the energy supplied from the solar source in order to meet the total demands of the system.

The utilization of storage systems for residential and commercial installations will be stimulated only if it results in a reduction in the cost of energy to the consumer. The initial investment is quite large, and unless these costs are subsidized in part by the utilities, which are using the system to reduce their generating capacity, or by the government through direct subsidies or tax incentives, it is questionable if storage systems will ever achieve sufficient economic attractiveness to make an impact on the residential and commercial markets.

Industrial Applications In 1919 Germany recognized the economic importance of energy savings in the industrial field. At that time the Institute of Fuel Economy was established in Dusseldorf. In 1931 the Institute of Fuel in Britain awarded its first Melchett Medal to the Director of the German Institute, Dr. Kurt Rummel. The presentation of this medal was in recognition of the significant theoretical and experimental advances made in the area of industrial waste heat recovery, particularly with respect to regenerators, by Dr. Rummel and his team at Dusseldorf. Many of their results are still in use today.

Industry is highly vulnerable to energy curtailment. Much of this energy is used in batch processes in which the energy demand varies in a timewise fashion during the period of operation. A large percentage of this energy is discharged to the surroundings as waste heat because of the intermittent nature of the industrial processes or because the temperatures have decreased to the point where the energy is of little value. As the cost of energy increases and energy supplies become scarcer, the establishment of energy management systems to match up timewise the available waste heat sources and the thermal energy demands of the system becomes an economic necessity. The successes of these systems are to a very large degree dependent on the thermal energy storage units employed. Even so, it is recognized that systems of waste heat recovery have been used for many years. One of the oldest, which embodies the utilization of stored heat, is in the pottery industry. Here pots are preheated by the waste products of combustion of the fuel before they enter the firing zone of the kiln. Subsequently, the air for the combustion is preheated by blowing it over the fired pots leaving the kiln.

1.2 SELECTION OF THE TYPE OF ENERGY TO BE STORED

Since energy appears in many different forms, the first step in the selection of the storage unit is to determine the type of energy to be stored. This depends to a certain extent on the form of the available energy and on the form in which the energy is to be retrieved from storage. A list of the several types of energy storage devices would include the storage of

1. Potential energy—The pumping of water to a higher elevation by off-peak electrical energy is one of the most common applications of this type of storage. The hydroelectric system thus uses pumped water to meet its peaking demands. Other examples include the storage of energy as compressed air and the compression of springs.
2. Chemical energy—The production of hydrogen for future use in a combustion process is the prime example of this type of storage. The hydrogen can be in the form of a liquid or a gas. Electrical storage in secondary batteries also falls into this classification.
3. Kinetic energy—The most common form of the storage of kinetic energy is the flywheel.

4. Electromagnetic fields—Energy is stored in capacitors or superconducting magnets.
5. Thermal energy—Many different types of energy devices fall within this classification. They include the storage of high-pressure steam, utilization of the sensible heat of a liquid or a solid, utilization of the heat of fusion or evaporation, reversible chemical heat absorption, heat of hydration, and heat of chemical change.

Each of these forms of energy storage has definite advantages and limitations. One must weigh very carefully all of the related factors before making a decision. The scope of this book will be restricted to thermal energy storage devices.

1.3 THERMAL ENERGY STORAGE UNITS

A detailed study of the technical and economic feasibility of thermal energy storage was presented by Glenn [10]. An assessment of five specific items was conducted with consideration of present energy use and future projection of energy supply, demand, and costs. The five areas were

1. Analysis of energy usage throughout the residential, commercial, and industrial sectors
2. Analysis of energy distribution by end use within the residential, commercial, and industrial sectors
3. Summary of the status of research and development related to thermal energy storage
4. Definition of barriers and economic considerations in implementation of thermal energy storage
5. Recognition and synthesis of applications of thermal energy storage

The results were then used to identify attractive areas of applications.

There are many factors that must be taken into consideration in the selection of the storage material to be used for a thermal storage unit. A set of desirable characteristics for the storage material would include the following:

1. High specific heat
2. High thermal diffusivity
3. High density
4. Reversible heating and cooling
5. Chemical and geometrical stability
6. Noncombustible, noncorrosive, and nontoxic
7. Low vapor pressure to reduce the cost of containment
8. Low cost—material and storage unit fabrication
9. Sufficient mechanical strength to be able to support compression load resulting from the stacking of the storage core

In addition to the above list, one must give serious consideration to the operating temperature range of the storage material. If the temperature range is very high, many materials may be eliminated from consideration.

Thermal energy storage employing steam accumulators has been used in Europe for many years. A very complete discussion of these systems has been presented by Goldstern [11]. The first units of this type of storage device were essentially steam accumulators, high-pressure steam or water storage tanks from which the fluid was expanded to a low pressure for use in industrial processes. The storage function has recently been taken over by "thermal storage boilers," which can be used for both steam generation and storage. Such systems have been described by Reay [12].

The heat of hydration can be used in the design of a thermal storage device. When a material such as calcium oxide comes in contact with water or water vapor, a chemical reaction occurs that produces a considerable amount of heat. Electrical or thermal energy can be utilized to drive the moisture from the calcium oxide; thus the combination of these two processes gives a thermal storage device. The major disadvantage of this type of system is the deterioration of the calcium oxide under continued heating and hydration and the removal of noncondensable gases from the system.

Many devices have been proposed for thermal energy storage that involve chemical changes. Gas hydrates, hydrogen zinc reactions, sodium sulfide, and ammonia absorption systems are examples of such applications. Some of these involve a considerable safety hazard. Silica gel has also been considered where the heat of absorption is utilized in the thermal storage process, but it has been found to be an extremely costly thermal storage device.

Storage units that utilize the latent heat of the storage material are receiving considerable attention particularly for solar heating and air conditioning applications. These units are attractive because they have a smaller temperature swing as one cycles from storage to retrieval, but the major advantage of these units is smaller size and lower weight per unit of storage capacity. The mean storage temperature can be controlled to a large degree by the selection of the storage material. The phase change utilized involves the liquid and solid phases. Lorsch et al. [13] presented the following criteria for phase-change storage materials:

1. The material should have a melting point slightly above the heating temperature or slightly below the cooling temperature. This permits the desired heat transfer to take place.
2. The material should have a large heat of fusion. The larger the heat of fusion, the less material is required to store a given amount of energy.
3. The material should have a congruent melting point. The material should melt completely so that the liquid and solid phases are identical in composition. Otherwise, the difference in densities between solid and liquid will cause segregation, which causes changes in the chemical composition of the material.
4. The material should not supercool. During the cooling of the liquid phase of the material, the melt should solidify at the thermodynamic melting point. This requires a large rate of nucleation and growth. Otherwise, the liquid may supercool and ultimately form a glass, and the stored energy will not be released.

A material for which no nucleating agent exists and which supercools could be used with a "cold finger," a region of material that is always kept below the melting point.

5. The material should be stable. Applications in buildings require long-term thermal energy storage materials. The desired lifetime of such a system is about 20 years. Therefore, stability is critical for materials, especially at high temperatures in the liquid state, since the diffusion of atoms is enhanced, and the rates of chemical reactions generally increase under these conditions.
6. The material should not interact with the container. This is an extension of criterion 5. There is a sufficiently wide choice of container materials, such as plastics, aluminum alloys, and ferrous alloys, so that this point should not constitute a serious limitation.
7. The material should not be dangerous. Since the possibility of accidental leakage is always present, it is preferable to choose a material that is nonflammable and nontoxic.
8. The material should be cheap and available. This criterion is not very well defined. The cost of a material that is presently available may be much higher than the cost of the same material if a sufficiently large demand were generated for that material.

The technical literature currently available indicates that difficulties have been encountered in obtaining a material that will satisfy a large number of the above criteria. In solar system applications Lorsch [13] has indicated that paraffin waxes and tetrahydrofuran hydrates appear to be the most promising materials. In addition to the above materials, serious consideration is being given to salt hydrates in the low-temperature storage range, 30-60°C (86-140°F): sodium hydroxide for storage in the 315°C (600°F) range; and lithium hydroxide, lithium hydride, and lithium flouride for high-temperature storage systems, 1000-1700°C (1832-3100°F).

It is well to note that latent heat storage systems require both a transport fluid and a storage material, thus a heat exchanger is needed. This presents a distinct disadvantage for these systems when compared with units that use one fluid for both the storage and the transport of energy.

There are two basic types of sensible heat storage units. One involves systems where the storage material also serves as the energy transporting fluid. The other involves storage units that require a separate fluid system to transport the energy from one location to another. This latter type of unit is somewhat more complex, since a heat transfer surface area and a temperature difference between the fluid and the surface of the storage material are required.

Hot water storage is used in most homes for domestic water systems and falls into the first category mentioned in the previous paragraph. In addition, hot water systems are commonly used for storage in solar energy systems. Its major advantages are a relatively high storage per unit volume, due mainly to its high specific heat, and the fact that it can be used for both energy storage and as the energy transporting fluid. This latter means that no heat exchanger is needed. There is little difficulty encountered in selecting and designing the container for the fluid storage material. These types of units store energy in the low-temperature range, 0–110°C (32–230°F). Because of the temperature range used, it is often necessary to use additives to lower the freezing point or to prevent the growth of biological materials in the system. The arrangement of the storage tank's inlet and outlet nozzles is very important, since one would ideally like to have thermal stratification present in the tank, with the hot water at the top and the cold water at the bottom.

There are at the present time two major groups of heat storage media utilizing the sensible heat of solids. One is comprised of firebricks formed from clay, olivine, chrome, Feolite, magnesite, and various mixtures of these ingredients. The other is composed of castable metals, which include gray cast iron and cast irons containing alloying ingredients such as silicon and aluminum. For the same heat storage, the firebricks as a group are approximately twice as bulky as the iron, but are only about one-half to one-third the cost. Another material that is attracting considerable interest is concrete. Its thermal characteristics are borderline but its cost and the fact that the unit can be fabricated on site are responsible for its being given further consideration as a storage material.

One of the most promising materials now available for storage is Feolite. This material was developed by the Electricity Council Research Centre at Capenhurst, United Kingdom. The development of Feolite originated from a search for materials exhibiting the factors that contribute to high-volume enthalpy and close atomic packing. This study revealed that there did not appear to be any material as good, or likely to be as readily available, as ferric oxide, Fe_2O_3. In fact, at ideal single-crystal density and 100% purity, it would have almost the same energy storage per unit volume as cast iron up to 700°C and could be used at temperatures higher than this. The commercial material Tenemax utilizes low-cost enriched iron ores and uses a virtually standard brick-making process for its manufacture.

At the present time and in the foreseeable future the most commonly used thermal energy storage units will be those employing sensible heat storage. In order to design these units and evaluate their economic feasibility it will be necessary to predict their transient response or performance. The performance of liquid storage systems in which the liquid serves as both storage and transport medium can be adequately predicted by the use of theories similar to those associated with automatic control systems. These relationships are well established and adequately described in textbooks and technical papers.

The transient response of solid sensible heat storage units is not as well understood, and some of the earlier work in this area is based on assumptions that may lead to the introduction of considerable errors when the units are used in certain systems or when certain fluid and storage material combinations are used. This book deals with the prediction of the transient response of solid sensible heat storage units. The process is transient in nature and the temperature and mass flow rates of the fluids entering the unit may be either arbitrary or periodic functions of time. If the fluid inlet temperature or the mass flow rate is a periodic function of time, the units are usually referred to as regenerators.

The temperature of the fluid leaving as well as the energy stored will thus be time dependent. It will be a function of the geometry of the storage unit and the material used, the flow rate and flow structure of the fluid, and the thermodynamic properties of the fluid. Three basic geometric configurations will be discussed. These involve one-fluid systems with slab or hollow cylindrical storage unit cross sections, one-fluid packed bed, and two-fluid slab and cylindrical cross-sectioned heat storage exchangers.

The parameters that affect the transient response of these units have been non-dimensionalized in order to generalize the solution obtained, and to allow them to be used for many different combinations of the independent variables. In many instances the results for the single-blow operating mode have been presented in the form of curves or tables. The method of superposition can be used with these results to obtain the response of the storage units that are operating under arbitrary timewise-varying inlet fluid temperatures and flow rates. In more complex situations, the response of the unit has been obtained through the use of computer programs based on the finite difference or numerical solution of the appropriate mathematical models.

REFERENCES

1. Committee on Advanced Storage Systems, *Criteria for Energy Storage R & D*, National Academy of Science, Washington, D.C., 1976.
2. "Heat Storage Using Sodium Hydroxide," *Engineer*, Dec. 2, 1966, p. 842.
3. "Cheap Heat for Industry," *Elec. Rev.*, Aug. 23, 1968, p. 677.
4. "Off Peak Heating Could Be Cheaper," *Engineering*, Nov. 22, 1968, p. 755.
5. M. G. Gibbs, "Design Criteria for Eight-Hour Storage Heating Systems," *Elec. Rev.*, Mar. 14, 1969, p. 379.
6. R. Polland, "Engineering Outline: Thermal Storage Electrical Space Heating," *Engineering*, Oct. 30, 1970, p. 473.
7. M. G. Gibbs, "Current Development in Storage Heaters," *Elec. Rev.*, Feb. 5, 1971, p. 189.
8. A. A. Field, "Storage Heating by Low Cost Electricity," *Heat and Piping*, Feb. 1971, p. 75.
9. "Off Peak Kw-Hr Are Used for Heat by Day," *Elec. World*, Mar. 1, 1971, p. 175.
10. D. R. Glenn, "Technical and Economic Feasibility of Thermal Energy Storage," NTIS-C00-2558-1, Feb. 1976.
11. W. Goldstern, *Steam Storage Installations*, Pergamon Press, Oxford, 1970.
12. D. A. Reay, *Industrial Energy Conservation*, Pergamon Press, Oxford, 1977.
13. H. G. Lorsch, K. W. Kauffman, and J. C. Denton, "Thermal Energy Storage for Heating and Air Conditioning," *Future Energy Production Systems*, vol. 1, Academic Press, New York, 1976, p. 69.

CHAPTER

TWO

SINGLE-BLOW OPERATING MODE: INFINITE FLUID HEAT CAPACITY AND SIMPLIFIED MODELS

2.1 INTRODUCTION

One common method for the storage or removal of energy from a solid substance is to pass a fluid over the material. This section will consider cases where the storage material has a geometric configuration that allows discrete fluid passages and continuous elements of storage material to be identified. If the storage material exists as relatively small particles, the storage unit is termed a packed bed. A discussion of the transient responses of packed beds will be presented in Chap. 12.

The flow rate and the inlet temperature of the energy transporting fluid may remain constant after an initial disturbance, experience arbitrary timewise variations, or vary in a periodic fashion. When the material is at a uniform temperature and the inlet fluid experiences a step change in temperature, the unit is said to be operating under "single-blow" conditions. An example of this type of unit would be the heating of air for a blowdown-type wind tunnel. The heat storage unit is heated over a long period of time and reaches a state of equilibrium at a uniform temperature. The unit is then exposed to a cold air stream flowing at a high mass flow rate for a relatively short period of time during which the temperature of the air passing through the heat storage unit increases.

The second mode of operation, arbitrary timewise variation in flow rate or inlet fluid temperature, is also transient in nature. Most heat storage units can be described by a mathematical model that is linear irrespective of timewise variations in inlet fluid temperature, fluid mass flow rate, or the initial temperature distribution in the unit. The performance of these storage units under time-varying operating conditions can be obtained by using the results from the single-blow problem and the method of superposition.

In the last mode of operation the performance of the unit, although time dependent, is repeated in a periodic fashion. These units are usually referred to as regenerators. The transient response of such units will be discussed in later chapters.

Two storage unit geometries will be considered. In one, the unit will be composed of a number of rectangular cross-sectioned channels for the flowing fluid, connected in parallel and separated by the heat storage material as shown in Fig. 2.1*a*. The second type of unit is a hollow cylinder of storage material. The fluid flows in contact with either the inside or the outside surface of the storage material. The surface that is not in contact with the fluid is considered to be adiabatic or perfectly insulated. The transient response of a heat storage unit composed of a series of circular holes in which

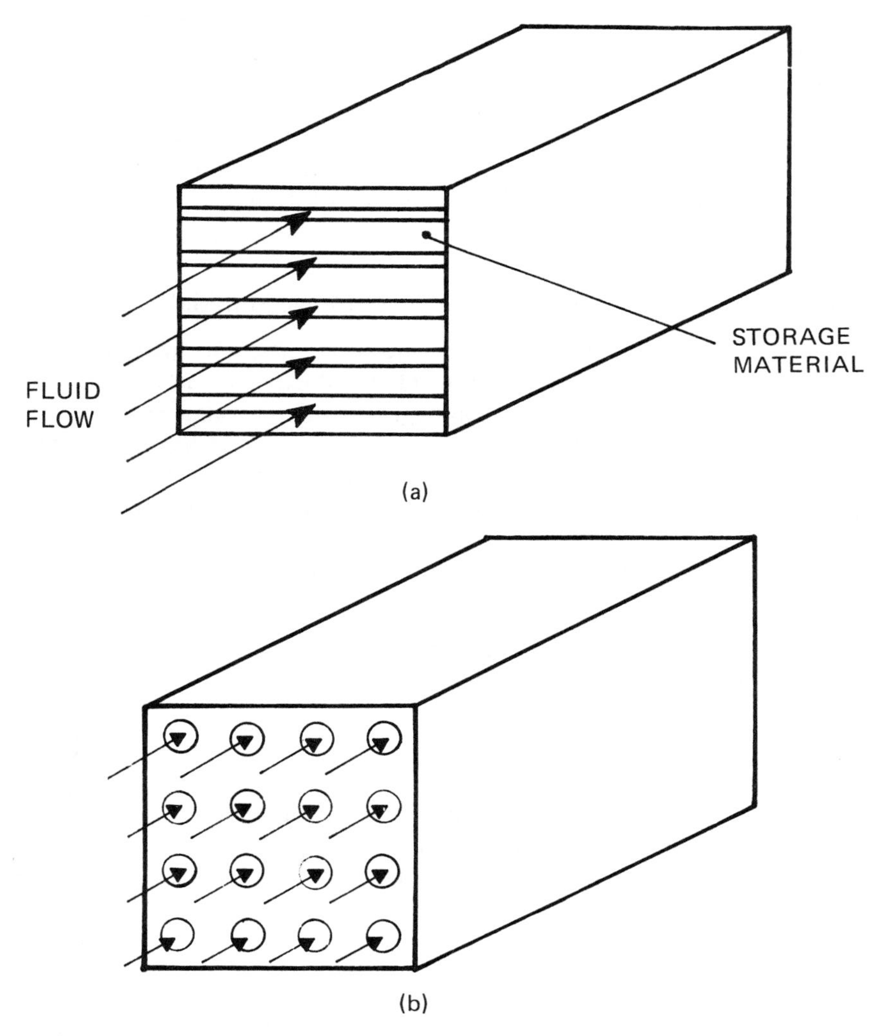

Figure 2.1 Storage unit configurations. (*a*) Rectangular flow passages. (*b*) Circular flow passages.

the fluid passes through the unit in a parallel fashion may be very closely approximated by the hollow cylinder analysis in which the fluid is in contact with the inner surface while the outer surface is insulated. A sketch of such a configuration is shown in Fig. 2.1*b*.

2.2 INFINITE FLUID HEAT CAPACITY

The simplest analysis of the performance of a thermal energy storage unit is one based on the assumption that the temperature of the fluid remains constant as it passes through the channels surrounded by the storage material. This may be interpreted to mean that the heat capacity of the fluid $\dot{m}_f c_f$ is infinite. This will be strictly true only when the fluid is undergoing a change in phase as it passes through the storage unit. There are many occasions, however, when the temperature change experienced by the fluid is small because of the high heat capacity of the fluid. This is usually the case when the energy transporting fluid is a liquid. Under these conditions the infinite heat capacity model may be used to obtain a very good approximation of the transient response of a heat storage unit. The conditions under which these solutions can be used will be discussed in detail in Sec. 3.5.

2.2.1 Negligible Temperature Gradients in the Storage Material

If the major resistance to the transfer of heat is offered by the convective film at the fluid-solid interface, the temperature gradients within the storage material may be very small; that is, the storage material is at an approximately uniform temperature. When these conditions are present, the response of the heat storage unit can be predicted by using a lumped parameter analysis in conjunction with the infinite fluid heat capacity model. The solution is independent of the geometric configuration of the unit.

The mathematical model for a storage unit operating under these conditions is obtained by performing an energy balance on the storage material:

$$\left\{\begin{array}{l}\text{Net rate of accumulation}\\ \text{of energy by the}\\ \text{storage material}\end{array}\right\} = \left\{\begin{array}{l}\text{net rate of}\\ \text{heat removed}\\ \text{from the fluid}\end{array}\right\} \tag{2.1}$$

or

$$\rho_m c_m V \frac{dt_m}{d\tau} = hA(t_{fi} - t_m) \tag{2.2}$$

The initial temperature of the storage material is t_o and the temperature of the fluid flowing through the unit remains constant at its inlet value, t_{fi}.

The following nondimensional quantities are introduced:

$$\text{Temperature:} \qquad T_m \equiv \frac{t_m - t_o}{t_{fi} - t_o}$$

$$\text{Time:} \qquad \eta \equiv \frac{hA\tau}{\rho_m c_m V} \tag{2.3}$$

The energy equation (2.2) thus becomes

$$\frac{dT_m}{d\eta} = 1.0 - T_m \tag{2.4}$$

with the initial conditions $\eta = 0$, $T_m = 0$.

The nondimensional temperature of the storage material is obtained by integrating Eq. (2.4) and using the initial condition to yield

$$T_m = 1.0 - e^{-\eta} \tag{2.5}$$

The total amount of heat that can be stored in the unit is

$$Q_{\max} = \rho_m c_m V(t_{fi} - t_o) \tag{2.6}$$

The actual amount of heat stored is

$$\begin{aligned} Q &= \rho_m c_m V \int_0^{\tau} \frac{\partial t_m}{\partial \tau}\, d\tau \\ &= \rho_m c_m V(t_{fi} - t_o) \int_0^{\eta} \frac{dT_m}{d\eta}\, d\eta \\ &= \rho_m c_m V(t_{fi} - t_o)(1.0 - e^{-\eta}) \end{aligned} \tag{2.7}$$

The nondimensional heat storage is defined as

$$\begin{aligned} Q^+ &\equiv \frac{Q}{Q_{\max}} \\ &= 1.0 - e^{-\eta} \end{aligned} \tag{2.8}$$

The rate at which heat can be removed from the storage unit may be calculated by using the same expressions that were used for the heat storage. The maximum and the actual heat storage under these conditions will be negative quantities, since t_{fi} is less than t_o.

2.2.2 Internal Temperature Gradients in Storage Material

There are many applications in which the thermal energy storage unit is operating under conditions in which the resistance to the transfer of heat offered by the convective film is of the same order of magnitude as that offered by the storage material. Under these conditions significant temperature gradients will be present in the storage material. The mathematical model for the transient response of these units will depend on the geometric configuration of the storage material and the initial temperature distribution within the storage unit. The governing differential equations can be obtained from an energy balance.

Flat Slab The analysis of the transient response of a heat storage unit composed of a number of rectangular cross-sectioned channels for the flowing fluid separated by the heat storage material, Fig. 2.1*a*, will be presented. If all the channels and storage sections are identical and if the fluid flow rates in the channels are equal, then surfaces of symmetry, or surfaces across which no heat flows, are present. It is thus possible to consider only one-half of the channel and storage material as shown in Fig. 2.2 in the prediction of the response of these units.

Since the heat capacity of the fluid is infinite, the differential equation for the storage material is

$$\rho_m c_m \frac{\partial t_m}{\partial \tau} = k_m \frac{\partial^2 t_m}{\partial y^2} \tag{2.9}$$

The initial conditions are

$$\tau = 0 \qquad t_m = t_o$$

and the boundary conditions are

$$y = 0 \qquad \frac{\partial t_m}{\partial y} = 0$$

$$y = w \qquad -k_m \frac{\partial t_m}{\partial y} = h(t_m - t_{fi}) \tag{2.10}$$

In addition to the nondimensional temperature and time introduced in Sec. 2.2.1, a nondimensional length $Y = y/w$ and the Biot number

$$\text{Bi} = \frac{hw}{k_m}$$

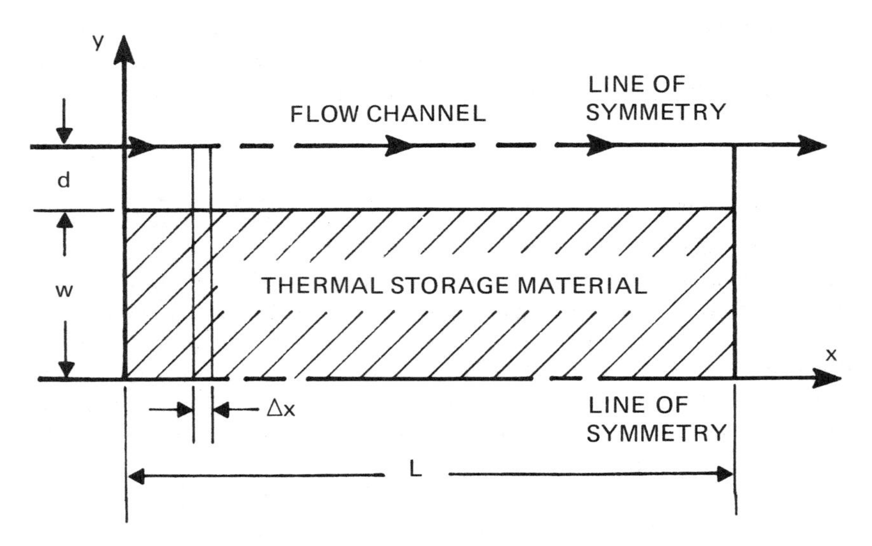

Figure 2.2 Cross section of storage unit.

are introduced. The nondimensional energy equation thus becomes

$$\frac{\partial T_m}{\partial \eta} = \frac{1}{\text{Bi}} \frac{\partial^2 T_m}{\partial Y^2} \tag{2.11}$$

The initial and boundary conditions are

$$\begin{aligned} &\eta = 0 \qquad && T_m = 0 \\ &Y = 0 && \frac{\partial T_m}{\partial Y} = 0 \\ &Y = 1 && \frac{\partial T_m}{\partial Y} = \text{Bi}\,(1 - T_m) \end{aligned} \tag{2.12}$$

An expression for the temperature distribution in the storage material has been presented by Gröber et al. [1]:

$$T_m = 1.0 - \sum_{j=1}^{\infty} 2 \frac{\sin M_j}{\sin M_j \cos M_j + M_j} \exp\left[-(M_j^2)\frac{\eta}{\text{Bi}}\right] \cos(M_j Y) \tag{2.13}$$

where $M_j \tan M_j = \text{Bi}$. The expression for the nondimensional heat storage is

$$Q^+ = \sum_{j=1}^{\infty} 2 \left[\frac{\sin^2 M_j}{M_j \sin M_j \cos M_j + (M_j^2)}\right] \left\{1.0 - \exp\left[-(M_j^2)\frac{\eta}{\text{Bi}}\right]\right\} \tag{2.14}$$

Hollow Cylindrical Cross Section The transient response of a storage unit composed of a hollow cylindrical cross section with one surface insulated and the other exposed to a fluid that has an infinite heat capacity can be predicted analytically. The mathematical model is based on an energy balance of the storage material. The differential energy equation becomes

$$\rho_m c_m \frac{\partial t_m}{\partial \tau} = k_m \left(\frac{\partial^2 t_m}{\partial r^2} + \frac{1}{r}\frac{\partial t_m}{\partial r}\right) \tag{2.15}$$

The initial conditions are $\tau = 0$, $t_m = t_o$.

The boundary conditions will depend on the surface that is in contact with the fluid. The two possibilities are shown in Fig. 2.3 with the appropriate boundary conditions given by

$$\begin{aligned} &\text{Internal flow:} \qquad && r = r_i \qquad && k_m \frac{\partial t_m}{\partial r} = h(t_m - t_{fi}) \\ & && r = r_o && \frac{\partial t_m}{\partial r} = 0 \\ &\text{External flow:} && r = r_i && \frac{\partial t_m}{\partial r} = 0 \\ & && r = r_o && k_m \frac{\partial t_m}{\partial r} = h(t_{fi} - t_m) \end{aligned} \tag{2.16}$$

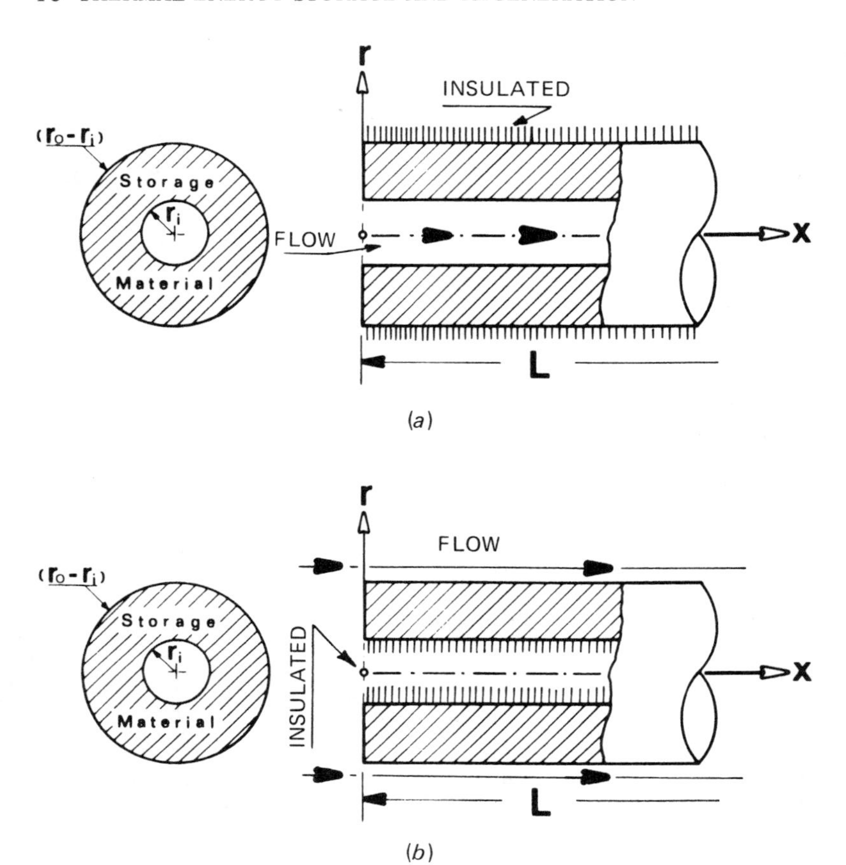

Figure 2.3 Hollow cylinder configuration, single-fluid case. (*a*) Internal flow. (*b*) External flow.

Solutions of these models in their dimensional form have been given by Ozisik [2].

$$t_m(\eta, \tau) = t_{fi} + (t_o - t_{fi}) \sum_{j=1}^{\infty} \exp(-\alpha\beta_j^2\tau) K_o(\beta_j, r) \int_{r_i}^{r_o} K_o(\beta_j, r') r' \, dr' \tag{2.17}$$

where $K_o(\beta_j, r) = \dfrac{R_o(\beta_j, r)}{\sqrt{N}}$

For internal flow,

$$R_o(\beta_j, r) = \frac{J_o(\beta_j r)}{\beta_j J_o'(\beta_j r_o)} - \frac{Y_o(\beta_j r)}{\beta_j Y_o'(\beta_j r_o)} \tag{2.18}$$

and $$N = \frac{r_o^2}{2} R_o^2(\beta_j, r_o) - \frac{r_i^2}{2}\left(\frac{h^2}{k_m^2 \beta_j^2} + 1\right) R_o^2(\beta_j, r_i)$$

with the eigenvalues β_j given by

$$\frac{hJ_o(\beta r_i) - \beta k_m J_o'(\beta r_i)}{J_o'(\beta r_o)} - \frac{hY_o(\beta r_i) - \beta k_m Y_o'(\beta r_i)}{Y_o'(\beta r_o)} = 0$$

For external flow,

$$R_o(\beta_j, r) = \frac{J_o(\beta_j r)}{k_m \beta_j J_o'(\beta_j r_o) + hJ_o(\beta_j r_o)} - \frac{Y_o(\beta_j r)}{k_m \beta_j Y_o'(\beta_j r_o) + hY_o(\beta_j r_o)} \tag{2.19}$$

and $$N = \frac{r_o^2}{2}\left(\frac{h^2}{k_m^2 \beta_j^2} + 1\right) R_o^2(\beta_j, r_o) - \frac{r_i^2}{2} R_o^2(\beta_j, r_i)$$

with the eigenvalues β_j given by

$$\frac{J_o'(\beta r_i)}{\beta k_m J_o'(\beta r_o) + hJ_o(\beta r_o)} - \frac{Y_o'(\beta r_i)}{\beta k_m Y_o'(\beta r_o) + hY_o(\beta r_o)} = 0$$

The heat stored in the unit may be obtained by using the expression for the temperature given by Eq. (2.17) and

$$Q = 2\pi L \int_0^\tau \int_{r_i}^{r_o} (t_m - t_o) r\, dr\, d\tau \tag{2.20}$$

where L is the length of the hollow cylinder. These expressions are not easily evaluated, since the determination of the eigenvalues β_j is difficult. It will be shown in Sec. 3.4 that if the Biot number, defined as

$$\text{Bi} = \frac{h(r_o - r_i)}{k_m}$$

is less than 2, the flat slab solution can be used with a high degree of accuracy for the calculations of the transient response of hollow cylindrical cross-sectioned storage units. This approximation has been found to be valid with radius ratios as large as 10.

2.3 SIMPLIFIED MODEL

In the simplified model the thermal conductivity of the storage material is considered to be infinite in the direction perpendicular to the flow and zero in the direction of flow. The temperature distribution within the storage material will be independent of the geometric cross section of the unit and thus the same expression can be used to determine the transient response of both a flat slab and hollow cylindrical cross-section unit. The temperature of the fluid and the storage material will be a function of the axial coordinate, x, only.

Anzelius [3], Nusselt [4], Hausen [5, 6], Schumann [7], and Larsen [8] have presented results for the transient response of a heat storage unit initially at a uniform temperature and exposed to a step change in the inlet fluid temperature. In the simplified analysis the following assumptions have been made:

1. Constant fluid and material properties
2. Infinite thermal conductivity for the solid in the y direction
3. Zero thermal conductivity for the solid in the x direction
4. Uniform heat transfer coefficient
5. Step change in inlet fluid temperature
6. Uniform initial temperature distribution in the storage material
7. No heat transfer through the sides of the storage unit
8. Constant fluid velocity

The transient response of the storage unit is governed by the one-dimensional conservation of energy equation for the moving fluid and the storage material. These equations can be derived by considering an incremental volume of length Δx. The energy entering the incremental volume is equal to the energy leaving plus the energy accumulated within the volume.

An energy balance for the fluid passing through the incremental volume yields

$$\frac{hA\Delta x}{L}(t_m - t_f) + \dot{m}_f c_f t_f \Big|_x = \dot{m}_f c_f t_f \Big|_{x+\Delta x} + S_f \Delta x \rho_f c_f \frac{\partial t_f}{\partial \tau}$$

The terms on the left-hand side of the equation represent the heat transferred from the storage material to the fluid and the energy content of the fluid entering the section. These terms are equated to the energy content of the fluid leaving the section and the rate of accumulation of energy by the fluid contained within the volume. Note that

$$t_f \Big|_{x+\Delta x} = t_f \Big|_x + \frac{\partial t_f}{\partial x}\Delta x$$

so the equation can be simplified to yield

$$\frac{hA}{L}(t_m - t_f) - \dot{m}_f c_f \frac{\partial t_f}{\partial x} = S_f \rho_f c_f \frac{\partial t_f}{\partial \tau} \tag{2.21}$$

In most practical applications the rate of accumulation of energy by the fluid contained within the volume can be neglected. The final form of the energy equation for the fluid is

$$\frac{\dot{m}_f c_f L}{hA}\frac{\partial t_f}{\partial x} = t_m - t_f \tag{2.22}$$

An energy balance on the solid contained within the incremental volume yields

$$\frac{hA}{L}\,\Delta x(t_f - t_m) = S_m \Delta x \rho_m c_m \frac{\partial t_m}{\partial \tau} \tag{2.23}$$

which can be rearranged to obtain

$$\frac{S_m \rho_m c_m L}{hA}\frac{\partial t_m}{\partial \tau} = t_f - t_m \tag{2.24}$$

The boundary and the initial conditions are

$$\begin{aligned} &\tau = 0 \qquad t = t_m = t_o \\ &\tau > 0 \qquad x = 0 \qquad t_f = t_{fi} \\ &\qquad\qquad x = 0 \qquad t_m = (t_o - t_{fi}) \exp\left(\frac{\tau h A}{S_m \rho_m c_m L}\right) + t_{fi} \end{aligned} \tag{2.25}$$

The following nondimensional variables are introduced:

Nondimensional distance: $\xi \equiv \dfrac{hAx}{\dot{m}_f c_f L}$

Nondimensional length: $\lambda \equiv \dfrac{hA}{\dot{m}_f c_f}$

Nondimensional time: $\eta \equiv \dfrac{hA(\tau - x/v)}{S_m L \rho_m c_m}$ (2.26)

Nondimensional fluid temperature: $T_f \equiv \dfrac{t_f - t_o}{t_{fi} - t_o}$

Nondimensional storage material temperature: $T_m \equiv \dfrac{t_m - t_o}{t_{fi} - t_o}$

The term x/v represents the time required by the fluid to pass from the entrance of the heat storage unit to the location x. This quantity is usually small when compared to the time scale of interest and can be neglected without introducing significant errors.

The mathematical model of the storage unit in nondimensional form is

Fluid: $$\frac{\partial T_f}{\partial \xi} = T_m - T_f \tag{2.27}$$

Storage material: $$\frac{\partial T_m}{\partial \eta} = T_f - T_m \tag{2.28}$$

Initial and boundary conditions are

$$\begin{aligned} &\xi = 0 \qquad T_f = 1.0 \qquad T_m = 1.0 - e^{-\eta} \\ &\eta = 0 \qquad\qquad\qquad T_m = 0 \end{aligned} \tag{2.29}$$

A complete evaluation of the numerous solutions to this problem was presented by Klinkenberg [9] and provides the source for the following relationships used to generate the results presented in Table 2.1 and Fig. 2.4.

$\eta < 2.0 \qquad \xi < 2.0$

$$T_f(\eta, \xi) = 1.0 - e^{-\eta-\xi} \sum_{N=1}^{N=\infty} \left(\frac{\xi^N}{N!} \sum_{k=0}^{k=N-1} \frac{\eta^k}{k!} \right) \tag{2.30}$$

$2.0 \leqslant \eta < 4.0 \qquad 2.0 \leqslant \xi < 4.0$

$$T_f(\eta, \xi) = 1.0 - \frac{1}{2}[1.0 + \text{erf}(\sqrt{\xi} - \sqrt{\eta})] - \frac{\xi^{1/4}}{\eta^{1/4} + \xi^{1/4}} e^{-\eta-\xi} I_o(2\sqrt{\eta\xi}) \tag{2.31}$$

$\eta \geqslant 4.0 \qquad \xi \geqslant 4.0$

$$T_f(\eta, \xi) = 1.0 - \frac{1}{2}\left[1.0 + \text{erf}\left(\sqrt{\xi} - \sqrt{\eta} - \frac{1}{8\sqrt{\xi}} - \frac{1}{8\sqrt{\eta}}\right)\right] \tag{2.32}$$

As pointed out by Larsen, the equations used in the description of the mathematical model of the heat storage unit are symmetric, so the value of the nondimensional temperature in the storage material can be obtained by using

$$T_m(A, B) = 1.0 - T_f(B, A) \tag{2.33}$$

Example 2.1 A thermal storage unit similar to that shown in Fig. 2.1*a* is composed of 12 channels 5.8 m (19.03 ft) long and 0.5 m (1.64 ft) wide. The thickness of the Feolite storage material between channels is 8 cm (3.15 in). The top and bottom channels have a storage material thickness of 4 cm (1.575 in) between the channel and the surrounding air. The entire unit is well insulated. The channels are 1.9 cm (0.748 in) high. Air flows through the channels at a mean velocity of 15.0 m/s (49.21 ft/s), and the convective heat transfer coefficient is 50.23 W/m² °C (8.8455 Btu/ft² h °F). The storage unit is initially at a temperature of 10°C (50°F), and the temperature of the hot gas entering the unit is 80°C (176°F). Determine the variation in the outlet fluid temperature during the first 3 h of operation and the temperature distribution in the storage material at the end of 3 h.

SOLUTION The physical properties of the Feolite and air are

Feolite @ 45.0°C (113°F):

$$\rho_m = 3900 \text{ kg/m}^3 \quad (243.48 \text{ lb}_m/\text{ft}^3)$$

$$c_m = 0.92 \text{ kJ/kg °C} \quad (0.2197 \text{ Btu/lb}_m \text{ °F})$$

Air @ 50°C (122°F):

$$\rho_f = 1.095 \text{ kg/m}^3 \quad (0.068 \text{ lb}_m/\text{ft}^3)$$

$$c_f = 1.011 \text{ kJ/kg °C} \quad (0.24 \text{ Btu/lb}_m \text{ °F}).$$

Table 2.1 Nondimensional fluid temperature, T_f

Nondimensional Distance, ξ

Nondimensional Time, η	1	2	3	4	5	6	7	8	9	10
0	0.3679	0.1353	0.0569	0.0207	0.0075	0.0027	0.0010	0.0004	0.0001	0.0000
1	0.6543	0.3943	0.2248	0.1233	0.0656	0.0340	0.0173	0.0087	0.0043	0.0021
2	0.8174	0.6035	0.4146	0.2700	0.1685	0.1017	0.0596	0.0341	0.0191	0.0105
3	0.9063	0.7531	0.5833	0.4269	0.2982	0.2003	0.1302	0.0823	0.0507	0.0306
4	0.9529	0.8520	0.7170	0.5702	0.4339	0.3174	0.2242	0.1537	0.1026	0.0669
5	0.9767	0.9140	0.8150	0.6919	0.5628	0.4401	0.3323	0.2432	0.1730	0.1200
6	0.9887	0.9513	0.8828	0.7871	0.6747	0.5574	0.4449	0.3441	0.2587	0.1894
7	0.9945	0.9730	0.9278	0.8573	0.7659	0.6615	0.5532	0.4487	0.3537	0.2717
8	0.9974	0.9853	0.9565	0.9070	0.8363	0.7488	0.6509	0.5497	0.4518	0.3618
9	0.9988	0.9921	0.9744	0.9408	0.8885	0.8185	0.7346	0.6421	0.5469	0.4544
10	0.9994	0.9958	0.9852	0.9631	0.9257	0.8720	0.8032	0.7226	0.6347	0.5445
11	0.9997	0.9978	0.9915	0.9774	0.9516	0.9118	0.8574	0.7899	0.7123	0.6283
12	0.9999	0.9989	0.9952	0.9864	0.9690	0.9404	0.8989	0.8444	0.7782	0.7033
13	0.9999	0.9994	0.9974	0.9920	0.9805	0.9605	0.9297	0.8870	0.8326	0.7679
14	1.0000	0.9997	0.9986	0.9953	0.9879	0.9742	0.9519	0.9194	0.8760	0.8219
15	1.0000	0.9999	0.9992	0.9973	0.9926	0.9835	0.9677	0.9436	0.9097	0.8658
16	1.0000	0.9999	0.9996	0.9985	0.9956	0.9895	0.9786	0.9611	0.9354	0.9006
17	1.0000	1.0000	0.9998	0.9991	0.9974	0.9935	0.9861	0.9736	0.9545	0.9275
18	1.0000	1.0000	0.9999	0.9995	0.9985	0.9960	0.9910	0.9823	0.9684	0.9480
19	1.0000	1.0000	0.9999	0.9997	0.9991	0.9975	0.9943	0.9883	0.9784	0.9632
20	1.0000	1.0000	1.0000	0.9999	0.9995	0.9985	0.9964	0.9924	0.9854	0.9743

Nondimensional Time, η	11	12	13	14	15	16	17	18	19	20
0	0.0000	0.0000	0.0000	0.0000	0.0000	0.0000	0.0000	0.0000	0.0000	0.0000
1	0.0010	0.0005	0.0002	0.0001	0.0000	0.0000	0.0000	0.0000	0.0000	0.0000
2	0.0057	0.0031	0.0016	0.0008	0.0004	0.0002	0.0001	0.0001	0.0000	0.0000
3	0.0181	0.0105	0.0060	0.0034	0.0019	0.0011	0.0006	0.0003	0.0002	0.0001
4	0.0427	0.0267	0.0165	0.0100	0.0060	0.0035	0.0020	0.0012	0.0007	0.0004
5	0.0814	0.0541	0.0353	0.0226	0.0143	0.0089	0.0054	0.0033	0.0020	0.0012
6	0.1355	0.0948	0.0650	0.0438	0.0290	0.0189	0.0122	0.0077	0.0048	0.0030
7	0.2037	0.1493	0.1072	0.0755	0.0523	0.0356	0.0239	0.0158	0.0103	0.0066
8	0.2828	0.2161	0.1617	0.1187	0.0855	0.0606	0.0422	0.0290	0.0196	0.0131
9	0.3686	0.2924	0.2271	0.1729	0.1292	0.0949	0.0686	0.0488	0.0342	0.0237
10	0.4566	0.3745	0.3009	0.2369	0.1831	0.1390	0.1038	0.0763	0.0553	0.0395
11	0.5424	0.4585	0.3797	0.3083	0.2458	0.1924	0.1482	0.1123	0.0838	0.0616
12	0.6228	0.5406	0.4602	0.3843	0.3151	0.2538	0.2010	0.1566	0.1202	0.0909
13	0.6953	0.6179	0.5391	0.4617	0.3884	0.3211	0.2611	0.2089	0.1646	0.1278
14	0.7585	0.6882	0.6136	0.5376	0.4630	0.3921	0.3266	0.2678	0.2162	0.1720
15	0.8121	0.7501	0.6819	0.6097	0.5364	0.4642	0.3954	0.3316	0.2739	0.2230
16	0.8563	0.8032	0.7425	0.6761	0.6062	0.5352	0.4653	0.3985	0.3362	0.2796
17	0.8919	0.8475	0.7950	0.7355	0.6708	0.6030	0.5342	0.4663	0.4013	0.3405
18	0.9199	0.8837	0.8393	0.7874	0.7291	0.6660	0.6001	0.5332	0.4672	0.4039
19	0.9415	0.9126	0.8760	0.8317	0.7804	0.7232	0.6616	0.5974	0.5323	0.4681
20	0.9579	0.9353	0.9056	0.8686	0.8245	0.7738	0.7177	0.6575	0.5949	0.5315

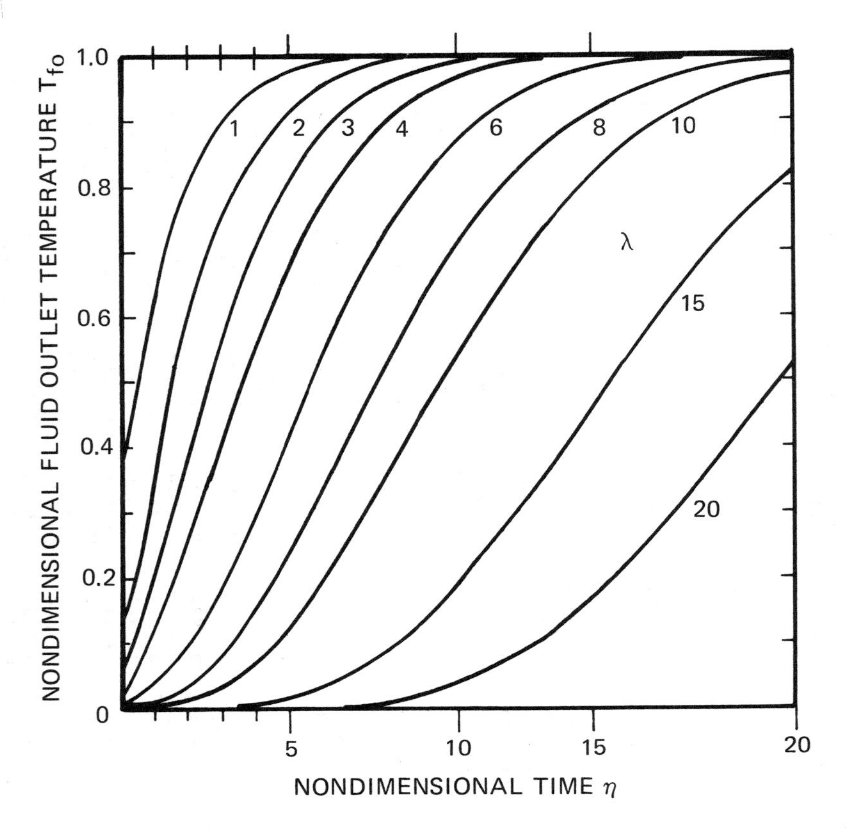

Figure 2.4 Nondimensional fluid outlet temperature.

Symmetry will be used so that the surface area, the cross-sectional area of the storage material, and the mass rate of flow of the air will be those associated with only one channel.

$$A = (2.0)(5.8)(0.5) = 5.8 \text{ m}^2 \quad (62.43 \text{ ft}^2)$$

$$\dot{m}_f = \rho_f v S_f = (1.095)(15.0)(0.019)(0.5) = 0.156 \text{ kg/s} \quad (0.344 \text{ lb}_m\text{/s})$$

$$S_m = (0.08)(0.5) = 0.04 \text{ m}^2 \quad (0.43 \text{ ft}^2)$$

The nondimensional length of the storage unit is

$$\lambda = \frac{hA}{\dot{m}_f c_f} = \frac{(50.23)(5.8)}{(0.156)(1011.0)} = 1.847$$

The nondimensional time is

$$\eta = \frac{hA\tau}{S_m L \rho_m c_m} = \frac{(50.23)(5.8)\tau}{(0.04)(5.8)(3900.0)(920.0)} = 3.5 \times 10^{-4}\tau$$

The nondimensional temperature of the fluid leaving the storage unit at the end of the 3-h period, $\eta = 3.78$, is 0.8537 or an exit fluid temperature of 69.76°C (157.57°F). The temperature of the solid material at the exit at 3 h is obtained by

$$T_m(3.78, 1.847) = 1.0 - T_f(1.847, 3.78)$$
$$= 0.7271$$

which gives

$$t_m\,|_{x=5.8\text{ m}} = 60.90°\text{C} \quad (141.62°\text{F})$$

The complete results are presented in Fig. 2.5.

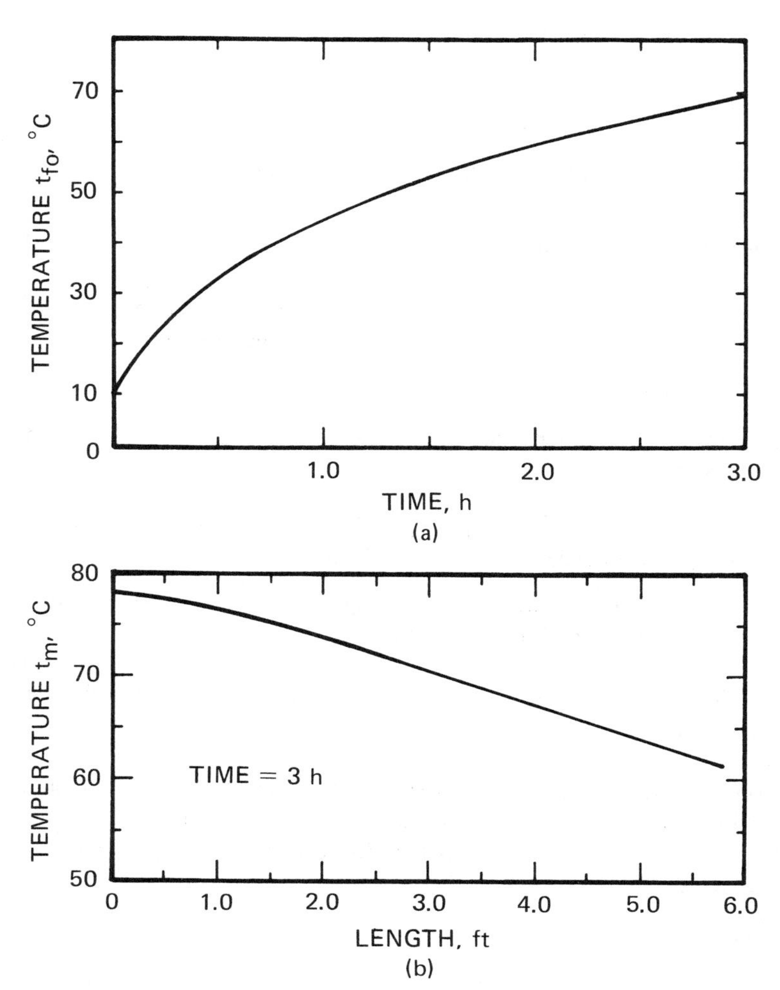

Figure 2.5 Transient response of storage unit. (*a*) Outlet fluid temperature. (*b*) Storage material temperature.

The amount of energy stored in the unit can be obtained in two different ways. The energy lost by the fluid as it passes through the channels in the heat storage unit may be obtained by using

$$Q = \dot{m}_f c_f \int_0^\tau (t_{fi} - t_{fo})\, d\tau$$

$$= \dot{m}_f c_f (t_{fi} - t_0) \int_0^\tau (1.0 - T_{fo})\, d\tau \tag{2.34}$$

The energy stored within the storage material is

$$Q = S_m L \rho_m c_m (\bar{t}_m - t_o) \tag{2.35}$$

where $\bar{t}_m$ is the average material temperature:

$$\bar{t}_m = \frac{1}{L} \int_0^L t_m\, dx$$

This may be written in terms of the nondimensional temperature as

$$Q = S_m \rho_m c_m (t_{fi} - t_o) \int_0^L T_m\, dx \tag{2.36}$$

The maximum possible heat storage is obtained when the mean temperature of the storage material is equal to the temperature of the fluid entering the unit,

$$Q_{\max} = S_m L \rho_m c_m (t_{fi} - t_o) \tag{2.37}$$

The expression for the nondimensional heat storage where Eq. (2.34) is used to find the actual heat storage is

$$Q^+ = \frac{\dot{m}_f c_f (t_{fi} - t_o)}{S_m L \rho_m c_m (t_{fi} - t_o)} \int_0^\tau (1 - T_{fo})\, d\tau = \frac{1}{\lambda} \int_0^\eta (1 - T_{fo})\, d\eta \tag{2.38}$$

where λ is the nondimensional length of the storage unit. Values of Q^+ are tabulated as a function of η and λ in Table 2.2 and Fig. 2.6.

Example 2.2 Determine the heat stored in the unit described in Example 2.1 during the 3-h operating period.

SOLUTION The nondimensional length of the unit was found to be $\lambda = 1.847$ and the nondimensional time $\eta = 3.5 \times 10^{-4} \tau$. After 3 h of operation, $\eta = 3.78$. The value of Q^+, obtained from Table 2.2, is 0.8638. The maximum possible heat stored per channel is

$$Q_{\max} = S_m L \rho_m c_m (t_{fi} - t_o)$$

$$= (0.04)(5.8)(3900.0)(920.0)(80.0 - 10.0)$$

$$= 5.827 \times 10^4 \text{ kJ per channel}$$

Table 2.2 Nondimensional heat storage

Nondimensional Length, λ

Nondimensional Time, η	1	2	3	4	5	6	7	8	9	10
1	0.4762	0.3662	0.2887	0.2335	0.1936	0.1641	0.1418	0.1246	0.1109	0.0999
2	0.7324	0.6142	0.5153	0.4349	0.3708	0.3200	0.2795	0.2471	0.2209	0.1994
3	0.8658	0.7727	0.6814	0.5977	0.5244	0.4618	0.4091	0.3651	0.3283	0.2974
4	0.9336	0.8696	0.7970	0.7225	0.6511	0.5855	0.5269	0.4756	0.4311	0.3927
5	0.9675	0.9269	0.8741	0.8142	0.7512	0.6890	0.6301	0.5759	0.5270	0.4835
6	0.9841	0.9597	0.9237	0.8788	0.8271	0.7724	0.7175	0.6643	0.6143	0.5682
7	0.9921	0.9782	0.9547	0.9227	0.8827	0.8373	0.7889	0.7398	0.6914	0.6452
8	0.9960	0.9883	0.9736	0.9518	0.9221	0.8862	0.8456	0.8023	0.7578	0.7136
9	0.9978	0.9938	0.9849	0.9706	0.9494	0.9220	0.8894	0.8527	0.8134	0.7728
10	0.9987	0.9967	0.9915	0.9824	0.9677	0.9476	0.9222	0.8923	0.8588	0.8228
11	0.999	0.9982	0.9953	0.9897	0.9798	0.9654	0.9463	0.9226	0.8949	0.8641
12	0.999	0.9990	0.9974	0.9941	0.9877	0.9776	0.9635	0.9453	0.9231	0.8974
13	0.999	0.999	0.998	0.9967	0.9926	0.9858	0.9757	0.9620	0.9446	0.9238
14	1.000	0.999	0.999	0.9983	0.9957	0.9911	0.9840	0.9740	0.9607	0.9442
15	1.000	0.999	0.999	0.999	0.9976	0.9946	0.9897	0.9825	0.9726	0.9597
16	1.000	1.000	0.999	0.999	0.9988	0.9968	0.9935	0.9884	0.9811	0.9713
17	1.000	1.000	1.000	1.000	0.999	0.9982	0.9960	0.9924	0.9872	0.9799
18	1.000	1.000	1.000	1.000	0.999	0.9991	0.9976	0.9951	0.9914	0.9861
19	1.000	1.000	1.000	1.000	1.000	0.9996	0.9986	0.9969	0.9943	0.9905
20	1.000	1.000	1.000	1.000	1.000	0.9999	0.9993	0.9981	0.9963	0.9936
	11	12	13	14	15	16	17	18	19	20
1	0.0909	0.0833	0.0769	0.0714	0.0667	0.0625	0.0588	0.0556	0.0526	0.0500
2	0.1815	0.1665	0.1538	0.1428	0.1333	0.1250	0.1176	0.1111	0.1053	0.1000
3	0.2714	0.2493	0.2304	0.2141	0.1999	0.1875	0.1765	0.1667	0.1579	0.1500
4	0.3597	0.3312	0.3066	0.2851	0.2663	0.2498	0.2352	0.2222	0.2105	0.2000
5	0.4451	0.4113	0.3815	0.3554	0.3324	0.3120	0.2938	0.2776	0.2631	0.2500
6	0.5262	0.4885	0.4547	0.4245	0.3976	0.3736	0.3521	0.3329	0.3155	0.2999
7	0.6018	0.5617	0.5251	0.4918	0.4616	0.4345	0.4099	0.3878	0.3678	0.3496
8	0.6707	0.6299	0.5917	0.5563	0.5238	0.4940	0.4669	0.4421	0.4196	0.3991
9	0.7320	0.6921	0.6538	0.6174	0.5833	0.5517	0.5225	0.4956	0.4709	0.4482
10	0.7854	0.7477	0.7104	0.6742	0.6397	0.6069	0.5763	0.5477	0.5212	0.4967
11	0.8309	0.7963	0.7612	0.7262	0.6921	0.6591	0.6277	0.5980	0.5702	0.5442
12	0.8688	0.8380	0.8058	0.7729	0.7401	0.7077	0.6763	0.6462	0.6175	0.5904
13	0.8997	0.8730	0.8443	0.8142	0.7833	0.7523	0.7216	0.6916	0.6627	0.6350
14	0.9245	0.9019	0.8768	0.8499	0.8216	0.7925	0.7631	0.7339	0.7053	0.6775
15	0.9439	0.9252	0.9039	0.8803	0.8549	0.8282	0.8007	0.7729	0.7451	0.7177
16	0.9589	0.9438	0.9260	0.9058	0.8835	0.8595	0.8342	0.8082	0.7817	0.7551
17	0.9703	0.9583	0.9437	0.9267	0.9075	0.8864	0.8637	0.8397	0.8149	0.7896
18	0.9788	0.9694	0.9577	0.9437	0.9275	0.9092	0.8891	0.8675	0.8447	0.8210
19	0.9851	0.9778	0.9686	0.9573	0.9438	0.9283	0.9108	0.8916	0.8710	0.8492
20	0.9896	0.9841	0.9770	0.9680	0.9570	0.9440	0.9290	0.9123	0.8939	0.8742

The value of the heat stored during the first 3 h of operation is

$$Q = Q^+ Q_{max}$$
$$= (0.8638)(5.827 \times 10^4)$$
$$= 5.033 \times 10^4 \text{ kJ per channel} \quad (4.70 \times 10^4 \text{ Btu per channel})$$

or a total of 6.04×10^5 kJ (5.72×10^5 Btu) for the complete unit. The complete transient response of the heat storage unit is given in Fig. 2.7.

In the evaluation of the performance of a thermal storage unit, the fraction of the available energy that is stored is of interest to the engineer since it can be thought of as an indication of the value of the energy recovered by the storage system. The available energy in the fluid stream is defined as

$$AE = \dot{m}_f c_f (t_{fi} - t_{env})\tau \tag{2.39}$$

where the temperature of the environment and the initial temperature of the unit are usually equal, $t_{env} = t_o$. The fraction of the available energy stored, A^+, is

$$A^+ \equiv \frac{Q}{AE} = \frac{\lambda}{\eta} Q^+ \tag{2.40}$$

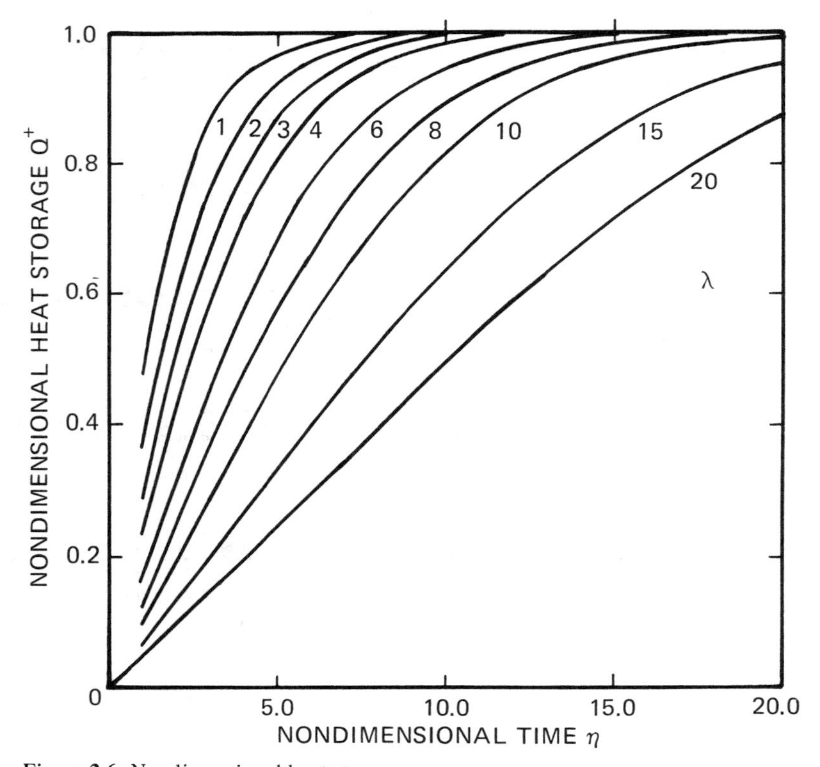

Figure 2.6 Nondimensional heat storage.

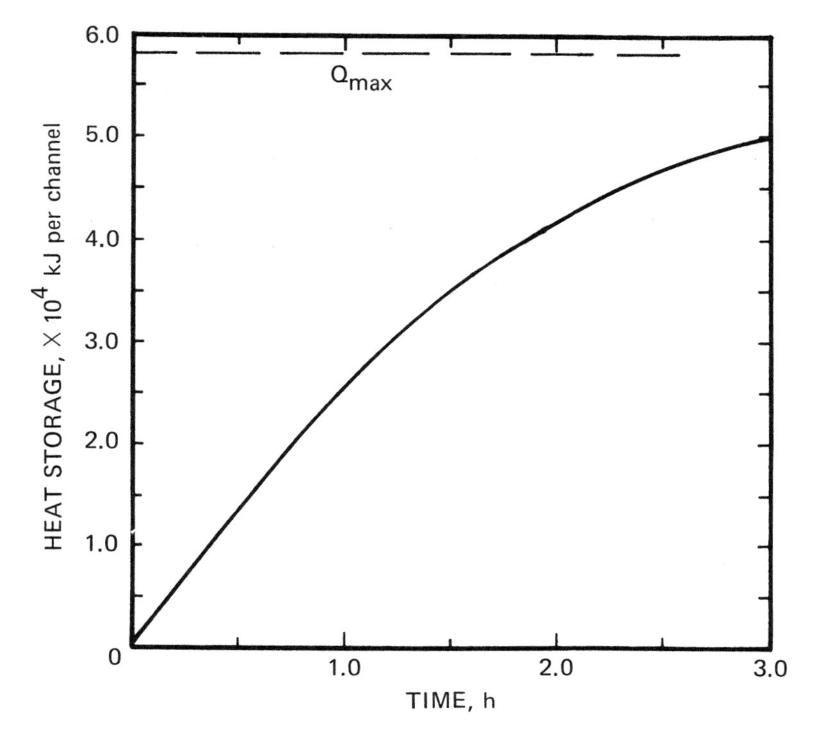

Figure 2.7 Heat storage.

For Example 2.2, the amount of available energy passing through the storage unit during 3 h of operation is

$$\begin{aligned} AE &= \dot{m}_f c_f (t_{fi} - t_{\text{env}})\tau \\ &= (0.156)(12.0)(1.011)(80.0 - 10.0)(3.0)(3600.0) \\ &= 1.43 \times 10^6 \text{ kJ} \quad (1.36 \times 10^6 \text{ Btu}) \end{aligned}$$

The fraction of that heat that was stored was

$$A^+ = \frac{Q}{AE} = \frac{6.04 \times 10^5}{1.43 \times 10^6} = 0.42$$

2.4 SIMPLIFIED MODEL EMPLOYING A MODIFIED HEAT TRANSFER COEFFICIENT

The simplified model assumes that the thermal conductivity in the direction perpendicular to the flow is infinite. The rate of heat transfer at the wall is $\delta Q = h\ \delta A(t_m - t_f)$, where t_m is the temperature of the storage material at a particular location. If the thermal conductivity is not infinite, a transverse temperature gradient will exist in the storage material. The corresponding boundary condition at the surface will be $\delta Q = h\,\delta A(t_{ms} - t_f)$, where t_{ms} is the temperature at the surface of the storage material.

In order to use the simplified model for situations in which the thermal conductivity in the direction perpendicular to the flow is finite, the mean temperature of the storage material for the cross section $\bar{t}_m$ is introduced. A modified heat transfer coefficient $\bar{h}$ is also introduced, which has a value selected to ensure the equality $\delta Q = \bar{h}\,\delta A(\bar{t}_m - t_f)$. The relationships between the convective film coefficient h and the modified film coefficient $\bar{h}$ have been developed by Hausen [10, 11] and by Smith and Willmott [12]:

$$\frac{1}{\bar{h}} = \frac{1}{h} + \frac{\Delta}{k} \tag{2.41}$$

where for

Slabs: $\Delta = \dfrac{w}{3}$

Solid cylinders: $\Delta = \dfrac{r_o}{4}$

Hollow cylinders: $\Delta = \dfrac{r_i}{2(r_o^2 - r_i^2)} \left[\dfrac{2r_o^4}{r_o^2 - r_i^2} \ln\left(\dfrac{r_o}{r_i}\right) - 2r_o^2 + r_i^2 \right]$

Solid spheres: $\Delta = \dfrac{r_o}{5}$

These expressions have been developed on the assumption that the mean solid temperature varied linearly with respect to time.

Although the mean temperature of the storage unit operating under single-blow conditions does not vary timewise in a linear fashion, the transient performance of a unit having finite conductivity in the direction perpendicular to the flow may be estimated using the modified heat transfer coefficient. The relationships between the modified and the actual value of η and λ are

$$\bar{\eta} = \frac{\eta}{1 + \mathrm{Bi}_\Delta}$$

and $$\bar{\lambda} = \frac{\lambda}{1 + \mathrm{Bi}_\Delta}$$

where $\mathrm{Bi}_\Delta = \dfrac{h\Delta}{k}$

Example 2.3 Determine the heat storage and the fluid outlet temperature of the heat storage unit described in Example 2.1 after 3 h of operation using the modified heat transfer coefficient.

SOLUTION The thermal conductivity of the Feolite is $k_m = 2.1$ W/m °C (1.213 Btu/ft h °F). The Biot number Bi_Δ is 0.319. The modified nondimensional length and time are

$$\bar{\eta} = \frac{3.78}{1 + 0.319} = 2.866$$

$$\bar{\lambda} = \frac{1.847}{1 + 0.319} = 1.400$$

The values of Q^+ and T_{fo} are

$$Q^+ = 0.814 \qquad T_{fo} = 0.8371$$

The total amount of heat stored per channel is

$$Q = Q^+ Q_{\text{max}} = (0.814)(5.827 \times 10^4) = 4.743 \times 10^4 \text{ kJ} \qquad (4.496 \times 10^4 \text{ Btu})$$

The temperature of the fluid leaving the storage unit is

$$t_{fo} = t_o + (t_{fi} - t_{fo})T_{fo} = 10.0 + (70.0)(0.8371) = 68.597°\text{C} \qquad (155.47°\text{F})$$

REFERENCES

1. H. Gröber, S. Erk and V. Grigull, *Fundamentals of Heat Transfer*, McGraw-Hill, New York, 1961, p. 43.
2. M. N. Ozisik, *Boundary Value Problems of Heat Conduction*, International Book Company, Scranton, Pa., 1968, p. 145.
3. A. Anzelius, "Über Erwärmung Vermittels Durchströmender Medien," *Z. Angew. Math. Mech.*, vol. 6, 1926, p. 291.
4. W. Nusselt, "Die Theorie des Windehitzers," *Z. Ver. Deutsch. Ing.*, vol. 71, 1927, p. 85.
5. H. Hausen, "Über den Wärmeaustausch in Regeneratoren," *Tech. Mech. Thermodynam.*, vol. 1, 1930, p. 219.
6. H. Hausen, "Über die Theorie des Wärmeaustausches in Regeneratoren," *Z. Angew. Math. Mech.*, vol. 9, 1929, p. 173.
7. T. E. W. Schumann, "Heat Transfer: A Liquid Flowing Through a Porous Prism," *J. Franklin Inst.*, vol. 208, 1929, p. 405.
8. F. W. Larsen, "Rapid Calculations of Temperature in a Regenerative Heat Exchanger Having Arbitrary Initial Solid and Entering Fluid Temperatures," *Int. J. Heat Mass Transfer*, vol. 10, 1967, p. 149.
9. A. Klinkenberg, "Heat Transfer in Cross-Flow Heat Exchangers and Packed Beds," *Ind. Eng. Chem.*, vol. 46, 1954, p. 2285.
10. H. Hausen, "Berechnung der Steintemperature in Winderhitzern," *Arch. Eisenuttes*, vol. 10, 1938/1939, p. 474.
11. H. Hausen, *Wärmeübertragung im Gegenstrom, Gleichstrom und Kreuzstrom*, 2d ed., Springer-Verlag, Berlin, 1976.
12. S. A. H. Smith and A. J. Willmott, Private communication, May 1979.

CHAPTER

THREE

SINGLE-BLOW OPERATING MODE: FINITE CONDUCTIVITY MODEL

3.1 INTRODUCTION

There are two principal thermal resistances encountered in the transfer of heat from a fluid to the interior of the storage material. One occurs at the surface of the storage material and is inversely proportional to the convective heat transfer coefficient h. The other thermal resistance is associated with the transfer of heat from the surface to the interior of the storage material. It is composed of a component in the axial or flow direction as well as a component in the transverse or direction perpendicular to the flow. Both of these resistances are inversely proportional to the thermal conductivity of the storage material. The major transfer of heat in the storage material is, however, in the transverse direction.

If the convective heat transfer resistance is very large compared to that offered by the resistance to the transfer of heat within the material, negligible temperature gradients will exist in the storage material in the direction normal to the flow. Under these conditions the simplified model described in Chap. 2 can be used to predict the transient response of the unit. If the two resistances are of equal magnitude or if the internal resistance is greater, temperature gradients will be present within the storage material. The mathematical model used to predict the transient response of such units must, therefore, include axial and transverse conduction effects within the material. This model will be referred to as the finite conductivity model.

The relative influence of the internal and external resistances may usually be associated with the value of the Biot number, $\mathrm{Bi} = hw/k_m$, where w is a characteristic length associated with the geometry of the storage unit. If the Biot number is small, temperature gradients within the storage material will be insignificant and the simplified model may be used. If the Biot number is large, a condition frequently encountered when the energy transporting fluid is a liquid or when the storage material has a low thermal conductivity, an accurate prediction of the transient response of the unit can be obtained only by using the finite conductivity model.

3.2 FINITE CONDUCTIVITY MODEL: SLAB CONFIGURATION

The finite conductivity model is dependent on the geometric configuration of the storage unit. In this section the analysis of a unit composed of a number of rectangular cross-sectioned channels for the energy transporting fluid separated by the heat storage material will be presented. A sketch of the unit is shown in Fig. 3.1. If all the channels and storage sections are identical and if the flows in the channels are equal, surfaces of symmetry or surfaces across which no heat flows are present. The analysis can thus be restricted to the consideration of a section composed of one-half of the channel width and the storage material separating the two channels as shown in Fig. 3.2.

In the analysis of this section the following assumptions have been made:

1. Constant fluid and material properties
2. Uniform heat transfer coefficient
3. Step change in inlet temperature
4. Uniform initial temperature distribution in the storage material
5. No heat transfer through the sides of the storage units
6. Constant fluid velocity

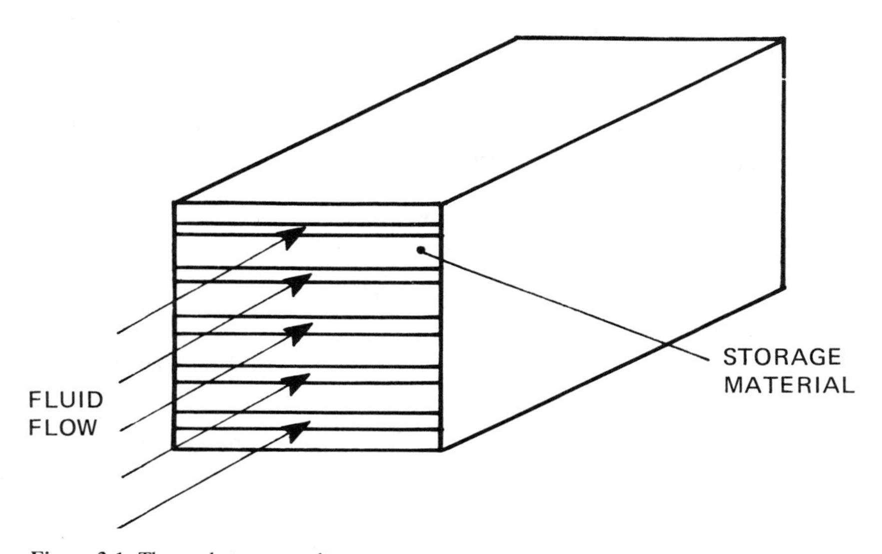

Figure 3.1 Thermal storage unit.

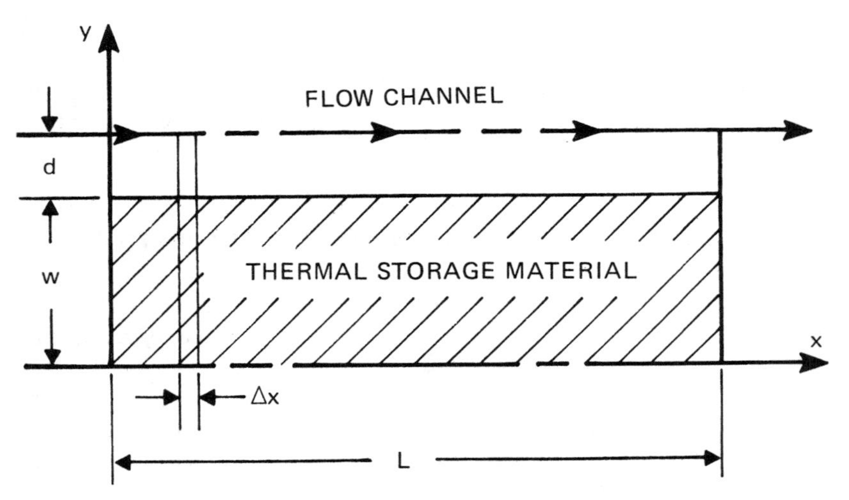

Figure 3.2 Cross section of storage unit.

The transient response of the storage unit is governed by the one-dimensional conservation of energy equation for the moving fluid and by the two-dimensional transient heat conduction equation for the storage material. The energy equation for the storage material is

$$\frac{1}{\alpha_m}\frac{\partial t_m}{\partial \tau} = \frac{\partial^2 t_m}{\partial x^2} + \frac{\partial^2 t_m}{\partial y^2} \tag{3.1}$$

whereas that for the fluid remains the same as for the simplified model,

$$\text{Fluid:} \qquad \frac{\dot{m}_f c_f L}{hA}\frac{\partial t_f}{\partial x} = t_w - t_f \tag{2.23}$$

The complete mathematical model is the same as that used for the simplified model in Sec. 2.2 except for those terms associated with the thermal conductivity of the storage material. The initial and boundary conditions are

$$\begin{aligned}
&\tau = 0 && t_f = t_m = t_o && && \\
&\tau > 0 && x = 0 && t_f = t_{fi} && \frac{\partial t_m}{\partial x} = 0 && 0 < y < w \\
& && x = L && && \frac{\partial t_m}{\partial x} = 0 && 0 < y < w \\
& && y = 0 && && \frac{\partial t_m}{\partial y} = 0 && 0 \leqslant x \leqslant L \\
& && y = w && && -k_m\frac{\partial t_m}{\partial y} = h(t_w - t_f) && 0 \leqslant x \leqslant L
\end{aligned} \tag{3.2}$$

The determination of the transient response of a storage unit described by this model has been presented by Schmidt and Szego [1]. The following nondimensional groups were introduced:

$$X \equiv \frac{x}{L} \qquad Y \equiv \frac{y}{w}$$

$$V^+ \equiv \frac{w}{L} \qquad \mathrm{Fo} \equiv \frac{\alpha \tau}{w^2} \qquad G^+ \equiv \frac{P k_m}{\dot{m}_f c_f} \tag{3.3}$$

$$\mathrm{Bi} \equiv \frac{hw}{k_m} \qquad T \equiv \frac{t - t_o}{t_{fi} - t_o}$$

The transformed equations, initial and boundary conditions, are

Fluid:
$$\frac{\partial T_f}{\partial X} + \frac{G^+}{V^+}(\mathrm{Bi})(T_f - T_w) = 0 \tag{3.4}$$

Storage material:
$$(V^+)^2 \frac{\partial^2 T_m}{\partial x^2} + \frac{\partial^2 T_m}{\partial y^2} = \frac{\partial T_m}{\partial \mathrm{Fo}} \tag{3.5}$$

with the initial conditions $\mathrm{Fo} = 0$, $T_m = T_f = 0$, and the boundary conditions

$$\begin{aligned}
&X = 0 \qquad T_f = 1 \qquad \frac{\partial T_m}{\partial x} = 0 \qquad 0 < Y < 1 \\
&X = 1 \qquad \frac{\partial T_m}{\partial x} = 0 \qquad 0 < Y < 1 \\
&Y = 1 \qquad \frac{\partial T_m}{\partial Y} = \mathrm{Bi}(T_f - T_w) \qquad 0 \leqslant X \leqslant 1 \\
&Y = 0 \qquad \frac{\partial T}{\partial Y} = 0 \qquad 0 \leqslant X \leqslant 1
\end{aligned} \tag{3.6}$$

The nondimensional parameters used in the simplified model are related to the new nondimensional parameters for the configuration under discussion by the following expressions:

Nondimensional length:
$$\xi = \mathrm{Bi}\,\frac{G^+}{V^+}\left(\frac{x}{L}\right) \tag{3.7}$$

Nondimensional unit length:
$$\lambda = \mathrm{Bi}\,\frac{G^+}{V^+} \tag{3.8}$$

Nondimensional time:
$$\eta = \mathrm{Bi}\ \mathrm{Fo} \tag{3.9}$$

The set of two equations, Eqs. (3.4) and (3.5), were solved simultaneously for the dependent variables T_f and T_m using a finite difference method. The dimensionless heat storage Q^+ was computed from the temperature distribution in the storage material.

For the finite difference method the thermal storage material was subdivided into six columns of nodes in the axial direction and four rows of nodes in the transverse direction. Initially a larger number of nodes was used, but it was verified that a finer grid configuration did not substantially change the accuracy of the results. Equation (3.5) was then rewritten in finite difference form for each of the 24 nodes. Prior to the formation of the difference equations, a transformed time domain γ was introduced:

$$\gamma \equiv 1 - e^{-B\,\mathrm{Fo}} \tag{3.10}$$

where B is a constant parameter. The differential equation for the storage material thus becomes

$$(V^+)^2 \frac{\partial^2 T_m}{\partial X^2} + \frac{\partial^2 T_m}{\partial Y^2} = B(1-\gamma)\frac{\partial T_m}{\partial \gamma} \tag{3.11}$$

The rationale for the use of the transformation is that the complete transient range must be covered by the solution. In order to keep discretization errors within reasonable limits, small steps in real time are required at the beginning of the process, whereas the step size can be increased as steady state is approached without introducing appreciable errors. With the time domain transformation, equal spacing in γ will yield the desired conditions. One hundred steps in γ ($\Delta\gamma = 0.01$) were taken, and the value of B was selected to obtain nearly steady-state conditions at the ninety-seventh step in $\Delta\gamma$. A backward difference method was used to approximate $\partial T_m/\partial\gamma$, whereas a central difference method was used for the approximation of the spacial derivatives.

Equations (3.4) and (3.5) are coupled, so the computational procedure involved an iterative scheme at each value of γ. At a particular value of γ corresponding to step n ($\gamma = n\,\Delta\gamma$), the temperature distribution in the fluid, T_f, is known and T_m at step $n + 1$ is initially approximated by using that fluid temperature distribution. An orthogonal polynomial of third order is fitted through the wall temperature T_w using the least-squares method proposed by Forsythe [2]. Substitution of this polynomial in Eq. (3.4) yields a first-order differential equation that is solved exactly and a new temperature distribution T_f is obtained. The finite difference solution is recalculated with this new fluid temperature distribution and the iterative process is repeated until convergence is obtained, that is, the maximum change in T_m is less than 10^{-7} and in T_f is less than 10^{-4}. The computations then proceed to step $n + 2$. The foregoing convergence criteria were shown to give good results at reasonable computational times. The curves to be presented summarize the results of more than 100 runs, so reasonable limits on computational times were a major concern.

The nondimensional fluid outlet temperatures for various operating conditions are presented in Figs. 3.3 through 3.9 [1]. The Biot number is varied from 0.03 to 30. A complete analysis of the results indicated that for the cases of practical interest, the effects of axial conduction could be neglected; that is, the first term on the right-hand side of Eq. (3.10) can be neglected. Thus it is possible to combine G^+ and V^+

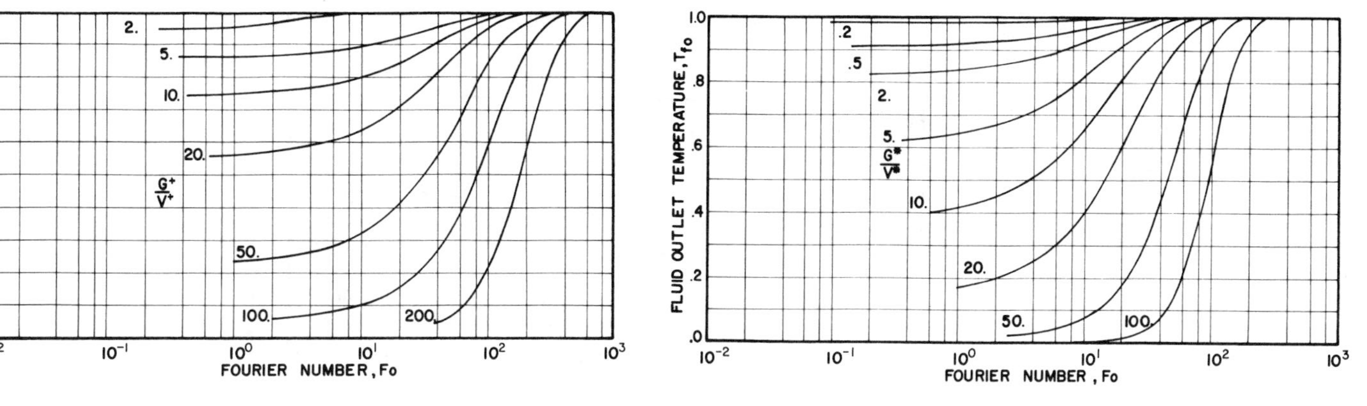

Figure 3.3 Nondimensional fluid outlet temperature, Bi = 0.03 [1].

Figure 3.4 Nondimensional fluid outlet temperature, Bi = 0.1 [1].

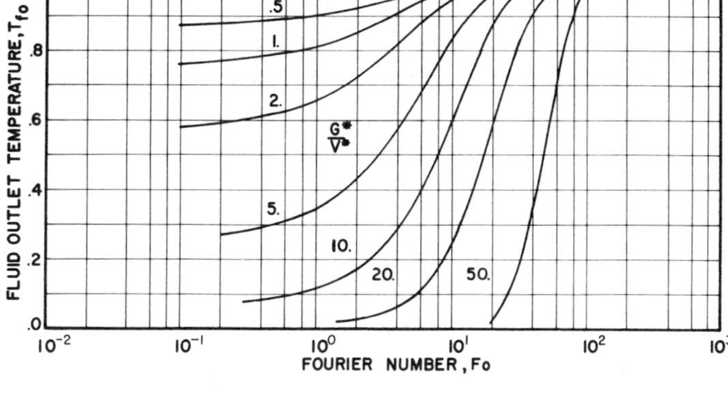

Figure 3.5 Nondimensional fluid outlet temperature, Bi = 0.3 [1].

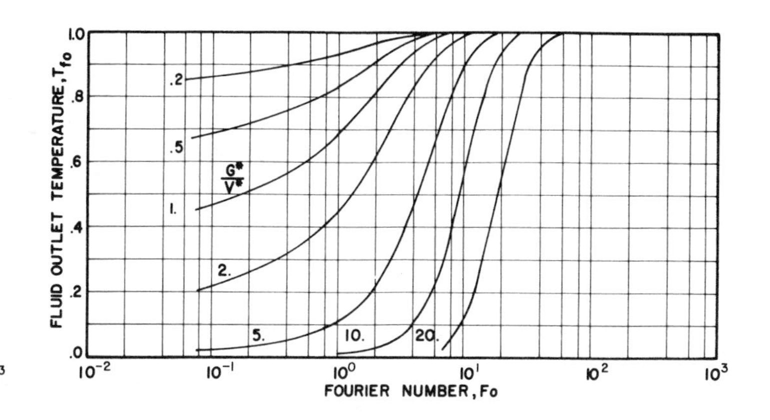

Figure 3.6 Nondimensional fluid outlet temperature, Bi = 1.0 [1].

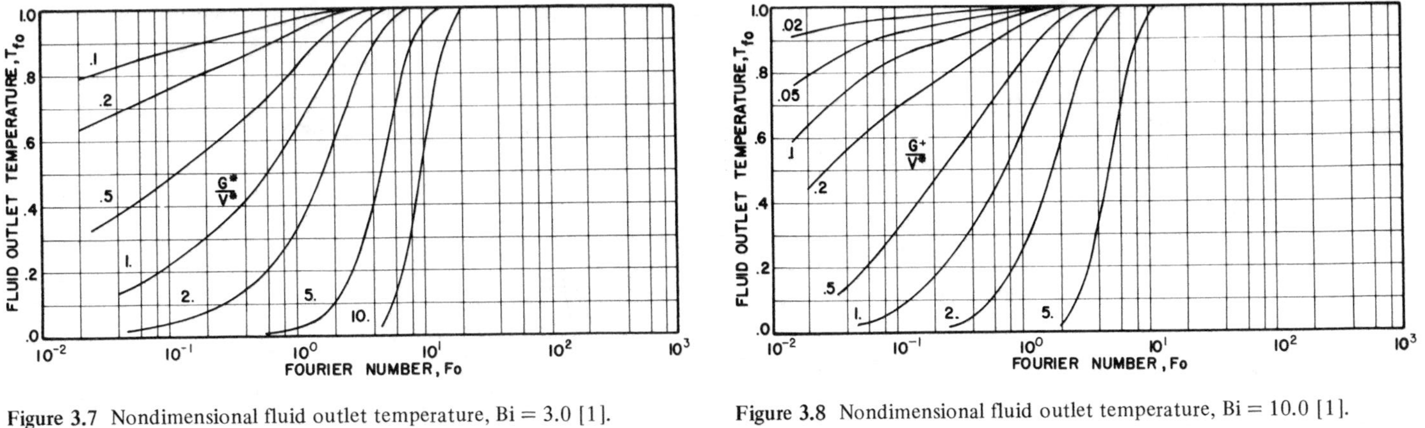

Figure 3.7 Nondimensional fluid outlet temperature, Bi = 3.0 [1].

Figure 3.8 Nondimensional fluid outlet temperature, Bi = 10.0 [1].

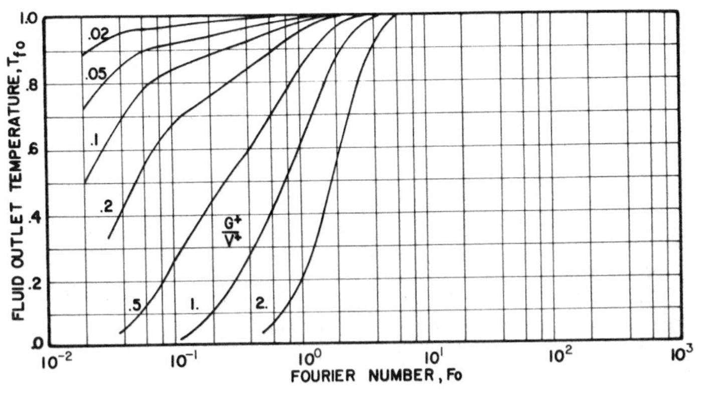

Figure 3.9 Nondimensional fluid outlet temperature, Bi = 30.0 [1].

into a single variable, namely, G^+/V^+, since these parameters appear in the mathematical model only as the indicated ratio.

The results are also tabulated in terms of the nondimensional variables used in the simplified model, η, T, and λ. The results for the nondimensional outlet fluid temperature are presented in Tables 3.1 through 3.5. The Biot number is varied from 0.1 to 1000. A comparison of the results presented in these tables with those tabulated in Table 2.1, in which the Biot number equals zero, allows one to evaluate the influence of the temperature gradients within the solid on the performance of the heat storage unit. If λ and η remain fixed, the nondimensional outlet fluid temperature decreases as the Biot number increases.

The nondimensional heat storage Q^+ was calculated using the finite conductivity model, and the results are presented in Figs. 3.10 through 3.14. They have also been tabulated as a function of η and λ and appear in Tables 3.6 through 3.10.

When the thermal conductivity of the storage unit is considered to be finite, the mathematical model describing the unit is no longer symmetric and the temperature distribution in the storage material cannot be obtained from the tabulated values of T_f. The temperature distribution in the storage material can be obtained only by using the finite difference program described by Schmidt and Szego [1].

The prediction of the response of a thermal storage unit under single-blow operation often requires double interpolation when curves or tabulated results are used. In order to simplify and improve the accuracy of the results obtained from this work, a computer program has been developed to perform the interpolation. This program is described by Schmidt and Szego [1] and will be referred to as HSSF.

Example 3.1 Determine the heat storage and the fluid outlet temperature of the heat storage unit described in Example 2.1 after 3 h of operation using the finite thermal conductivity model.

SOLUTION The thermal conductivity of the Feolite is $k_m = 2.1$ W/m °C (1.213 Btu/ft h °F), and the Biot number is

$$\text{Bi} = \frac{hw}{k_m} = \frac{(50.23)(0.04)}{2.1} = 0.957$$

The values of the nondimensional length and time calculated in Example 2.1 are

$$\eta = 3.78 \quad \text{and} \quad \lambda = 1.847$$

or

$$\text{Fo} = 3.95 \quad \text{and} \quad \frac{G^+}{V^+} = 1.93$$

By using HSSF or the results tabulated in Tables 3.1 through 3.10, it is found that

$$Q^+ = 0.811 \qquad T_f = 0.836$$

Table 3.1 Nondimensional temperature T_{fo}, Biot number 0.1

Nondimensional Time, η	Nondimensional Storage Unit Length, λ									
	1	2	3	4	5	6	7	8	9	10
1	0.6587	0.4058	0.2311	0.1177	0.0000	0.0000	0.0000	0.0000	0.0000	0.0000
2	0.8163	0.6106	0.4352	0.3002	0.1811	0.0967	0.0000	0.0000	0.0000	0.0000
3	0.9037	0.7551	0.5973	0.4534	0.3348	0.2482	0.1668	0.0756	0.0000	0.0000
4	0.9503	0.8500	0.7239	0.5914	0.4664	0.3556	0.2672	0.2109	0.1491	0.0696
5	0.9701	0.9078	0.8142	0.7021	0.5869	0.4731	0.3721	0.2821	0.2136	0.1799
6	0.9852	0.9414	0.8749	0.7876	0.6847	0.5818	0.4789	0.3802	0.2957	0.2212
7	0.9952	0.9621	0.9253	0.8488	0.7660	0.6704	0.5755	0.4797	0.3836	0.3009
8	1.0000	0.9793	0.9400	0.9016	0.8275	0.7475	0.6570	0.5679	0.4772	0.3877
9	1.0000	0.9915	0.9610	0.9205	0.8817	0.8098	0.7308	0.6435	0.5569	0.4717
10	1.0000	1.0000	0.9776	0.9435	0.9036	0.8650	0.7948	0.7152	0.6296	0.5429
11	1.0000	1.0000	0.9900	0.9627	0.9276	0.8943	0.8512	0.7821	0.7004	0.6150
12	1.0000	1.0000	1.0000	0.9782	0.9485	0.9136	0.8867	0.8399	0.7715	0.6862
13	1.0000	1.0000	1.0000	0.9947	0.9660	0.9355	0.9011	0.8783	0.8311	0.7631
14	1.0000	1.0000	1.0000	1.0000	0.9802	0.9544	0.9238	0.8900	0.8690	0.8249
15	1.0000	1.0000	1.0000	1.0000	1.0000	0.9703	0.9438	0.9134	0.8802	0.8583
16	1.0000	1.0000	1.0000	1.0000	1.0000	0.9908	0.9610	0.9343	0.9043	0.8715
17	1.0000	1.0000	1.0000	1.0000	1.0000	1.0000	0.9762	0.9525	0.9259	0.8963
18	1.0000	1.0000	1.0000	1.0000	1.0000	1.0000	1.0000	0.9682	0.9450	0.9186
19	1.0000	1.0000	1.0000	1.0000	1.0000	1.0000	1.0000	1.0000	0.9616	0.9385
20	1.0000	1.0000	1.0000	1.0000	1.0000	1.0000	1.0000	1.0000	1.0000	0.9561

Table 3.2 Nondimensional temperature T_{fo}, Biot number 1.0

Nondimensional Time, η	Nondimensional Storage Unit Length, λ									
	1	2	3	4	5	6	7	8	9	10
1	0.6902	0.4529	0.2745	0.1599	0.1045	0.0625	0.0353	0.0175	0.0059	0.0000
2	0.8145	0.6247	0.4606	0.3342	0.2352	0.1495	0.0620	0.0211	0.0000	0.0000
3	0.8904	0.7479	0.5865	0.4508	0.3505	0.2664	0.1986	0.1410	0.0902	0.0023
4	0.9366	0.8334	0.7025	0.5691	0.4523	0.3537	0.2772	0.2166	0.1612	0.1227
5	0.9628	0.8942	0.7926	0.6782	0.5631	0.4557	0.3658	0.2898	0.2284	0.1764
6	0.9744	0.9309	0.8609	0.7663	0.6621	0.5592	0.4609	0.3781	0.3052	0.2439
7	0.9846	0.9505	0.9059	0.8355	0.7470	0.6521	0.5574	0.4705	0.3897	0.3222
8	0.9923	0.9665	0.9300	0.8839	0.8151	0.7321	0.6450	0.5569	0.4768	0.4012
9	0.9975	0.9794	0.9507	0.9174	0.8649	0.7977	0.7202	0.6394	0.5570	0.4826
10	1.0000	0.9893	0.9678	0.9346	0.9062	0.8481	0.7830	0.7104	0.6351	0.5584
11	1.0000	0.9971	0.9816	0.9544	0.9194	0.8950	0.8332	0.7704	0.7021	0.6316
12	1.0000	1.0000	0.9941	0.9709	0.9410	0.9052	0.8801	0.8199	0.7595	0.6949
13	1.0000	1.0000	1.0000	0.9841	0.9594	0.9279	0.8920	0.8663	0.8078	0.7499
14	1.0000	1.0000	1.0000	1.0000	0.9748	0.9476	0.9153	0.8797	0.8534	0.7968
15	1.0000	1.0000	1.0000	1.0000	0.9974	0.9646	0.9359	0.9033	0.8688	0.8414
16	1.0000	1.0000	1.0000	1.0000	1.0000	0.9821	0.9539	0.9244	0.8919	0.8643
17	1.0000	1.0000	1.0000	1.0000	1.0000	1.0000	0.9693	0.9432	0.9133	0.8812
18	1.0000	1.0000	1.0000	1.0000	1.0000	1.0000	1.0000	0.9596	0.9326	0.9027
19	1.0000	1.0000	1.0000	1.0000	1.0000	1.0000	1.0000	0.9837	0.9497	0.9222
20	1.0000	1.0000	1.0000	1.0000	1.0000	1.0000	1.0000	1.0000	0.9646	0.9398
	11	12	13	14	15	16	17	18	19	20
1	0.0186	0.0133	0.0091	0.0058	0.0030	0.0008	0.0000	0.0000	0.0000	0.0000
2	0.0266	0.0188	0.0128	0.0080	0.0042	0.0011	0.0000	0.0000	0.0000	0.0000
3	0.0358	0.0227	0.0134	0.0067	0.0039	0.0013	0.0000	0.0000	0.0000	0.0000
4	0.0704	0.0495	0.0324	0.0164	0.0074	0.0008	0.0000	0.0000	0.0000	0.0000
5	0.0920	0.0638	0.0538	0.0379	0.0252	0.0149	0.0000	0.0000	0.0000	0.0000
6	0.1753	0.1082	0.0688	0.0417	0.0355	0.0268	0.0169	0.0000	0.0000	0.0000
7	0.2598	0.2015	0.1491	0.0869	0.0518	0.0271	0.0109	0.0000	0.0000	0.0000
8	0.3424	0.2813	0.2248	0.1745	0.1289	0.0738	0.0392	0.0164	0.0000	0.0000
9	0.4200	0.3572	0.2994	0.2463	0.1971	0.1530	0.1129	0.0657	0.0000	0.0000
10	0.4935	0.4292	0.3701	0.3154	0.2649	0.2179	0.1746	0.1356	0.0999	0.0000
11	0.5640	0.4978	0.4371	0.3810	0.3294	0.2812	0.2366	0.1943	0.1561	0.1212
12	0.6303	0.5657	0.5017	0.4442	0.3906	0.3417	0.2957	0.2528	0.2130	0.1746
13	0.6886	0.6275	0.5670	0.5051	0.4503	0.3992	0.3520	0.3085	0.2674	0.2289
14	0.7405	0.6829	0.6250	0.5676	0.5082	0.4558	0.4068	0.3612	0.3200	0.2804
15	0.7825	0.7327	0.6778	0.6227	0.5682	0.5123	0.4607	0.4137	0.3697	0.3295
16	0.8245	0.7736	0.7257	0.6732	0.6206	0.5686	0.5162	0.4651	0.4198	0.3774
17	0.8623	0.8145	0.7654	0.7192	0.6690	0.6187	0.5690	0.5198	0.4690	0.4253
18	0.8680	0.8516	0.8052	0.7579	0.7133	0.6651	0.6168	0.5692	0.5225	0.4725
19	0.8890	0.8588	0.8417	0.7967	0.7510	0.7079	0.6615	0.6151	0.5693	0.5244
20	0.9082	0.8796	0.8527	0.8324	0.7888	0.7446	0.7028	0.6581	0.6134	0.5694

Table 3.3 Nondimensional temperature T_{fo}, Biot number 10

Nondimensional Time, η	Nondimensional Storage Unit Length, λ									
	1	2	3	4	5	6	7	8	9	10
1	0.8368	0.6850	0.5606	0.4386	0.3300	0.2416	0.1714	0.1169	0.0753	0.0000
2	0.8811	0.7662	0.6592	0.5679	0.4872	0.4049	0.3303	0.2638	0.2056	0.1476
3	0.9039	0.8102	0.7199	0.6345	0.5543	0.4804	0.4137	0.3538	0.3072	0.2543
4	0.9206	0.8408	0.7626	0.6896	0.6167	0.5474	0.4826	0.4241	0.3704	0.3218
5	0.9341	0.8661	0.7983	0.7305	0.6644	0.6019	0.5420	0.4836	0.4291	0.3792
6	0.9452	0.8878	0.8279	0.7673	0.7076	0.6484	0.5910	0.5360	0.4838	0.4331
7	0.9549	0.9053	0.8531	0.7999	0.7450	0.6903	0.6368	0.5842	0.5334	0.4847
8	0.9623	0.9209	0.8750	0.8268	0.7784	0.7281	0.6776	0.6278	0.5795	0.5322
9	0.9688	0.9325	0.8943	0.8514	0.8064	0.7603	0.7148	0.6678	0.6212	0.5756
10	0.9744	0.9436	0.9087	0.8727	0.8324	0.7901	0.7467	0.7034	0.6599	0.6160
11	0.9792	0.9532	0.9228	0.8891	0.8536	0.8168	0.7766	0.7354	0.6939	0.6532
12	0.9818	0.9613	0.9351	0.9053	0.8725	0.8377	0.8035	0.7652	0.7259	0.6863
13	0.9853	0.9658	0.9457	0.9195	0.8902	0.8584	0.8247	0.7920	0.7553	0.7177
14	0.9883	0.9718	0.9514	0.9318	0.9058	0.8770	0.8460	0.8133	0.7809	0.7466
15	0.9903	0.9772	0.9594	0.9385	0.9193	0.8935	0.8652	0.8350	0.8032	0.7710
16	0.9905	0.9821	0.9667	0.9480	0.9267	0.9080	0.8825	0.8546	0.8251	0.7942
17	0.9919	0.9813	0.9733	0.9567	0.9374	0.9159	0.8977	0.8723	0.8450	0.8161
18	0.9932	0.9839	0.9726	0.9647	0.9473	0.9275	0.9059	0.8881	0.8630	0.8361
19	0.9944	0.9862	0.9761	0.9663	0.9563	0.9383	0.9182	0.8966	0.8792	0.8544
20	0.9954	0.9884	0.9793	0.9685	0.9611	0.9482	0.9297	0.9095	0.8883	0.8708
	11	**12**	**13**	**14**	**15**	**16**	**17**	**18**	**19**	**20**
1	0.0603	0.0440	0.0312	0.0211	0.0132	0.0070	0.0021	0.0000	0.0000	0.0000
2	0.1241	0.0946	0.0707	0.0513	0.0355	0.0227	0.0123	0.0039	0.0000	0.0000
3	0.1977	0.1575	0.1233	0.0944	0.0709	0.0511	0.0347	0.0210	0.0000	0.0000
4	0.2879	0.2359	0.1882	0.1514	0.1193	0.0915	0.0676	0.0472	0.0300	0.0000
5	0.3528	0.3128	0.2745	0.2378	0.2028	0.1538	0.1178	0.0908	0.0672	0.0468
6	0.3832	0.3436	0.3222	0.2923	0.2606	0.2300	0.2004	0.1719	0.1444	0.0953
7	0.4290	0.3857	0.3466	0.3116	0.2806	0.2717	0.2462	0.2206	0.1955	0.1712
8	0.4822	0.4386	0.3919	0.3513	0.3162	0.2847	0.2567	0.2319	0.2313	0.2099
9	0.5269	0.4844	0.4435	0.4043	0.3634	0.3226	0.2906	0.2618	0.2361	0.2134
10	0.5666	0.5256	0.4867	0.4482	0.4111	0.3755	0.3402	0.2984	0.2688	0.2421
11	0.6049	0.5643	0.5251	0.4882	0.4526	0.4175	0.3837	0.3511	0.3197	0.2788
12	0.6404	0.6011	0.5627	0.5254	0.4894	0.4561	0.4234	0.3912	0.3601	0.3300
13	0.6733	0.6354	0.5981	0.5615	0.5259	0.4915	0.4584	0.4284	0.3980	0.3683
14	0.7043	0.6677	0.6313	0.5957	0.5606	0.5265	0.4935	0.4617	0.4313	0.4043
15	0.7330	0.6979	0.6628	0.6280	0.5937	0.5601	0.5273	0.4956	0.4650	0.4355
16	0.7579	0.7260	0.6922	0.6585	0.6251	0.5922	0.5598	0.5282	0.4976	0.4680
17	0.7817	0.7503	0.7197	0.6872	0.6548	0.6226	0.5909	0.5598	0.5291	0.4996
18	0.8040	0.7742	0.7437	0.7138	0.6827	0.6514	0.6204	0.5898	0.5598	0.5302
19	0.8246	0.7964	0.7673	0.7377	0.7085	0.6787	0.6485	0.6185	0.5889	0.5599
20	0.8433	0.8169	0.7893	0.7611	0.7323	0.7036	0.6750	0.6457	0.6167	0.5881

Table 3.4 Nondimensional temperature T_{fo}, Biot number 50

Nondimensional Storage Unit Length, λ

Nondimensional Time, η	1	2	3	4	5	6	7	8	9	10
1	0.9071	0.8137	0.7248	0.6425	0.5673	0.4991	0.4375	0.3819	0.3319	0.2868
2	0.9390	0.8786	0.8648	0.8061	0.7467	0.6884	0.6323	0.5789	0.5287	0.4817
3	0.9567	0.9109	0.8639	0.8170	0.7708	0.7260	0.6829	0.6424	0.6535	0.6110
4	0.9608	0.9314	0.8936	0.8544	0.8147	0.7751	0.7361	0.6979	0.6610	0.6253
5	0.9653	0.9301	0.8947	0.8596	0.8274	0.8086	0.7741	0.7397	0.7057	0.6724
6	0.9691	0.9375	0.9055	0.8733	0.8412	0.8093	0.7778	0.7470	0.7212	0.7064
7	0.9706	0.9432	0.9141	0.8846	0.8550	0.8252	0.7956	0.7664	0.7375	0.7093
8	0.9722	0.9445	0.9169	0.8902	0.8661	0.8385	0.8108	0.7832	0.7558	0.7287
9	0.9739	0.9480	0.9221	0.8963	0.8706	0.8450	0.8201	0.7976	0.7718	0.7460
10	0.9755	0.9511	0.9268	0.9025	0.8783	0.8541	0.8301	0.8061	0.7822	0.7588
11	0.9766	0.9537	0.9309	0.9080	0.8852	0.8624	0.8396	0.8169	0.7943	0.7717
12	0.9778	0.9556	0.9335	0.9121	0.8913	0.8697	0.8482	0.8267	0.8052	0.7837
13	0.9790	0.9579	0.9368	0.9157	0.8947	0.8741	0.8556	0.8353	0.8149	0.7946
14	0.9801	0.9600	0.9399	0.9198	0.8996	0.8795	0.8595	0.8397	0.8223	0.8042
15	0.9811	0.9620	0.9428	0.9236	0.9043	0.8850	0.8657	0.8466	0.8275	0.8086
16	0.9818	0.9636	0.9455	0.9271	0.9087	0.8901	0.8716	0.8531	0.8347	0.8164
17	0.9826	0.9651	0.9475	0.9302	0.9128	0.8950	0.8772	0.8593	0.8415	0.8238
18	0.9835	0.9667	0.9499	0.9329	0.9159	0.8993	0.8824	0.8652	0.8480	0.8309
19	0.9842	0.9683	0.9522	0.9359	0.9195	0.9031	0.8866	0.8706	0.8542	0.8376
20	0.9850	0.9698	0.9543	0.9387	0.9230	0.9072	0.8912	0.8753	0.8594	0.8439

Nondimensional Time, η	11	12	13	14	15	16	17	18	19	20
1	0.2460	0.2092	0.1760	0.1458	0.1185	0.0000	0.0000	0.0000	0.0000	0.0000
2	0.4379	0.3971	0.3593	0.3242	0.2916	0.2615	0.2336	0.2078	0.1838	0.1616
3	0.5699	0.5304	0.4927	0.4569	0.4231	0.3911	0.3609	0.3325	0.3058	0.2806
4	0.5910	0.5583	0.5271	0.5049	0.5128	0.4824	0.4531	0.4250	0.3980	0.3723
5	0.6399	0.6084	0.5780	0.5487	0.5206	0.4937	0.4680	0.4435	0.4201	0.4104
6	0.6764	0.6470	0.6181	0.5900	0.5628	0.5364	0.5110	0.4865	0.4629	0.4403
7	0.6817	0.6548	0.6343	0.6214	0.5957	0.5705	0.5459	0.5221	0.4989	0.4765
8	0.7020	0.6759	0.6504	0.6256	0.6015	0.5782	0.5619	0.5503	0.5281	0.5064
9	0.7204	0.6952	0.6704	0.6461	0.6224	0.5993	0.5768	0.5551	0.5340	0.5137
10	0.7368	0.7127	0.6888	0.6652	0.6420	0.6192	0.5970	0.5754	0.5544	0.5340
11	0.7492	0.7268	0.7047	0.6828	0.6603	0.6381	0.6163	0.5949	0.5740	0.5537
12	0.7623	0.7410	0.7198	0.6987	0.6776	0.6566	0.6347	0.6136	0.5930	0.5729
13	0.7742	0.7539	0.7337	0.7135	0.6933	0.6733	0.6534	0.6336	0.6135	0.5919
14	0.7849	0.7655	0.7462	0.7269	0.7077	0.6885	0.6694	0.6504	0.6315	0.6127
15	0.7915	0.7758	0.7574	0.7390	0.7206	0.7023	0.6840	0.6657	0.6475	0.6294
16	0.7982	0.7801	0.7630	0.7489	0.7322	0.7147	0.6972	0.6797	0.6622	0.6447
17	0.8061	0.7885	0.7711	0.7539	0.7368	0.7234	0.7090	0.6922	0.6755	0.6587
18	0.8137	0.7967	0.7797	0.7628	0.7461	0.7296	0.7132	0.6996	0.6874	0.6714
19	0.8210	0.8044	0.7879	0.7715	0.7552	0.7390	0.7229	0.7070	0.6913	0.6773
20	0.8280	0.8119	0.7958	0.7798	0.7639	0.7481	0.7324	0.7168	0.7013	0.6860

Table 3.5 Nondimensional temperature T_{fo}, Biot number 1000

Nondimensional Storage Unit Length, λ

Nondimensional Time, η	1	2	3	4	5	6	7	8	9	10
1	0.9824	0.9652	0.9484	0.9320	0.9160	0.9004	0.8851	0.8702	0.8556	0.8413
2	0.9834	0.9673	0.9515	0.9360	0.9209	0.9061	0.8917	0.8775	0.8637	0.8501
3	0.9845	0.9693	0.9544	0.9399	0.9256	0.9117	0.8980	0.8846	0.8715	0.8586
4	0.9855	0.9713	0.9573	0.9437	0.9303	0.9171	0.9042	0.8915	0.8791	0.8669
5	0.9865	0.9732	0.9601	0.9473	0.9347	0.9224	0.9102	0.8983	0.8865	0.8750
6	0.9874	0.9750	0.9628	0.9508	0.9390	0.9274	0.9160	0.9048	0.8937	0.8828
7	0.9883	0.9768	0.9655	0.9543	0.9432	0.9324	0.9216	0.9111	0.9006	0.8904
8	0.9892	0.9785	0.9680	0.9576	0.9473	0.9371	0.9271	0.9172	0.9074	0.8977
9	0.9901	0.9802	0.9704	0.9608	0.9512	0.9417	0.9324	0.9231	0.9139	0.9048
10	0.9909	0.9818	0.9728	0.9639	0.9550	0.9462	0.9374	0.9288	0.9202	0.9116
11	0.9917	0.9834	0.9751	0.9668	0.9586	0.9505	0.9424	0.9343	0.9262	0.9182
12	0.9924	0.9848	0.9773	0.9697	0.9621	0.9546	0.9471	0.9396	0.9321	0.9246
13	0.9932	0.9863	0.9794	0.9725	0.9655	0.9586	0.9516	0.9447	0.9377	0.9308
14	0.9938	0.9876	0.9814	0.9751	0.9687	0.9624	0.9560	0.9496	0.9431	0.9366
15	0.9945	0.9890	0.9833	0.9776	0.9718	0.9660	0.9602	0.9542	0.9483	0.9423
16	0.9952	0.9902	0.9852	0.9800	0.9748	0.9695	0.9642	0.9587	0.9532	0.9477
17	0.9958	0.9914	0.9869	0.9823	0.9776	0.9728	0.9680	0.9630	0.9580	0.9529
18	0.9963	0.9925	0.9886	0.9845	0.9803	0.9760	0.9716	0.9671	0.9625	0.9578
19	0.9969	0.9936	0.9902	0.9866	0.9829	0.9790	0.9750	0.9710	0.9668	0.9625
20	0.9974	0.9946	0.9916	0.9885	0.9853	0.9819	0.9783	0.9747	0.9709	0.9670
	11	12	13	14	15	16	17	18	19	20
1	0.8274	0.8137	0.8004	0.7873	0.7745	0.7620	0.7497	0.7377	0.7259	0.7144
2	0.8368	0.8238	0.8111	0.7986	0.7864	0.7744	0.7626	0.7511	0.7398	0.7287
3	0.8460	0.8336	0.8215	0.8096	0.7979	0.7865	0.7752	0.7642	0.7533	0.7427
4	0.8549	0.8432	0.8316	0.8203	0.8092	0.7982	0.7875	0.7769	0.7665	0.7563
5	0.8636	0.8525	0.8415	0.8307	0.8201	0.8096	0.7994	0.7892	0.7793	0.7695
6	0.8720	0.8615	0.8511	0.8408	0.8307	0.8207	0.8109	0.8013	0.7917	0.7823
7	0.8802	0.8702	0.8603	0.8506	0.8410	0.8315	0.8221	0.8129	0.8038	0.7948
8	0.8881	0.8787	0.8693	0.8601	0.8510	0.8419	0.8330	0.8242	0.8155	0.8069
9	0.8958	0.8868	0.8780	0.8693	0.8606	0.8521	0.8436	0.8352	0.8269	0.8187
10	0.9032	0.8948	0.8864	0.8782	0.8700	0.8619	0.8538	0.8458	0.8379	0.8301
11	0.9103	0.9024	0.8946	0.8868	0.8790	0.8713	0.8637	0.8561	0.8485	0.8411
12	0.9172	0.9098	0.9024	0.8950	0.8877	0.8804	0.8732	0.8660	0.8588	0.8517
13	0.9238	0.9169	0.9099	0.9030	0.8961	0.8893	0.8824	0.8756	0.8688	0.8620
14	0.9302	0.9237	0.9172	0.9107	0.9042	0.8977	0.8913	0.8848	0.8783	0.8719
15	0.9363	0.9302	0.9242	0.9181	0.9120	0.9059	0.8998	0.8936	0.8875	0.8814
16	0.9421	0.9365	0.9309	0.9252	0.9195	0.9137	0.9080	0.9022	0.8964	0.8906
17	0.9477	0.9425	0.9373	0.9320	0.9266	0.9212	0.9158	0.9103	0.9048	0.8993
18	0.9531	0.9483	0.9434	0.9384	0.9334	0.9284	0.9233	0.9182	0.9130	0.9078
19	0.9582	0.9537	0.9492	0.9446	0.9400	0.9352	0.9305	0.9256	0.9207	0.9158
20	0.9630	0.9589	0.9548	0.9505	0.9462	0.9418	0.9373	0.9327	0.9281	0.9235

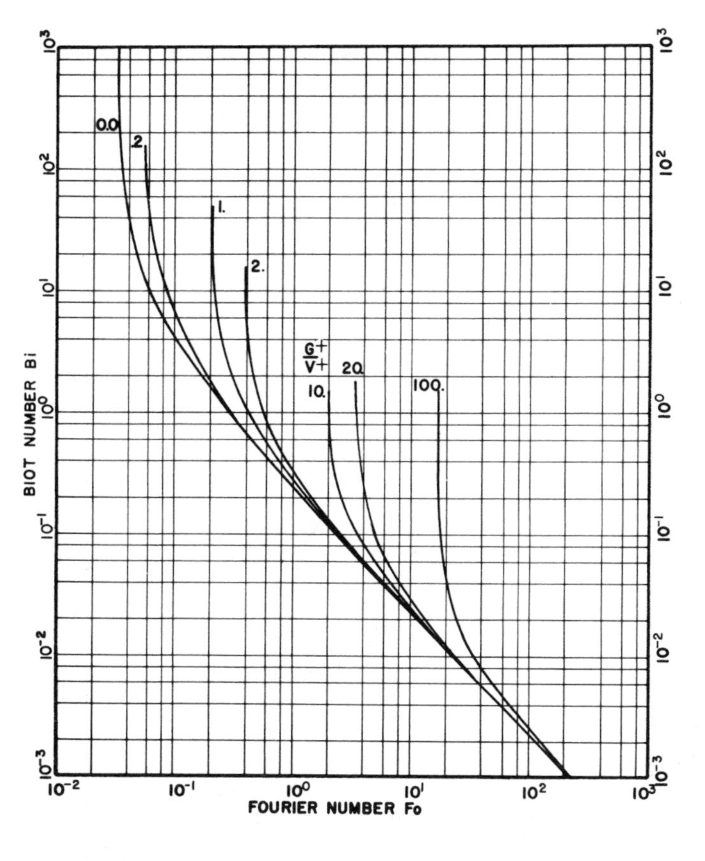

Figure 3.10 Twenty percent heat storage [1].

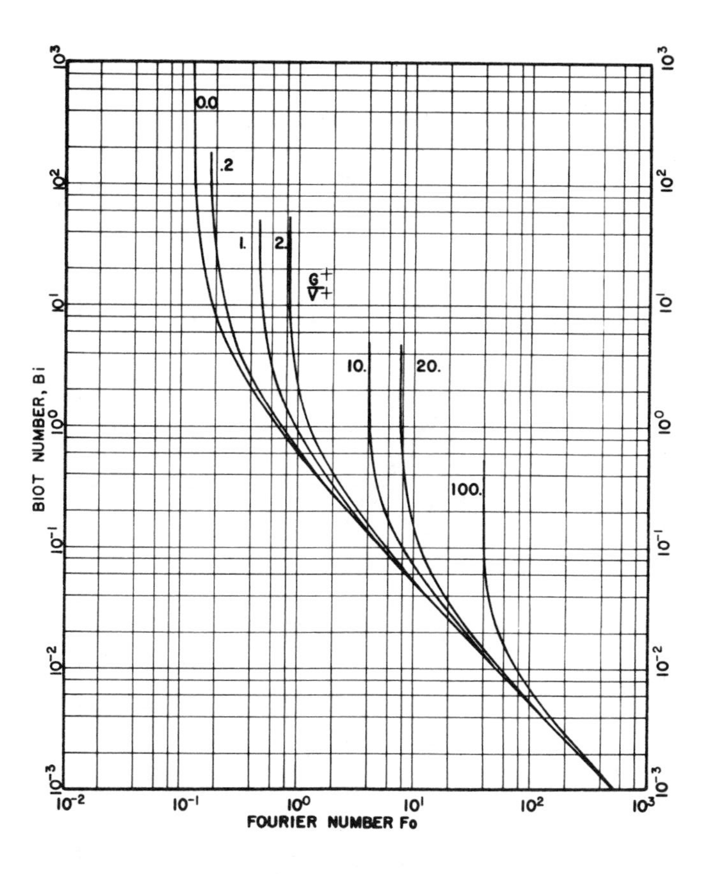

Figure 3.11 Forty percent heat storage [1].

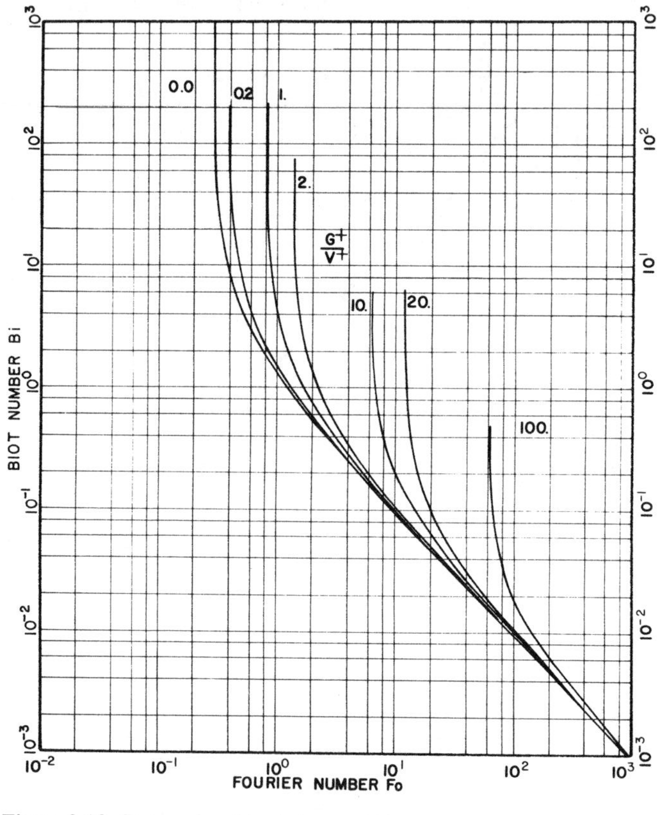

Figure 3.12 Sixty percent heat storage [1].

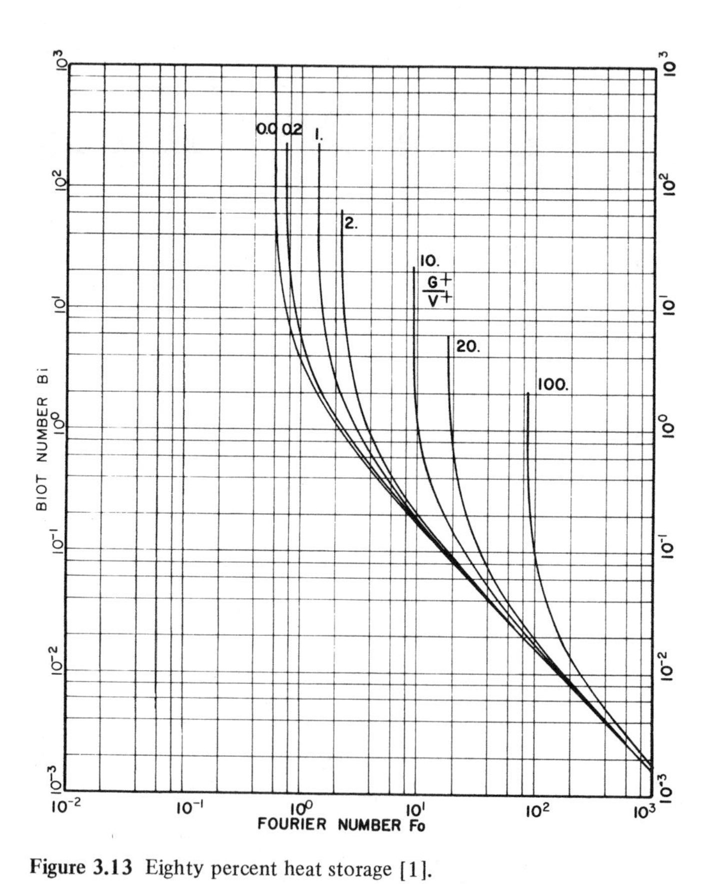

Figure 3.13 Eighty percent heat storage [1].

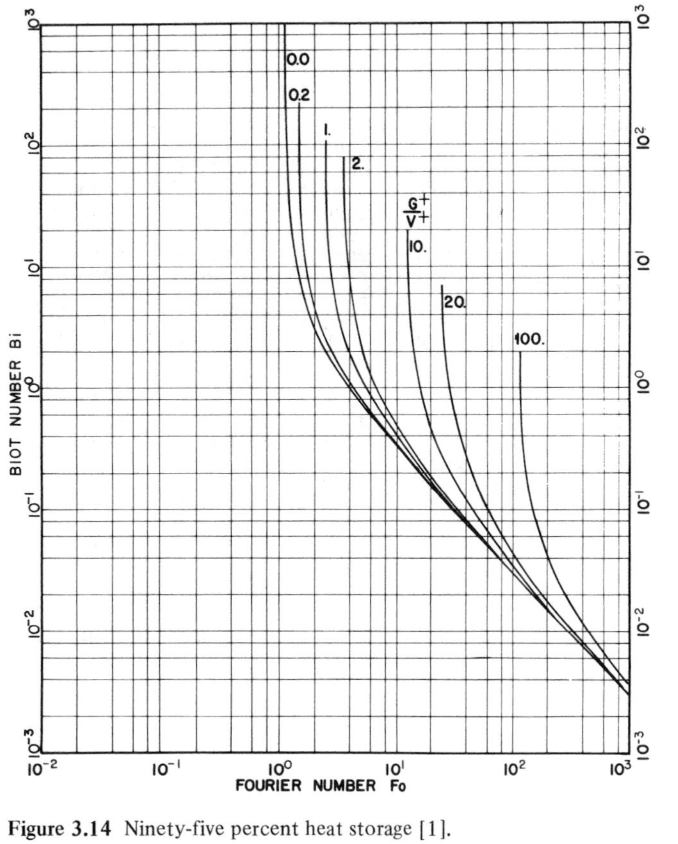

Figure 3.14 Ninety-five percent heat storage [1].

Table 3.6 Nondimensional heat storage, Biot number 0.1

Nondimensional Time, η	Nondimensional Storage Unit Length, λ									
	1	2	3	4	5	6	7	8	9	10
1	0.4688	0.3626	0.2910	0.2392	0.2000	0.1667	0.1429	0.1250	0.1111	0.1000
2	0.7235	0.6055	0.5118	0.4378	0.3781	0.3286	0.2857	0.2500	0.2222	0.2000
3	0.8591	0.7619	0.6722	0.5931	0.5248	0.4667	0.4171	0.3732	0.3333	0.3000
4	0.9299	0.8591	0.7843	0.7120	0.6446	0.5832	0.5280	0.4794	0.4366	0.3981
5	0.9674	0.9181	0.8607	0.7996	0.7391	0.6806	0.6254	0.5741	0.5268	0.4843
6	0.9893	0.9528	0.9114	0.8631	0.8114	0.7591	0.7076	0.6576	0.6100	0.5649
7	0.9986	0.9771	0.9444	0.9078	0.8661	0.8211	0.7750	0.7289	0.6834	0.6389
8	1.0000	0.9915	0.9675	0.9388	0.9062	0.8695	0.8297	0.7883	0.7465	0.7047
9	1.0000	0.9986	0.9838	0.9603	0.9351	0.9061	0.8733	0.8376	0.8000	0.7616
10	1.0000	1.0000	0.9939	0.9772	0.9552	0.9330	0.9071	0.8776	0.8452	0.8107
11	1.0000	1.0000	0.9992	0.9889	0.9720	0.9519	0.9322	0.9090	0.8824	0.8528
12	1.0000	1.0000	1.0000	0.9962	0.9844	0.9683	0.9499	0.9325	0.9117	0.8877
13	1.0000	1.0000	1.0000	1.0000	0.9929	0.9808	0.9658	0.9493	0.9337	0.9151
14	1.0000	1.0000	1.0000	1.0000	0.9982	0.9900	0.9783	0.9645	0.9495	0.9356
15	1.0000	1.0000	1.0000	1.0000	1.0000	0.9962	0.9877	0.9767	0.9641	0.9505
16	1.0000	1.0000	1.0000	1.0000	1.0000	1.0000	0.9945	0.9862	0.9761	0.9647
17	1.0000	1.0000	1.0000	1.0000	1.0000	1.0000	0.9990	0.9933	0.9855	0.9762
18	1.0000	1.0000	1.0000	1.0000	1.0000	1.0000	1.0000	0.9982	0.9926	0.9855
19	1.0000	1.0000	1.0000	1.0000	1.0000	1.0000	1.0000	1.0000	0.9978	0.9926
20	1.0000	1.0000	1.0000	1.0000	1.0000	1.0000	1.0000	1.0000	1.0000	0.9978

Table 3.7 Nondimensional heat storage, Biot number 1.0

Nondimensional Storage Unit Length, λ

Nondimensional Time, η	1	2	3	4	5	6	7	8	9	10
1	0.4110	0.3307	0.2646	0.2185	0.1844	0.1589	0.1391	0.1234	0.1107	0.1000
2	0.6535	0.5587	0.4720	0.4066	0.3546	0.3121	0.2762	0.2462	0.2212	0.2000
3	0.7979	0.7137	0.6312	0.5577	0.4952	0.4431	0.3989	0.3613	0.3287	0.3000
4	0.8831	0.8167	0.7491	0.6801	0.6157	0.5581	0.5076	0.4638	0.4254	0.3915
5	0.9333	0.8838	0.8325	0.7737	0.7143	0.6573	0.6048	0.5573	0.5148	0.4769
6	0.9629	0.9274	0.8897	0.8427	0.7916	0.7394	0.6886	0.6406	0.5964	0.5559
7	0.9832	0.9547	0.9287	0.8921	0.8503	0.8050	0.7585	0.7127	0.6689	0.6277
8	0.9945	0.9753	0.9539	0.9273	0.8938	0.8561	0.8154	0.7735	0.7318	0.6916
9	0.9993	0.9887	0.9737	0.9508	0.9260	0.8951	0.8606	0.8236	0.7855	0.7473
10	1.0000	0.9964	0.9871	0.9700	0.9483	0.9247	0.8959	0.8642	0.8302	0.7952
11	1.0000	0.9998	0.9955	0.9838	0.9665	0.9460	0.9234	0.8964	0.8670	0.8356
12	1.0000	1.0000	0.9998	0.9930	0.9804	0.9632	0.9438	0.9221	0.8967	0.8692
13	1.0000	1.0000	1.0000	0.9986	0.9903	0.9770	0.9601	0.9416	0.9208	0.8968
14	1.0000	1.0000	1.0000	1.0000	0.9968	0.9874	0.9738	0.9571	0.9396	0.9195
15	1.0000	1.0000	1.0000	1.0000	1.0000	0.9946	0.9844	0.9707	0.9543	0.9376
16	1.0000	1.0000	1.0000	1.0000	1.0000	0.9993	0.9922	0.9814	0.9676	0.9517
17	1.0000	1.0000	1.0000	1.0000	1.0000	1.0000	0.9977	0.9896	0.9784	0.9647
18	1.0000	1.0000	1.0000	1.0000	1.0000	1.0000	1.0000	0.9957	0.9869	0.9755
19	1.0000	1.0000	1.0000	1.0000	1.0000	1.0000	1.0000	0.9998	0.9935	0.9842
20	1.0000	1.0000	1.0000	1.0000	1.0000	1.0000	1.0000	1.0000	0.9982	0.9911

Nondimensional Time, η	11	12	13	14	15	16	17	18	19	20
1	0.0896	0.0825	0.0764	0.0711	0.0665	0.0625	0.0588	0.0556	0.0526	0.0500
2	0.1784	0.1645	0.1524	0.1420	0.1329	0.1249	0.1176	0.1111	0.1053	0.1000
3	0.2668	0.2461	0.2284	0.2129	0.1993	0.1873	0.1765	0.1667	0.1579	0.1500
4	0.3528	0.3266	0.3037	0.2835	0.2657	0.2499	0.2353	0.2222	0.2105	0.2000
5	0.4364	0.4049	0.3773	0.3529	0.3313	0.3120	0.2941	0.2778	0.2632	0.2500
6	0.5159	0.4814	0.4499	0.4214	0.3958	0.3732	0.3528	0.3333	0.3158	0.3000
7	0.5870	0.5516	0.5189	0.4884	0.4601	0.4341	0.4105	0.3889	0.3684	0.3500
8	0.6505	0.6148	0.5814	0.5502	0.5210	0.4937	0.4680	0.4442	0.4211	0.4000
9	0.7067	0.6716	0.6382	0.6066	0.5768	0.5488	0.5225	0.4978	0.4737	0.4500
10	0.7560	0.7221	0.6893	0.6579	0.6281	0.5998	0.5729	0.5475	0.5236	0.5000
11	0.7989	0.7668	0.7352	0.7045	0.6749	0.6466	0.6196	0.5939	0.5695	0.5463
12	0.8354	0.8058	0.7760	0.7464	0.7176	0.6896	0.6628	0.6370	0.6124	0.5889
13	0.8663	0.8394	0.8118	0.7839	0.7562	0.7290	0.7025	0.6770	0.6524	0.6287
14	0.8922	0.8681	0.8429	0.8170	0.7909	0.7648	0.7390	0.7139	0.6895	0.6660
15	0.9139	0.8924	0.8696	0.8459	0.8216	0.7970	0.7723	0.7479	0.7240	0.7007
16	0.9317	0.9130	0.8925	0.8710	0.8486	0.8257	0.8025	0.7791	0.7559	0.7331
17	0.9459	0.9301	0.9121	0.8927	0.8723	0.8511	0.8293	0.8073	0.7851	0.7630
18	0.9584	0.9440	0.9286	0.9114	0.8928	0.8734	0.8532	0.8326	0.8117	0.7906
19	0.9694	0.9559	0.9422	0.9273	0.9107	0.8930	0.8745	0.8552	0.8356	0.8156
20	0.9786	0.9668	0.9536	0.9405	0.9260	0.9101	0.8931	0.8754	0.8570	0.8382

Table 3.8 Nondimensional heat storage, Biot number 10

Nondimensional Time, η	Nondimensional Storage Unit Length, λ									
	1	2	3	4	5	6	7	8	9	10
1	0.2478	0.2219	0.1991	0.1794	0.1617	0.1460	0.1323	0.1202	0.1097	0.1000
2	0.3854	0.3546	0.3264	0.3008	0.2783	0.2576	0.2387	0.2214	0.2055	0.1909
3	0.4928	0.4598	0.4292	0.4007	0.3744	0.3500	0.3275	0.3067	0.2878	0.2703
4	0.5806	0.5467	0.5148	0.4850	0.4571	0.4308	0.4061	0.3830	0.3614	0.3414
5	0.6531	0.6197	0.5878	0.5575	0.5287	0.5014	0.4757	0.4512	0.4280	0.4061
6	0.7131	0.6810	0.6501	0.6202	0.5914	0.5639	0.5376	0.5124	0.4884	0.4655
7	0.7629	0.7327	0.7030	0.6742	0.6461	0.6189	0.5927	0.5673	0.5430	0.5196
8	0.8039	0.7760	0.7483	0.7207	0.6937	0.6673	0.6416	0.6165	0.5922	0.5687
9	0.8384	0.8125	0.7866	0.7609	0.7352	0.7099	0.6849	0.6605	0.6366	0.6132
10	0.8667	0.8434	0.8194	0.7953	0.7713	0.7473	0.7234	0.6997	0.6765	0.6536
11	0.8898	0.8691	0.8474	0.8252	0.8025	0.7800	0.7574	0.7348	0.7123	0.6901
12	0.9092	0.8904	0.8711	0.8508	0.8299	0.8087	0.7874	0.7660	0.7446	0.7232
13	0.9256	0.9086	0.8909	0.8727	0.8536	0.8340	0.8139	0.7936	0.7733	0.7529
14	0.9388	0.9241	0.9079	0.8912	0.8740	0.8560	0.8374	0.8183	0.7990	0.7797
15	0.9490	0.9369	0.9228	0.9073	0.8914	0.8751	0.8580	0.8403	0.8222	0.8037
16	0.9591	0.9470	0.9351	0.9215	0.9068	0.8916	0.8760	0.8597	0.8428	0.8255
17	0.9679	0.9563	0.9450	0.9334	0.9203	0.9062	0.8917	0.8767	0.8611	0.8450
18	0.9753	0.9650	0.9538	0.9432	0.9318	0.9192	0.9056	0.8917	0.8773	0.8623
19	0.9816	0.9725	0.9623	0.9514	0.9415	0.9304	0.9181	0.9050	0.8916	0.8778
20	0.9866	0.9788	0.9697	0.9598	0.9494	0.9398	0.9290	0.9171	0.9044	0.8915
	11	12	13	14	15	16	17	18	19	20
1	0.0880	0.0814	0.0756	0.0706	0.0662	0.0622	0.0587	0.0556	0.0526	0.0500
2	0.1706	0.1590	0.1487	0.1395	0.1313	0.1239	0.1172	0.1111	0.1053	0.1000
3	0.2469	0.2319	0.2183	0.2058	0.1945	0.1841	0.1747	0.1660	0.1579	0.1500
4	0.3160	0.2992	0.2833	0.2686	0.2549	0.2423	0.2306	0.2197	0.2097	0.2000
5	0.3776	0.3592	0.3421	0.3261	0.3113	0.2973	0.2840	0.2715	0.2598	0.2489
6	0.4352	0.4147	0.3958	0.3785	0.3624	0.3473	0.3332	0.3199	0.3074	0.2954
7	0.4892	0.4677	0.4474	0.4284	0.4106	0.3940	0.3788	0.3645	0.3510	0.3383
8	0.5385	0.5168	0.4961	0.4762	0.4574	0.4396	0.4228	0.4070	0.3923	0.3787
9	0.5835	0.5616	0.5407	0.5207	0.5016	0.4832	0.4656	0.4489	0.4330	0.4180
10	0.6247	0.6028	0.5818	0.5616	0.5423	0.5238	0.5061	0.4889	0.4724	0.4567
11	0.6624	0.6407	0.6198	0.5996	0.5802	0.5615	0.5435	0.5263	0.5097	0.4937
12	0.6966	0.6755	0.6549	0.6348	0.6154	0.5967	0.5786	0.5612	0.5445	0.5283
13	0.7278	0.7073	0.6871	0.6674	0.6482	0.6296	0.6115	0.5940	0.5771	0.5608
14	0.7561	0.7363	0.7167	0.6975	0.6787	0.6602	0.6423	0.6248	0.6079	0.5915
15	0.7817	0.7627	0.7439	0.7252	0.7068	0.6888	0.6711	0.6538	0.6369	0.6205
16	0.8047	0.7867	0.7687	0.7507	0.7329	0.7153	0.6979	0.6809	0.6642	0.6480
17	0.8257	0.8085	0.7913	0.7741	0.7569	0.7398	0.7229	0.7062	0.6898	0.6738
18	0.8445	0.8283	0.8119	0.7954	0.7789	0.7625	0.7461	0.7298	0.7138	0.6980
19	0.8613	0.8462	0.8307	0.8150	0.7992	0.7834	0.7676	0.7518	0.7362	0.7207
20	0.8764	0.8623	0.8477	0.8329	0.8178	0.8027	0.7875	0.7723	0.7571	0.7420

Table 3.9 Nondimensional heat storage, Biot number 50

Nondimensional Storage Unit Length, λ

Nondimensional Time, η	1	2	3	4	5	6	7	8	9	10
1	0.1502	0.1392	0.1294	0.1206	0.1128	0.1057	0.0994	0.0937	0.0885	0.0838
2	0.2133	0.2029	0.1949	0.1874	0.1797	0.1721	0.1648	0.1578	0.1511	0.1448
3	0.2649	0.2550	0.2454	0.2361	0.2271	0.2185	0.2104	0.2026	0.1959	0.1897
4	0.3028	0.2939	0.2854	0.2768	0.2683	0.2599	0.2517	0.2437	0.2361	0.2287
5	0.3398	0.3302	0.3207	0.3115	0.3025	0.2943	0.2865	0.2787	0.2711	0.2637
6	0.3725	0.3632	0.3539	0.3448	0.3358	0.3271	0.3185	0.3102	0.3021	0.2946
7	0.4021	0.3929	0.3839	0.3750	0.3662	0.3575	0.3490	0.3406	0.3324	0.3244
8	0.4309	0.4213	0.4119	0.4028	0.3940	0.3855	0.3770	0.3687	0.3605	0.3525
9	0.4578	0.4482	0.4387	0.4296	0.4206	0.4118	0.4032	0.3949	0.3868	0.3787
10	0.4831	0.4734	0.4639	0.4547	0.4457	0.4369	0.4282	0.4198	0.4115	0.4034
11	0.5069	0.4971	0.4876	0.4784	0.4693	0.4605	0.4518	0.4433	0.4350	0.4269
12	0.5297	0.5198	0.5102	0.5007	0.4916	0.4828	0.4741	0.4656	0.4573	0.4491
13	0.5513	0.5414	0.5318	0.5223	0.5131	0.5040	0.4952	0.4867	0.4784	0.4702
14	0.5717	0.5620	0.5523	0.5429	0.5336	0.5246	0.5157	0.5069	0.4984	0.4902
15	0.5912	0.5814	0.5719	0.5625	0.5532	0.5442	0.5353	0.5265	0.5180	0.5096
16	0.6097	0.6000	0.5905	0.5811	0.5720	0.5629	0.5540	0.5453	0.5367	0.5283
17	0.6275	0.6178	0.6083	0.5989	0.5898	0.5808	0.5720	0.5633	0.5547	0.5463
18	0.6445	0.6348	0.6254	0.6161	0.6069	0.5979	0.5891	0.5805	0.5720	0.5636
19	0.6606	0.6511	0.6417	0.6325	0.6234	0.6144	0.6056	0.5970	0.5885	0.5801
20	0.6760	0.6666	0.6573	0.6481	0.6391	0.6302	0.6215	0.6129	0.6044	0.5961
	11	12	13	14	15	16	17	18	19	20
1	0.0795	0.0756	0.0720	0.0687	0.0657	0.0625	0.0588	0.0556	0.0526	0.0500
2	0.1389	0.1333	0.1281	0.1231	0.1185	0.1141	0.1099	0.1061	0.1024	0.0989
3	0.1835	0.1776	0.1719	0.1664	0.1611	0.1560	0.1511	0.1465	0.1420	0.1378
4	0.2216	0.2147	0.2082	0.2019	0.1963	0.1910	0.1858	0.1808	0.1760	0.1713
5	0.2564	0.2494	0.2426	0.2360	0.2296	0.2234	0.2175	0.2118	0.2063	0.2010
6	0.2874	0.2803	0.2734	0.2667	0.2601	0.2537	0.2475	0.2415	0.2357	0.2301
7	0.3166	0.3090	0.3017	0.2948	0.2881	0.2816	0.2752	0.2690	0.2630	0.2571
8	0.3446	0.3369	0.3294	0.3220	0.3149	0.3079	0.3011	0.2947	0.2886	0.2825
9	0.3708	0.3631	0.3555	0.3480	0.3407	0.3336	0.3266	0.3199	0.3132	0.3068
10	0.3955	0.3877	0.3801	0.3726	0.3652	0.3580	0.3509	0.3440	0.3372	0.3306
11	0.4189	0.4111	0.4034	0.3959	0.3885	0.3812	0.3741	0.3670	0.3602	0.3534
12	0.4411	0.4332	0.4255	0.4180	0.4106	0.4033	0.3961	0.3890	0.3821	0.3752
13	0.4622	0.4543	0.4466	0.4390	0.4315	0.4242	0.4170	0.4099	0.4030	0.3961
14	0.4822	0.4743	0.4665	0.4589	0.4515	0.4441	0.4369	0.4298	0.4228	0.4160
15	0.5013	0.4934	0.4856	0.4780	0.4705	0.4632	0.4559	0.4488	0.4418	0.4349
16	0.5201	0.5119	0.5040	0.4963	0.4887	0.4814	0.4741	0.4670	0.4600	0.4531
17	0.5380	0.5299	0.5219	0.5141	0.5064	0.4988	0.4916	0.4844	0.4774	0.4705
18	0.5553	0.5472	0.5392	0.5313	0.5236	0.5160	0.5086	0.5012	0.4941	0.4872
19	0.5719	0.5638	0.5558	0.5480	0.5402	0.5326	0.5251	0.5178	0.5105	0.5034
20	0.5879	0.5798	0.5718	0.5640	0.5563	0.5487	0.5412	0.5338	0.5265	0.5194

Table 3.10 Nondimensional heat storage, Biot number 1000

Nondimensional Time, η	Nondimensional Storage Unit Length, λ									
	1	2	3	4	5	6	7	8	9	10
1	0.0182	0.0179	0.0177	0.0175	0.0173	0.0171	0.0169	0.0167	0.0165	0.0163
2	0.0352	0.0348	0.0344	0.0340	0.0336	0.0332	0.0328	0.0325	0.0321	0.0317
3	0.0513	0.0507	0.0501	0.0495	0.0489	0.0484	0.0478	0.0473	0.0468	0.0463
4	0.0663	0.0655	0.0648	0.0641	0.0633	0.0627	0.0620	0.0613	0.0607	0.0600
5	0.0803	0.0794	0.0785	0.0777	0.0768	0.0760	0.0752	0.0744	0.0737	0.0729
6	0.0934	0.0924	0.0914	0.0904	0.0895	0.0885	0.0876	0.0868	0.0859	0.0850
7	0.1055	0.1044	0.1033	0.1023	0.1012	0.1002	0.0992	0.0983	0.0973	0.0964
8	0.1167	0.1156	0.1144	0.1133	0.1122	0.1111	0.1100	0.1090	0.1080	0.1070
9	0.1271	0.1259	0.1247	0.1235	0.1223	0.1212	0.1201	0.1190	0.1179	0.1168
10	0.1366	0.1354	0.1341	0.1329	0.1317	0.1305	0.1294	0.1282	0.1271	0.1260
11	0.1454	0.1441	0.1428	0.1416	0.1403	0.1391	0.1380	0.1368	0.1357	0.1345
12	0.1533	0.1520	0.1507	0.1495	0.1483	0.1470	0.1459	0.1447	0.1435	0.1424
13	0.1605	0.1592	0.1580	0.1567	0.1555	0.1543	0.1531	0.1519	0.1508	0.1496
14	0.1670	0.1658	0.1645	0.1633	0.1621	0.1609	0.1597	0.1585	0.1574	0.1562
15	0.1728	0.1716	0.1704	0.1692	0.1680	0.1668	0.1657	0.1645	0.1634	0.1623
16	0.1780	0.1768	0.1756	0.1745	0.1733	0.1722	0.1711	0.1700	0.1689	0.1678
17	0.1826	0.1814	0.1803	0.1792	0.1781	0.1770	0.1759	0.1749	0.1738	0.1728
18	0.1865	0.1854	0.1844	0.1833	0.1823	0.1813	0.1802	0.1792	0.1782	0.1772
19	0.1899	0.1889	0.1879	0.1869	0.1860	0.1850	0.1840	0.1831	0.1821	0.1812
20	0.1928	0.1919	0.1910	0.1901	0.1892	0.1883	0.1874	0.1865	0.1856	0.1847
	11	12	13	14	15	16	17	18	19	20
1	0.0161	0.0160	0.0158	0.0156	0.0154	0.0153	0.0151	0.0150	0.0148	0.0146
2	0.0314	0.0311	0.0307	0.0304	0.0301	0.0298	0.0295	0.0292	0.0289	0.0286
3	0.0458	0.0453	0.0448	0.0444	0.0439	0.0435	0.0430	0.0426	0.0422	0.0418
4	0.0594	0.0588	0.0582	0.0576	0.0570	0.0565	0.0559	0.0554	0.0548	0.0543
5	0.0722	0.0715	0.0708	0.0701	0.0694	0.0687	0.0681	0.0674	0.0668	0.0662
6	0.0842	0.0834	0.0826	0.0818	0.0810	0.0803	0.0795	0.0788	0.0781	0.0774
7	0.0955	0.0946	0.0937	0.0928	0.0920	0.0911	0.0903	0.0895	0.0887	0.0879
8	0.1060	0.1050	0.1041	0.1031	0.1022	0.1013	0.1004	0.0996	0.0987	0.0979
9	0.1158	0.1148	0.1138	0.1128	0.1118	0.1109	0.1100	0.1090	0.1081	0.1072
10	0.1249	0.1239	0.1228	0.1218	0.1208	0.1198	0.1189	0.1179	0.1170	0.1160
11	0.1334	0.1323	0.1313	0.1302	0.1292	0.1282	0.1272	0.1262	0.1252	0.1242
12	0.1413	0.1402	0.1391	0.1380	0.1370	0.1359	0.1349	0.1339	0.1329	0.1319
13	0.1485	0.1474	0.1463	0.1452	0.1442	0.1431	0.1421	0.1411	0.1401	0.1391
14	0.1551	0.1540	0.1529	0.1519	0.1508	0.1498	0.1487	0.1477	0.1467	0.1457
15	0.1612	0.1601	0.1590	0.1580	0.1569	0.1559	0.1549	0.1539	0.1529	0.1519
16	0.1667	0.1657	0.1646	0.1636	0.1626	0.1615	0.1605	0.1595	0.1586	0.1576
17	0.1717	0.1707	0.1697	0.1687	0.1677	0.1667	0.1657	0.1648	0.1638	0.1629
18	0.1762	0.1752	0.1743	0.1733	0.1723	0.1714	0.1705	0.1695	0.1686	0.1677
19	0.1803	0.1793	0.1784	0.1775	0.1766	0.1757	0.1748	0.1739	0.1730	0.1721
20	0.1838	0.1830	0.1821	0.1812	0.1804	0.1795	0.1786	0.1778	0.1769	0.1761

The total amount of heat stored per channel is

$$Q = Q^{+}Q_{\max} = (0.811)(5.827 \times 10^4)$$
$$= 4.726 \times 10^4 \text{ kJ per channel} \quad (4.479 \times 10^4 \text{ Btu per channel})$$

The temperature of the fluid leaving is

$$t_{fo} = t_o + (t_{fi} - t_o)T_f$$
$$= 10.0 + (70.0)(0.836)$$
$$= 68.52^\circ\text{C} \quad (155.34^\circ\text{F})$$

The internal thermal resistance offered by the storage material decreases the amount of energy stored, from 5.033×10^4 kJ to 4.726×10^4 kJ. Thus, a 6% error was introduced in the calculation of the total heat stored through the use of the simplified model. The temperature of the fluid leaving the unit was 68.52°C as compared to a value of 69.76°C obtained from the simplified model. The magnitude of these differences demonstrates that, if the Biot number is much larger than 0.1, the finite conductivity model must be used to obtain an accurate prediction of the transient response of the thermal energy storage unit.

3.3 FINITE CONDUCTIVITY MODEL: HOLLOW CYLINDER

The prediction of the transient response of heat storage or removal units in which the storage material has an annular cross section is complicated by the presence of the two surfaces and the necessity of considering not only the thickness of the annulus, $r_o - r_i$, but also the ratio of the radii, r_o/r_i. Two configurations will be considered. In one, Fig. 3.15*a*, the fluid flows on the inside and the outer surface is considered to be insulated. In the second case, Fig. 3.15*b*, the fluid flows over the outer surface while the inner surface is insulated.

The results obtained from the solution of the above-noted problems can be used to predict the performance of several types of heat storage units. As noted earlier, the transient response of a storage unit composed of a group of cylindrical holes in the storage material through which the fluid flows in parallel, Fig. 3.16*a*, may be approximated by considering it to consist of a series of annuli with their outer surfaces insulated and the fluid flows in contact with the inner surfaces as shown in Fig. 3.15*a*. In a heat storage exchanger built in a shell-and-tube configuration, a possible mode of operation would be when the inner fluid is still and the outer fluid flows in a direction parallel to the axis of the annulus. Such a configuration is shown in Fig. 3.16*b*. The inner boundary is considered to be essentially adiabatic. A complementary case would be where the outer fluid is stationary and the fluid flows through the tube. If both fluids are flowing simultaneously, the techniques described in Chap. 11 must be used to determine the transient response of the unit.

The differential equations that describe the hollow cylindrical heat storage unit employ the same assumptions listed in Sec. 3.2:

Fluid: $$\frac{\dot{m}_f c_f L}{hA}\frac{\partial t_f}{\partial x} = t_m - t_f \qquad (3.12)$$

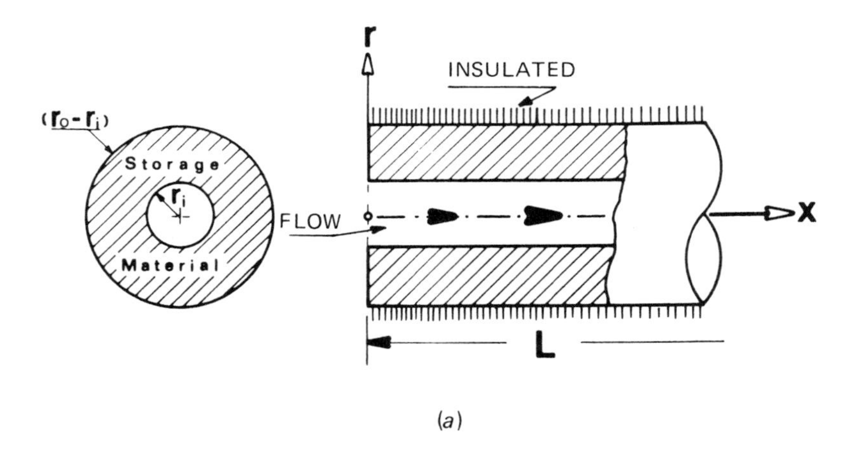

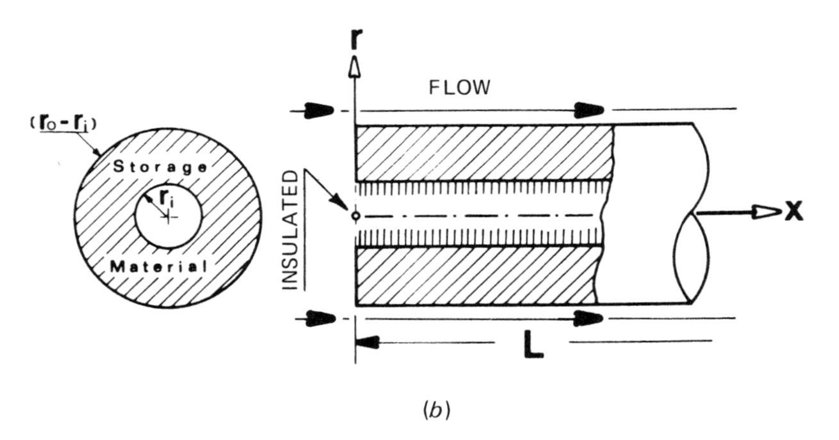

Figure 3.15 Hollow cylinder configuration, single-fluid case. (*a*) Internal flow. (*b*) External flow.

$$\text{Storage material:} \qquad \frac{1}{\alpha}\frac{\partial t_m}{\partial \tau} = \frac{\partial^2 t_m}{\partial r^2} + \frac{1}{r}\frac{\partial t_m}{\partial r} + \frac{\partial^2 t_m}{\partial x^2} \tag{3.13}$$

The initial conditions are

$$\tau = 0 \qquad t_f = t_m = t_o$$

and the boundary conditions are

$$\tau > 0 \qquad x = 0 \qquad t_f = t_{fi} \qquad \frac{\partial t_m}{\partial x} = 0 \qquad r_i < r < r_o$$

$$x = L \qquad \frac{\partial t_m}{\partial x} = 0 \qquad r_i < r < r_o$$

For a fluid in contact with the inner surface,

$$
\begin{aligned}
r &= r_i \qquad h(t_m - t_f) = k_m \frac{\partial t_m}{\partial r} \\
r &= r_o \qquad \frac{\partial t_m}{\partial r} = 0
\end{aligned}
\tag{3.14}
$$

For a fluid in contact with the outer surface,

$$
\begin{aligned}
r &= r_i \qquad \frac{\partial t_m}{\partial r} = 0 \\
r &= r_o \qquad h(t_f - t_m) = k_m \frac{\partial t_m}{\partial r}
\end{aligned}
$$

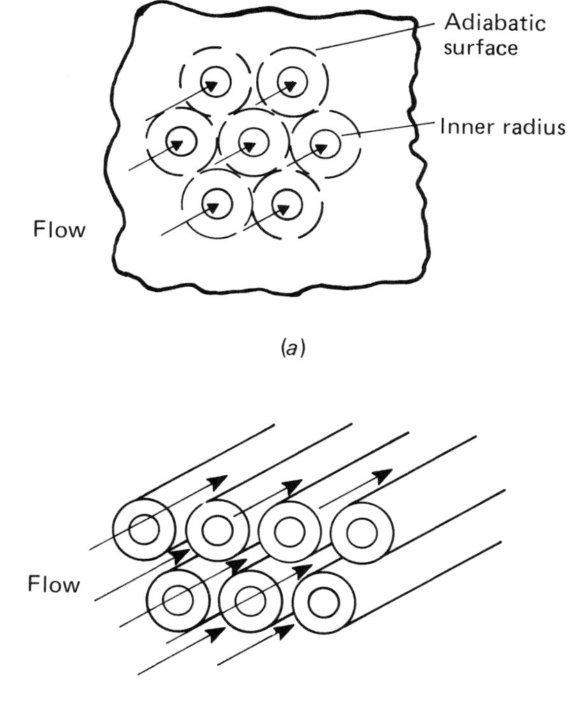

Figure 3.16 Heat storage unit configuration. (*a*) Circular holes in solid storage material. (*b*) Flow outside annuli.

These equations can be placed in nondimensional form with the following variables:

$$\mathrm{Bi} \equiv \frac{h(r_o - r_i)}{k_m} \qquad R \equiv \frac{r}{r_o - r_i}$$

$$V^+ \equiv \frac{r_o - r_i}{L} \qquad U^+ \equiv \frac{r_w}{r_a}$$

$$T_f \equiv \frac{t_f - t_o}{t_{fi} - t_o} \qquad T_m \equiv \frac{t_m - t_o}{t_{fi} - t_o} \tag{3.15}$$

$$\xi \equiv \frac{h(2\pi r_w)x}{\dot{m}_f c_f} \qquad \lambda \equiv \frac{h(2\pi r_w)L}{\dot{m}_f c_f}$$

$$\eta \equiv \frac{h(2\pi r_w)L\tau}{\pi(r_o^2 - r_i^2)L\rho_m c_m} = \frac{h\tau}{(r_o - r_i)\rho_m c_m} \frac{2U^+}{1 + U^+}$$

The radius of the surface in contact with the energy transporting fluid is denoted by r_w.

The nondimensional equations become

$$\text{Moving fluid:} \qquad \frac{\partial T_f}{\partial \xi} = T_m - T_f \tag{3.16}$$

$$\text{Storage material:} \qquad \frac{\partial T_m}{\partial \eta} = C^* \frac{1 + U^+}{2U^+} \frac{\partial^2 T_m}{\partial \xi^2} + \frac{1 + U^+}{2U^+} \times \frac{1}{\mathrm{Bi}} \left(\frac{\partial^2 T_m}{\partial R^2} + \frac{1}{R} \frac{\partial T_m}{\partial R} \right) \tag{3.17}$$

where $C^* = \frac{\lambda^2}{\mathrm{Bi}} (V^+)^2$

The initial conditions are

$$\eta = 0 \qquad T_m = T_f = 0$$

and the boundary conditions are

$$\begin{aligned}
&\xi = 0 && T_f = 1 && \frac{\partial T_m}{\partial \xi} = 0 \\
&\xi = \lambda && && \frac{\partial T_m}{\partial \xi} = 0 \\
&R = \frac{1}{|1 - U^+|} && && \frac{\partial T_m}{\partial R} = 0 \\
&R = \frac{U^+}{|1 - U^+|} && && \frac{\partial T_m}{\partial R} = \pm \mathrm{Bi}(T_f - T_m)
\end{aligned} \tag{3.18}$$

where + indicates that the outer surface is in contact with the fluid and − indicates that the inner surface is in contact with the fluid.

The solution of the mathematical model was obtained using a finite difference technique similar to that described in [1]. In order to reduce the discretization errors, the spacial derivatives were approximated by finite difference expressions using a modification of the method originally proposed by De Allen and Southwell [3]. All results presented in this paper were obtained with a value of $V^+ = 0.01$ in order to minimize the effects of axial conduction. The computer program, of course, imposes no restrictions on the value of V^+. The nondimensional fluid outlet temperature presented in Figs. 3.17 through 3.19 were obtained by Schmidt and Szego [4].

The amount of heat stored is determined directly from the temperature distribution in the storage material. The expression used is

$$Q = \rho_m c_m V(\bar{t}_m - t_o) \tag{3.19}$$

where V is the volume of the storage material and $\bar{t}_m$ is the mean temperature of the storage material. The maximum amount of heat is stored when the storage material reaches a uniform temperature equal to the inlet fluid temperature. The nondimensional heat storage Q^+ is the ratio of the actual heat stored to the maximum possible heat storage and can be expressed as

$$Q^+ = \frac{Q}{Q_{\max}} = \frac{\bar{t}_m - t_o}{t_{fi} - t_o} \tag{3.20}$$

Values of Q^+ for the hollow cylindrical heat storage unit are presented in Figs. 3.20 through 3.22. They appear in tabulated form in Tables 3.11 through 3.16.

Example 3.2 Air at 80°C (176°F) passes through a hollow concrete cylinder 10 m (32.808 ft) long with a mean velocity of 5 m/s (16.4 ft/s). The cylinder has an inside diameter of 5.0 cm (1.97 in) and an outside diameter of 25.0 cm (9.84 in). The outer surface is considered to be adiabatic and the convective film coefficient on the inside surface is 23.4 W/m^2°C (4.12 Btu/ft^{2}h °F). If the initial temperature of the concrete is 10°C (50°F), determine the temperature of the fluid leaving the storage unit and the amount of heat stored in the unit after 20 h of operation.

SOLUTION The thermal properties needed are

Air @ 50.0°C

$\rho_f = 1.106$ kg/m^3 (0.069 lb$_m$/ft^3)

$c_f = 1.007$ kJ/kg °C (0.24 Btu/lb$_m$ °F)

Concrete @ 60.0°C

$\rho_m = 2100.0$ kg/m^3 (131.7 lb$_m$/ft^3)

$c_m = 0.878$ kJ/kg °C (0.210 Btu/lb$_m$ °F)

$k_m = 1.17$ W/m °C (0.676 Btu/ft h °F)

$\alpha_m = 5.966 \times 10^{-7}$ m^2/s (0.023 ft^2/h)

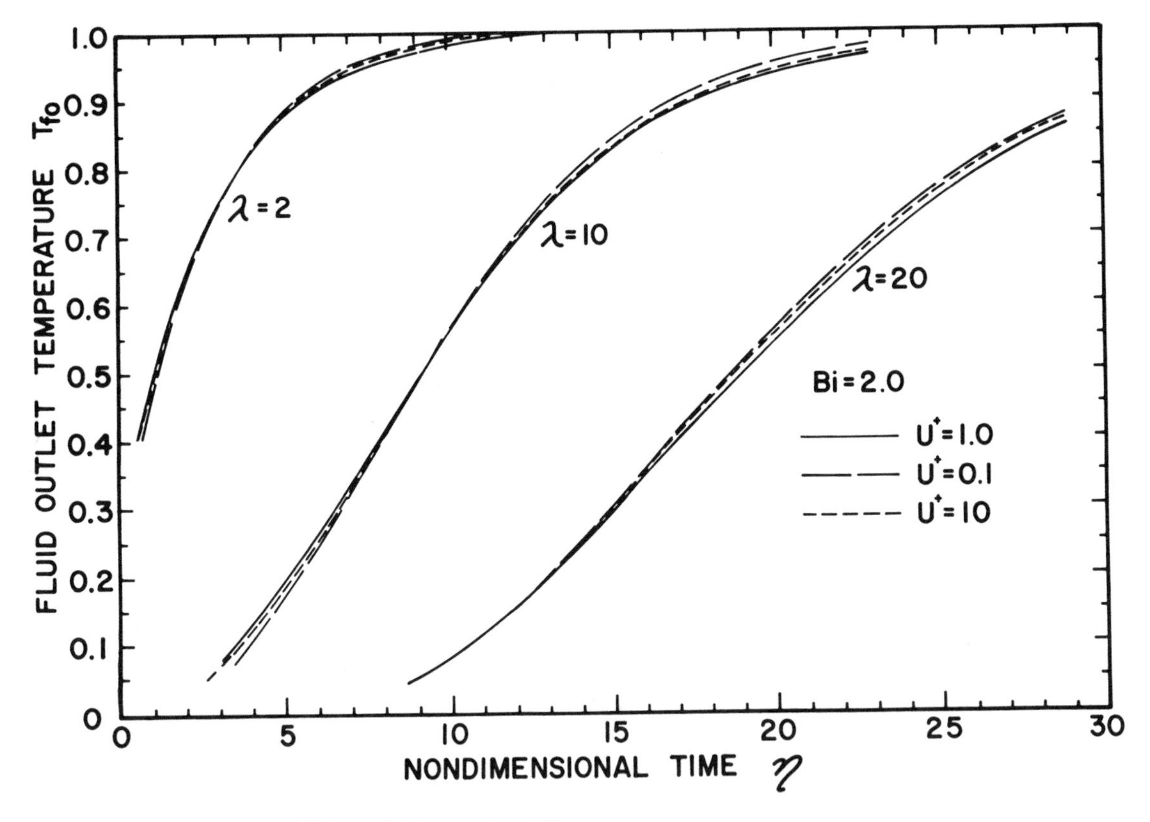

Figure 3.17 Nondimensional fluid outlet temperature [4].

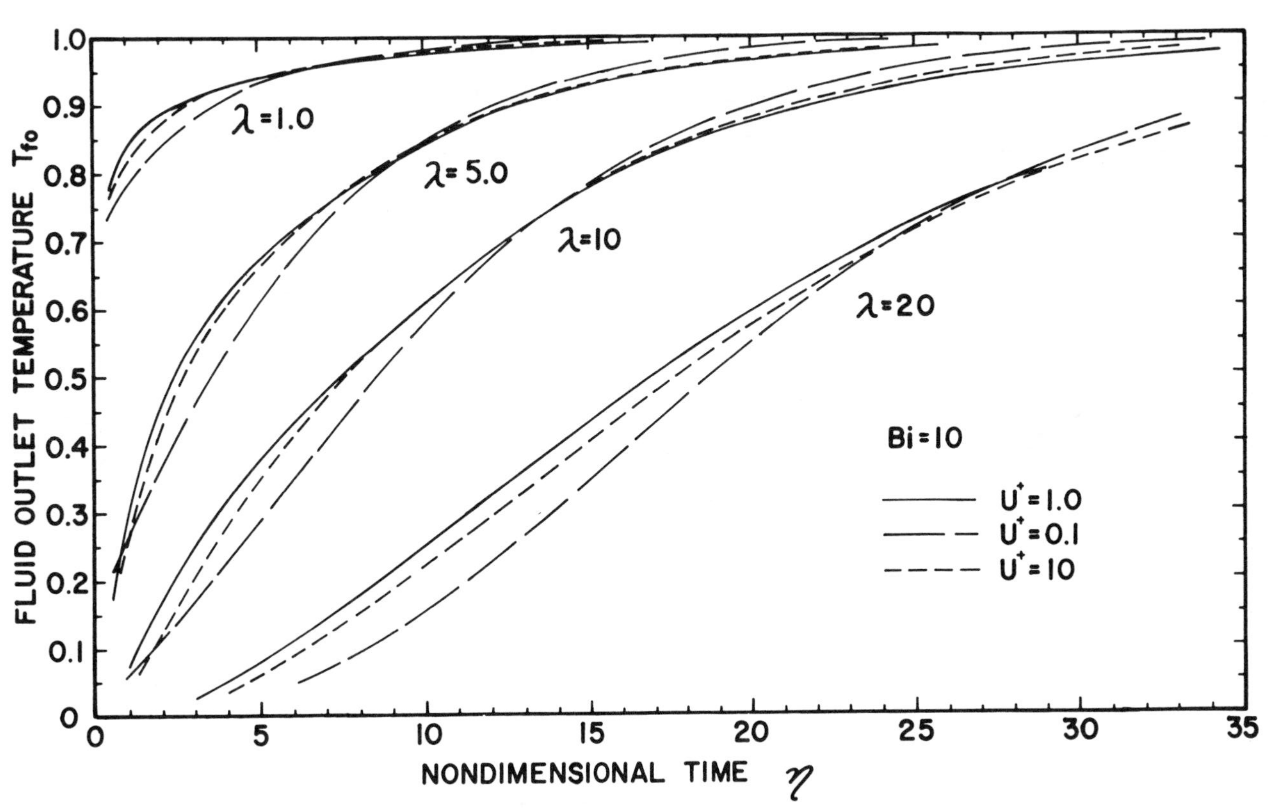

Figure 3.18 Nondimensional fluid outlet temperature [4].

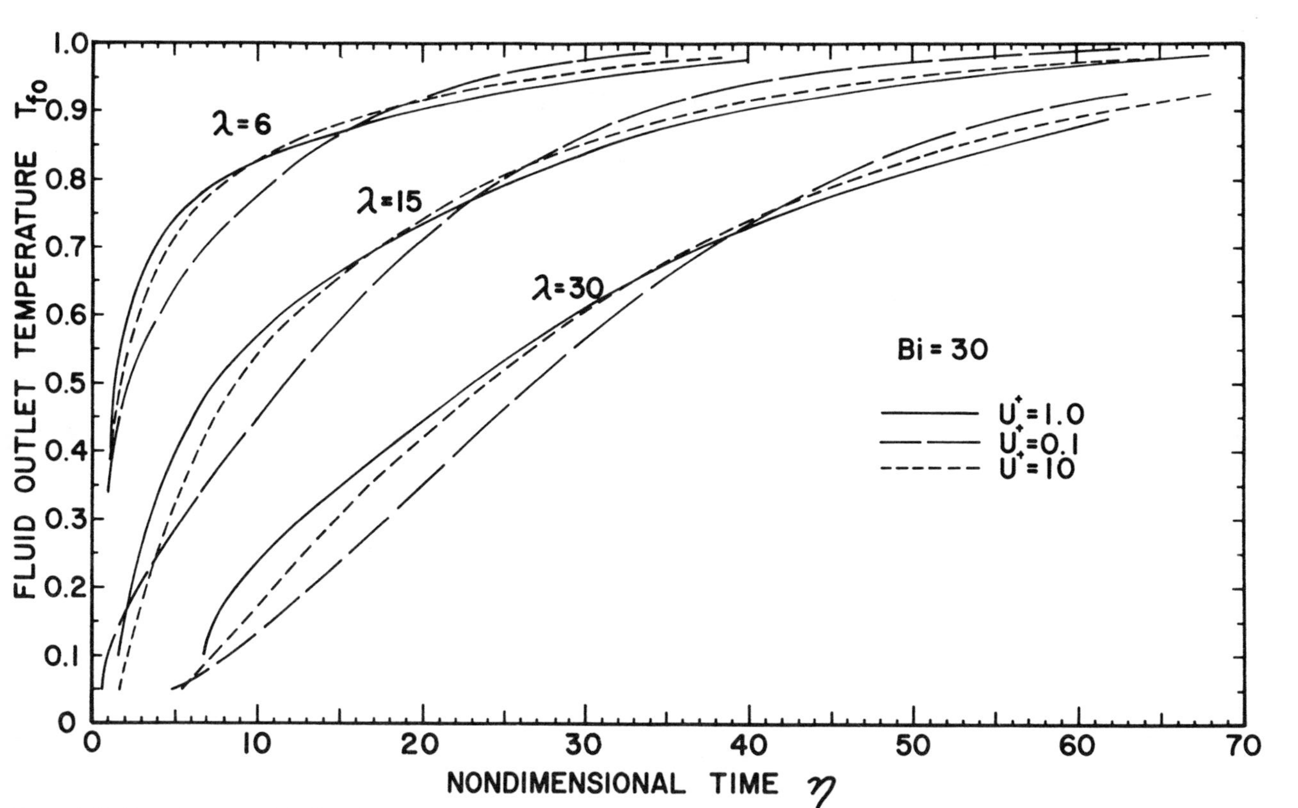

Figure 3.19 Nondimensional fluid outlet temperature [4].

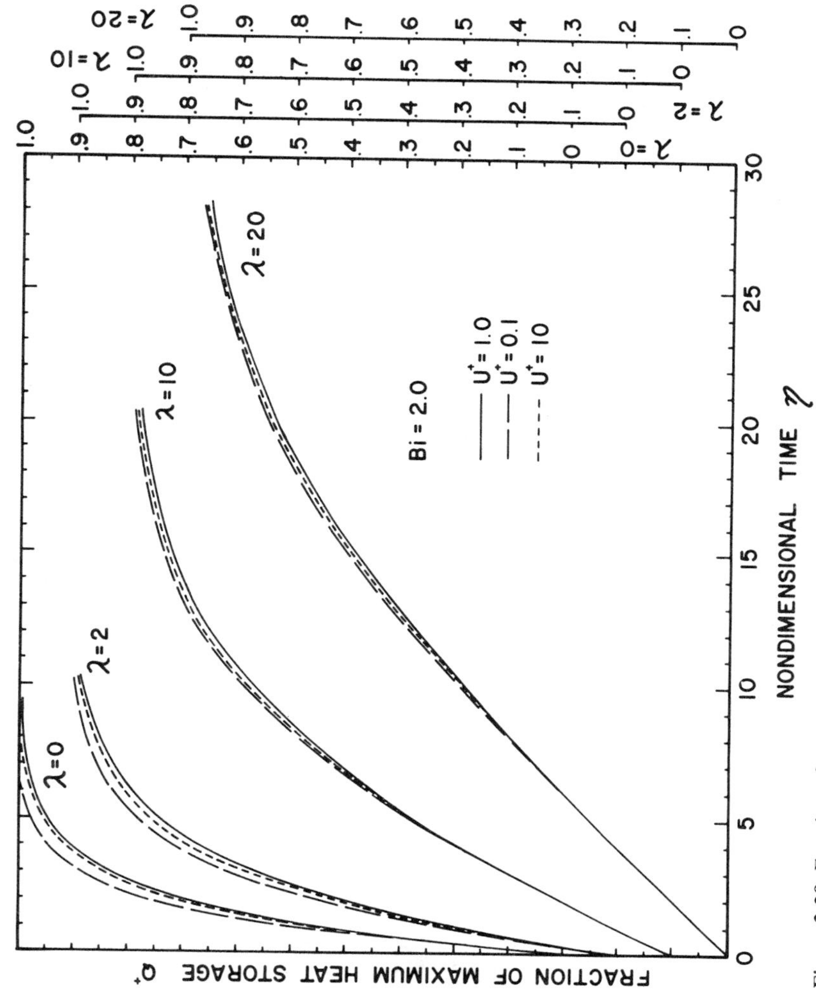

Figure 3.20 Fraction of maximum heat storage [4].

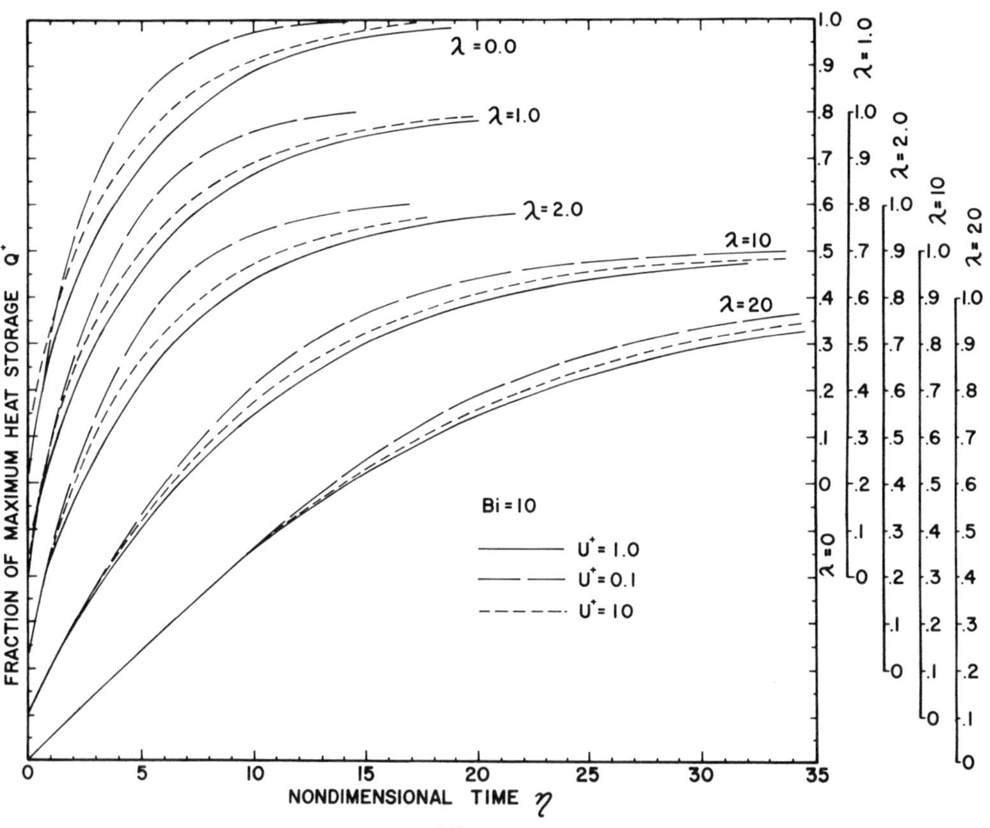

Figure 3.21 Fraction of maximum heat storage [4].

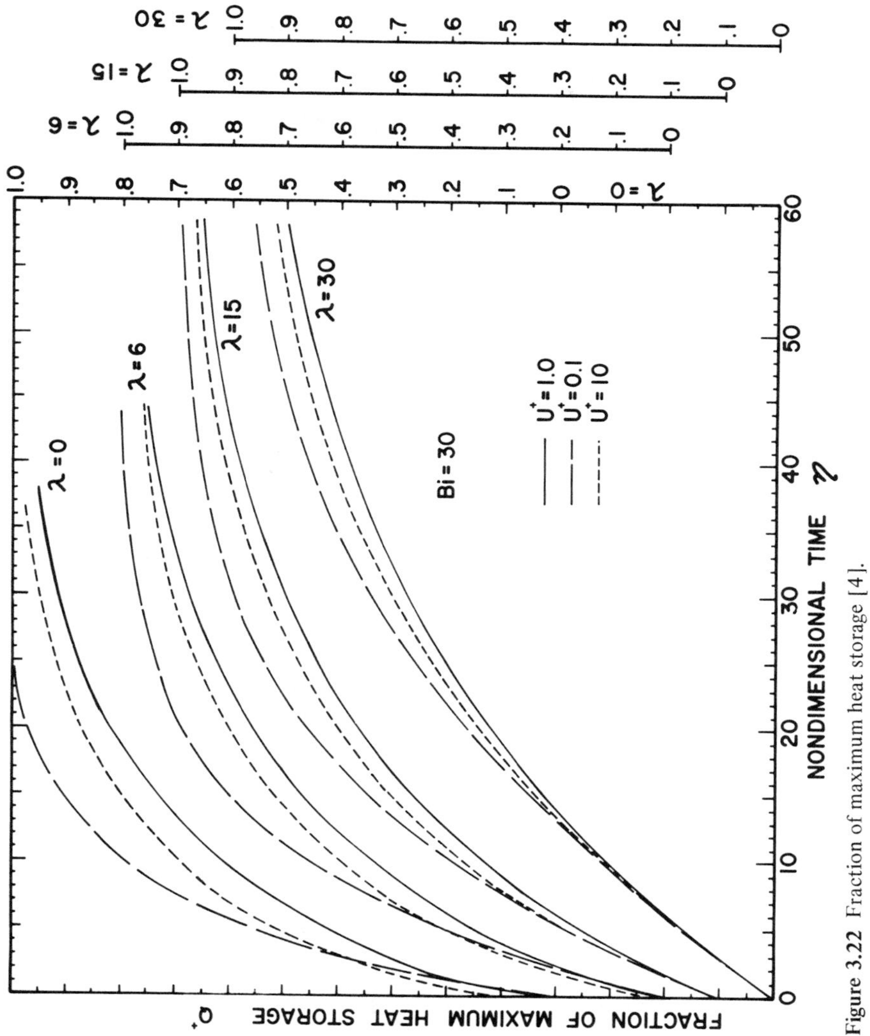

Figure 3.22 Fraction of maximum heat storage [4].

Table 3.11 Nondimensional heat storage, hollow cylinder, $U^{+} = 0.1$

Nondimensional Time, η	Nondimensional Length, λ										
	0	2	4	6	8	10	12	14	16	18	20
					*** BIOT= 2.0 ***						
3.00	0.889	0.713	0.560	0.446	0.362	0.290	0.245	0.212	0.187	0.167	0.150
6.00	0.994	0.931	0.844	0.742	0.644	0.551	0.480	0.423	0.375	0.333	0.300
9.00	1.000	0.989	0.950	0.897	0.827	0.750	0.675	0.607	0.548	0.496	0.450
12.00	1.000	1.000	0.991	0.962	0.924	0.876	0.817	0.756	0.696	0.641	0.590
15.00	1.000	1.000	1.000	0.992	0.970	0.942	0.904	0.860	0.810	0.759	0.710
18.00	1.000	1.000	1.000	1.000	0.994	0.978	0.953	0.923	0.887	0.848	0.805
21.00	1.000	1.000	1.000	1.000	1.000	0.998	0.983	0.962	0.937	0.907	0.874
24.00	1.000	1.000	1.000	1.000	1.000	1.000	0.999	0.987	0.968	0.947	0.922
27.00	1.000	1.000	1.000	1.000	1.000	1.000	1.000	1.000	0.989	0.974	0.954
30.00	1.000	1.000	1.000	1.000	1.000	1.000	1.000	1.000	1.000	0.992	0.978
					*** BIOT= 6.0 ***						
3.00	0.761	0.620	0.507	0.408	0.334	0.280	0.239	0.208	0.184	0.165	0.149
6.00	0.943	0.865	0.775	0.680	0.586	0.514	0.455	0.404	0.361	0.325	0.295
9.00	0.992	0.954	0.905	0.842	0.774	0.701	0.625	0.562	0.514	0.472	0.434
12.00	1.000	0.990	0.961	0.924	0.871	0.827	0.772	0.714	0.652	0.596	0.552
15.00	1.000	1.000	0.990	0.969	0.924	0.885	0.850	0.814	0.771	0.722	0.671
18.00	1.000	1.000	1.000	0.993	0.977	0.932	0.895	0.866	0.838	0.811	0.770
21.00	1.000	1.000	1.000	1.000	0.997	0.984	0.941	0.905	0.877	0.855	0.831
24.00	1.000	1.000	1.000	1.000	1.000	1.000	0.989	0.952	0.918	0.889	0.868
27.00	1.000	1.000	1.000	1.000	1.000	1.000	1.000	0.994	0.963	0.931	0.903
30.00	1.000	1.000	1.000	1.000	1.000	1.000	1.000	1.000	0.998	0.982	0.944
					*** BIOT= 10.0 ***						
5.00	0.833	0.736	0.646	0.566	0.498	0.428	0.375	0.332	0.297	0.268	0.243
10.00	0.975	0.938	0.890	0.834	0.777	0.711	0.647	0.588	0.540	0.499	0.462
15.00	1.000	0.991	0.970	0.944	0.911	0.867	0.826	0.782	0.734	0.686	0.640
20.00	1.000	1.000	0.997	0.985	0.968	0.937	0.901	0.869	0.843	0.817	0.787
25.00	1.000	1.000	1.000	1.000	0.994	0.990	0.967	0.936	0.897	0.867	0.847
30.00	1.000	1.000	1.000	1.000	1.000	1.000	1.000	1.000	0.973	0.939	0.896
35.00	1.000	1.000	1.000	1.000	1.000	1.000	1.000	1.000	1.000	1.000	0.982
40.00	1.000	1.000	1.000	1.000	1.000	1.000	1.000	1.000	1.000	1.000	1.000
45.00	1.000	1.000	1.000	1.000	1.000	1.000	1.000	1.000	1.000	1.000	1.000
50.00	1.000	1.000	1.000	1.000	1.000	1.000	1.000	1.000	1.000	1.000	1.000

Nondimensional Time, η	Nondimensional Length, λ										
	0	2	4	6	8	10	12	14	16	18	20
					*** BIOT= 15.0 ***						
5.00	0.750	0.661	0.581	0.513	0.455	0.407	0.366	0.332	0.290	0.262	0.239
10.00	0.935	0.889	0.836	0.781	0.727	0.676	0.628	0.583	0.525	0.486	0.450
15.00	0.989	0.967	0.942	0.910	0.874	0.834	0.794	0.754	0.704	0.659	0.616
20.00	1.000	0.995	0.983	0.965	0.945	0.921	0.893	0.863	0.826	0.793	0.757
25.00	1.000	1.000	0.998	0.991	0.979	0.964	0.947	0.928	0.894	0.869	0.845
30.00	1.000	1.000	1.000	1.000	0.996	0.988	0.977	0.963	0.939	0.911	0.885
35.00	1.000	1.000	1.000	1.000	1.000	0.999	0.994	0.986	0.985	0.962	0.939
40.00	1.000	1.000	1.000	1.000	1.000	1.000	1.000	0.998	1.000	1.000	0.993
45.00	1.000	1.000	1.000	1.000	1.000	1.000	1.000	1.000	1.000	1.000	1.000
50.00	1.000	1.000	1.000	1.000	1.000	1.000	1.000	1.000	1.000	1.000	1.000
					*** BIOT= 20.0 ***						
10.00	0.889	0.839	0.786	0.734	0.683	0.635	0.591	0.550	0.513	0.479	0.438
20.00	0.994	0.981	0.962	0.943	0.920	0.892	0.863	0.832	0.801	0.770	0.731
30.00	1.000	1.000	0.999	0.994	0.986	0.974	0.961	0.947	0.931	0.912	0.883
40.00	1.000	1.000	1.000	1.000	1.000	0.999	0.995	0.989	0.981	0.971	0.965
50.00	1.000	1.000	1.000	1.000	1.000	1.000	1.000	1.000	1.000	0.997	1.000
60.00	1.000	1.000	1.000	1.000	1.000	1.000	1.000	1.000	1.000	1.000	1.000
70.00	1.000	1.000	1.000	1.000	1.000	1.000	1.000	1.000	1.000	1.000	1.000
80.00	1.000	1.000	1.000	1.000	1.000	1.000	1.000	1.000	1.000	1.000	1.000
90.00	1.000	1.000	1.000	1.000	1.000	1.000	1.000	1.000	1.000	1.000	1.000
100.00	1.000	1.000	1.000	1.000	1.000	1.000	1.000	1.000	1.000	1.000	1.000
					*** BIOT= 30.0 ***						
10.00	0.799	0.749	0.700	0.653	0.608	0.567	0.530	0.495	0.464	0.436	0.410
20.00	0.958	0.939	0.916	0.890	0.862	0.832	0.802	0.772	0.743	0.714	0.686
30.00	0.997	0.991	0.982	0.970	0.956	0.943	0.927	0.909	0.889	0.869	0.848
40.00	1.000	1.000	1.000	0.997	0.992	0.985	0.976	0.966	0.955	0.945	0.933
50.00	1.000	1.000	1.000	1.000	1.000	1.000	0.997	0.993	0.988	0.981	0.973
60.00	1.000	1.000	1.000	1.000	1.000	1.000	1.000	1.000	1.000	0.998	0.994
70.00	1.000	1.000	1.000	1.000	1.000	1.000	1.000	1.000	1.000	1.000	1.000
80.00	1.000	1.000	1.000	1.000	1.000	1.000	1.000	1.000	1.000	1.000	1.000
90.00	1.000	1.000	1.000	1.000	1.000	1.000	1.000	1.000	1.000	1.000	1.000
100.00	1.000	1.000	1.000	1.000	1.000	1.000	1.000	1.000	1.000	1.000	1.000

Table 3.12 Nondimensional heat storage, hollow cylinder, $U^+ = 0.2$

Nondimensional Time, η	Nondimensional Length, λ										
	0	2	4	6	8	10	12	14	16	18	20
	*** BIOT= 2.0 ***										
3.00	0.853	0.684	0.545	0.438	0.360	0.291	0.248	0.214	0.187	0.167	0.150
6.00	0.975	0.902	0.811	0.715	0.627	0.542	0.476	0.421	0.375	0.333	0.300
9.00	0.996	0.968	0.920	0.862	0.795	0.722	0.655	0.594	0.540	0.493	0.450
12.00	1.000	0.990	0.967	0.930	0.887	0.836	0.782	0.727	0.675	0.625	0.580
15.00	1.000	0.996	0.987	0.967	0.937	0.902	0.861	0.819	0.776	0.731	0.687
18.00	1.000	0.999	0.993	0.985	0.968	0.943	0.912	0.879	0.844	0.807	0.771
21.00	1.000	1.000	0.997	0.991	0.983	0.970	0.947	0.920	0.892	0.862	0.830
24.00	1.000	1.000	0.999	0.996	0.989	0.982	0.970	0.950	0.927	0.902	0.875
27.00	1.000	1.000	1.000	0.998	0.994	0.988	0.980	0.970	0.952	0.932	0.909
30.00	1.000	1.000	1.000	1.000	0.997	0.993	0.987	0.979	0.970	0.954	0.936
	*** BIOT= 6.0 ***										
3.00	0.698	0.574	0.478	0.391	0.324	0.276	0.239	0.210	0.186	0.167	0.150
6.00	0.902	0.816	0.732	0.646	0.566	0.500	0.444	0.396	0.356	0.323	0.295
9.00	0.967	0.921	0.864	0.792	0.729	0.668	0.606	0.553	0.507	0.466	0.429
12.00	0.991	0.965	0.930	0.889	0.830	0.774	0.728	0.681	0.631	0.585	0.546
15.00	0.997	0.986	0.964	0.936	0.905	0.860	0.806	0.765	0.728	0.690	0.648
18.00	0.999	0.994	0.983	0.964	0.940	0.915	0.881	0.835	0.792	0.760	0.729
21.00	1.000	0.997	0.991	0.981	0.964	0.944	0.921	0.894	0.858	0.819	0.785
24.00	1.000	0.999	0.995	0.988	0.974	0.960	0.947	0.925	0.903	0.874	0.840
27.00	1.000	1.000	0.998	0.995	0.980	0.971	0.959	0.949	0.929	0.909	0.885
30.00	1.000	1.000	0.999	1.000	0.990	0.982	0.970	0.955	0.935	0.933	0.914
	*** BIOT= 10.0 ***										
5.00	0.763	0.674	0.596	0.528	0.470	0.409	0.362	0.323	0.291	0.265	0.243
10.00	0.939	0.891	0.837	0.783	0.731	0.672	0.618	0.569	0.525	0.486	0.451
15.00	0.986	0.964	0.936	0.903	0.867	0.818	0.773	0.733	0.696	0.656	0.618
20.00	0.997	0.990	0.976	0.956	0.933	0.909	0.878	0.839	0.799	0.767	0.738
25.00	1.000	0.997	0.991	0.982	0.968	0.951	0.931	0.912	0.888	0.856	0.821
30.00	1.000	0.999	0.996	0.992	0.986	0.975	0.962	0.948	0.930	0.914	0.894
35.00	1.000	1.000	0.999	0.996	0.993	0.987	0.978	0.959	0.939	0.946	0.930
40.00	1.000	1.000	1.000	0.999	0.996	0.994	0.985	0.970	0.960	0.950	0.947
45.00	1.000	1.000	1.000	1.000	0.999	0.998	0.992	0.980	0.971	0.961	0.963
50.00	1.000	1.000	1.000	1.000	1.000	1.000	0.999	0.990	0.980	0.973	0.968

Nondimensional Time, η	Nondimensional Length, λ										
	0	2	4	6	8	10	12	14	16	18	20
	*** BIOT= 15.0 ***										
5.00	0.667	0.592	0.527	0.471	0.424	0.383	0.348	0.318	0.282	0.257	0.237
10.00	0.877	0.826	0.774	0.723	0.675	0.630	0.588	0.550	0.502	0.466	0.434
15.00	0.956	0.927	0.892	0.856	0.818	0.781	0.744	0.709	0.664	0.627	0.592
20.00	0.986	0.970	0.950	0.926	0.900	0.872	0.843	0.813	0.770	0.740	0.711
25.00	0.996	0.989	0.978	0.963	0.945	0.926	0.905	0.882	0.851	0.820	0.789
30.00	0.999	0.996	0.991	0.983	0.971	0.957	0.941	0.925	0.908	0.887	0.862
35.00	1.000	0.999	0.996	0.992	0.986	0.977	0.965	0.952	0.939	0.925	0.910
40.00	1.000	1.000	0.998	0.996	0.992	0.988	0.981	0.971	0.961	0.950	0.937
45.00	1.000	1.000	1.000	0.998	0.996	0.993	0.989	0.984	0.975	0.967	0.958
50.00	1.000	1.000	1.000	0.999	0.998	0.996	0.993	0.990	0.985	0.980	0.970
	*** BIOT= 20.0 ***										
10.00	0.817	0.766	0.717	0.670	0.626	0.585	0.548	0.514	0.483	0.454	0.418
20.00	0.965	0.943	0.918	0.891	0.864	0.835	0.806	0.777	0.749	0.721	0.686
30.00	0.995	0.989	0.980	0.967	0.952	0.936	0.919	0.901	0.882	0.862	0.833
40.00	1.000	0.998	0.995	0.992	0.986	0.978	0.968	0.957	0.944	0.932	0.919
50.00	1.000	1.000	0.999	0.997	0.995	0.992	0.989	0.984	0.977	0.969	0.960
60.00	1.000	1.000	1.000	1.000	0.999	0.997	0.995	0.993	0.990	0.987	0.981
70.00	1.000	1.000	1.000	1.000	1.000	1.000	0.998	0.997	0.995	0.993	0.991
80.00	1.000	1.000	1.000	1.000	1.000	1.000	1.000	0.999	0.998	0.997	0.997
90.00	1.000	1.000	1.000	1.000	1.000	1.000	1.000	1.000	1.000	0.999	1.000
100.00	1.000	1.000	1.000	1.000	1.000	1.000	1.000	1.000	1.000	1.000	1.000
	*** BIOT= 30.0 ***										
10.00	0.709	0.664	0.622	0.583	0.547	0.514	0.484	0.457	0.431	0.408	0.387
20.00	0.902	0.876	0.850	0.822	0.795	0.767	0.739	0.712	0.686	0.661	0.637
30.00	0.972	0.958	0.943	0.926	0.907	0.888	0.869	0.849	0.829	0.809	0.789
40.00	0.994	0.988	0.981	0.972	0.961	0.949	0.937	0.923	0.909	0.894	0.880
50.00	0.999	0.997	0.994	0.991	0.986	0.979	0.971	0.962	0.953	0.942	0.932
60.00	1.000	0.999	0.998	0.996	0.994	0.992	0.988	0.983	0.978	0.971	0.963
70.00	1.000	1.000	1.000	0.999	0.998	0.996	0.994	0.992	0.990	0.986	0.982
80.00	1.000	1.000	1.000	1.000	0.999	0.999	0.998	0.996	0.994	0.993	0.991
90.00	1.000	1.000	1.000	1.000	1.000	1.000	0.999	0.998	0.997	0.996	0.994
100.00	1.000	1.000	1.000	1.000	1.000	1.000	1.000	1.000	0.999	0.998	0.997

Table 3.13 Nondimensional heat storage, hollow cylinder, $U^+ = 0.5$

Nondimensional Time, η	Nondimensional Length, λ: 0	2	4	6	8	10	12	14	16	18	20
*** BIOT= 2.0 ***											
3.00	0.835	0.669	0.534	0.431	0.354	0.285	0.243	0.211	0.186	0.166	0.150
6.00	0.984	0.906	0.810	0.712	0.622	0.531	0.465	0.411	0.367	0.330	0.300
9.00	1.000	0.986	0.939	0.871	0.798	0.720	0.649	0.585	0.529	0.482	0.441
12.00	1.000	1.000	0.994	0.956	0.906	0.846	0.785	0.726	0.669	0.616	0.569
15.00	1.000	1.000	1.000	0.999	0.967	0.928	0.880	0.829	0.778	0.729	0.681
18.00	1.000	1.000	1.000	1.000	1.000	0.978	0.943	0.904	0.860	0.816	0.773
21.00	1.000	1.000	1.000	1.000	1.000	1.000	0.985	0.954	0.920	0.884	0.845
24.00	1.000	1.000	1.000	1.000	1.000	1.000	1.000	0.990	0.962	0.933	0.901
27.00	1.000	1.000	1.000	1.000	1.000	1.000	1.000	1.000	0.994	0.969	0.943
30.00	1.000	1.000	1.000	1.000	1.000	1.000	1.000	1.000	1.000	0.998	0.975
*** BIOT= 6.0 ***											
3.00	0.644	0.540	0.455	0.378	0.316	0.269	0.232	0.204	0.181	0.163	0.147
6.00	0.866	0.786	0.707	0.624	0.549	0.486	0.435	0.392	0.353	0.319	0.290
9.00	0.953	0.905	0.849	0.784	0.717	0.652	0.593	0.540	0.495	0.457	0.424
12.00	0.990	0.963	0.928	0.883	0.830	0.776	0.722	0.669	0.619	0.574	0.534
15.00	1.000	0.993	0.971	0.943	0.907	0.861	0.815	0.770	0.724	0.680	0.637
18.00	1.000	1.000	0.995	0.974	0.953	0.924	0.884	0.844	0.804	0.765	0.726
21.00	1.000	1.000	1.000	0.997	0.978	0.961	0.937	0.901	0.866	0.831	0.796
24.00	1.000	1.000	1.000	1.000	0.999	0.982	0.967	0.946	0.915	0.883	0.852
27.00	1.000	1.000	1.000	1.000	1.000	1.000	0.985	0.972	0.953	0.926	0.897
30.00	1.000	1.000	1.000	1.000	1.000	1.000	1.000	0.989	0.977	0.959	0.934
*** BIOT= 10.0 ***											
5.00	0.699	0.626	0.561	0.503	0.452	0.397	0.354	0.316	0.285	0.259	0.237
10.00	0.903	0.857	0.807	0.756	0.706	0.649	0.598	0.552	0.510	0.475	0.444
15.00	0.973	0.950	0.922	0.888	0.852	0.809	0.765	0.722	0.681	0.641	0.604
20.00	0.998	0.989	0.975	0.955	0.932	0.905	0.873	0.838	0.803	0.768	0.734
25.00	1.000	1.000	0.999	0.989	0.975	0.957	0.940	0.916	0.887	0.858	0.829
30.00	1.000	1.000	1.000	1.000	0.999	0.987	0.973	0.960	0.945	0.923	0.898
35.00	1.000	1.000	1.000	1.000	1.000	1.000	0.998	0.984	0.974	0.962	0.947
40.00	1.000	1.000	1.000	1.000	1.000	1.000	1.000	1.000	0.993	0.984	0.974
45.00	1.000	1.000	1.000	1.000	1.000	1.000	1.000	1.000	1.000	1.000	0.992
50.00	1.000	1.000	1.000	1.000	1.000	1.000	1.000	1.000	1.000	1.000	1.000

Nondimensional Time, η	Nondimensional Length, λ										
	0	2	4	6	8	10	12	14	16	18	20
*** BIOT= 15.0 ***											
5.00	0.597	0.540	0.489	0.444	0.404	0.369	0.338	0.310	0.275	0.252	0.231
10.00	0.819	0.774	0.728	0.683	0.641	0.602	0.566	0.531	0.485	0.453	0.425
15.00	0.921	0.891	0.859	0.825	0.789	0.754	0.719	0.686	0.642	0.607	0.574
20.00	0.967	0.951	0.931	0.907	0.881	0.854	0.825	0.796	0.759	0.727	0.696
25.00	0.991	0.981	0.969	0.954	0.937	0.917	0.895	0.872	0.844	0.817	0.790
30.00	1.000	0.997	0.990	0.982	0.970	0.957	0.942	0.924	0.904	0.882	0.859
35.00	1.000	1.000	1.000	0.997	0.991	0.982	0.971	0.959	0.945	0.930	0.911
40.00	1.000	1.000	1.000	1.000	1.000	0.997	0.990	0.982	0.968	0.958	0.948
45.00	1.000	1.000	1.000	1.000	1.000	1.000	1.000	0.997	0.987	0.977	0.968
50.00	1.000	1.000	1.000	1.000	1.000	1.000	1.000	1.000	1.000	0.994	0.984
*** BIOT= 20.0 ***											
10.00	0.745	0.703	0.662	0.624	0.588	0.554	0.522	0.493	0.465	0.439	0.407
20.00	0.928	0.907	0.883	0.858	0.832	0.805	0.778	0.751	0.724	0.698	0.663
30.00	0.984	0.975	0.965	0.953	0.939	0.923	0.906	0.887	0.868	0.848	0.824
40.00	1.000	0.999	0.995	0.990	0.984	0.976	0.967	0.957	0.945	0.932	0.918
50.00	1.000	1.000	1.000	1.000	1.000	1.000	0.996	0.991	0.984	0.977	0.965
60.00	1.000	1.000	1.000	1.000	1.000	1.000	1.000	1.000	1.000	1.000	0.994
70.00	1.000	1.000	1.000	1.000	1.000	1.000	1.000	1.000	1.000	1.000	1.000
80.00	1.000	1.000	1.000	1.000	1.000	1.000	1.000	1.000	1.000	1.000	1.000
90.00	1.000	1.000	1.000	1.000	1.000	1.000	1.000	1.000	1.000	1.000	1.000
100.00	1.000	1.000	1.000	1.000	1.000	1.000	1.000	1.000	1.000	1.000	1.000
*** BIOT= 30.0 ***											
10.00	0.629	0.596	0.564	0.535	0.506	0.480	0.455	0.432	0.411	0.391	0.372
20.00	0.843	0.819	0.795	0.771	0.746	0.722	0.698	0.675	0.652	0.629	0.608
30.00	0.935	0.921	0.906	0.890	0.873	0.856	0.838	0.819	0.800	0.782	0.763
40.00	0.974	0.967	0.959	0.950	0.941	0.929	0.917	0.904	0.890	0.876	0.862
50.00	0.994	0.990	0.985	0.980	0.975	0.969	0.961	0.953	0.944	0.935	0.924
60.00	1.000	1.000	0.999	0.996	0.993	0.990	0.986	0.981	0.976	0.969	0.962
70.00	1.000	1.000	1.000	1.000	1.000	1.000	0.999	0.996	0.994	0.990	0.986
80.00	1.000	1.000	1.000	1.000	1.000	1.000	1.000	1.000	1.000	1.000	0.999
90.00	1.000	1.000	1.000	1.000	1.000	1.000	1.000	1.000	1.000	1.000	1.000
100.00	1.000	1.000	1.000	1.000	1.000	1.000	1.000	1.000	1.000	1.000	1.000

Table 3.14 Nondimensional heat storage, hollow cylinder, $U^+ = 2.0$

Nondimensional Time, η	Nondimensional Length, λ										
	0	2	4	6	8	10	12	14	16	18	20
					*** BIOT= 2.0 ***						
3.00	0.831	0.670	0.533	0.431	0.355	0.288	0.245	0.213	0.187	0.167	0.150
6.00	0.976	0.903	0.810	0.710	0.619	0.534	0.468	0.413	0.369	0.333	0.300
9.00	1.000	0.980	0.936	0.874	0.800	0.724	0.652	0.589	0.533	0.485	0.444
12.00	1.000	1.000	0.987	0.955	0.909	0.855	0.793	0.732	0.674	0.621	0.574
15.00	1.000	1.000	1.000	0.993	0.967	0.934	0.890	0.841	0.788	0.737	0.688
18.00	1.000	1.000	1.000	1.000	0.997	0.979	0.951	0.915	0.874	0.830	0.784
21.00	1.000	1.000	1.000	1.000	1.000	1.000	0.986	0.962	0.933	0.898	0.861
24.00	1.000	1.000	1.000	1.000	1.000	1.000	1.000	0.992	0.971	0.946	0.917
27.00	1.000	1.000	1.000	1.000	1.000	1.000	1.000	1.000	0.996	0.978	0.957
30.00	1.000	1.000	1.000	1.000	1.000	1.000	1.000	1.000	1.000	0.999	0.984
					*** BIOT= 6.0 ***						
3.00	0.647	0.548	0.462	0.383	0.322	0.273	0.235	0.206	0.183	0.164	0.149
6.00	0.858	0.783	0.708	0.627	0.551	0.488	0.434	0.389	0.353	0.321	0.293
9.00	0.945	0.898	0.845	0.785	0.720	0.654	0.593	0.542	0.497	0.456	0.420
12.00	0.983	0.955	0.921	0.881	0.833	0.780	0.725	0.671	0.620	0.575	0.536
15.00	0.999	0.985	0.962	0.937	0.905	0.866	0.822	0.775	0.729	0.683	0.639
18.00	1.000	0.999	0.987	0.970	0.949	0.922	0.890	0.853	0.813	0.771	0.731
21.00	1.000	1.000	0.999	0.990	0.976	0.957	0.935	0.908	0.876	0.842	0.804
24.00	1.000	1.000	1.000	1.000	0.993	0.981	0.964	0.946	0.922	0.895	0.864
27.00	1.000	1.000	1.000	1.000	1.000	0.995	0.985	0.970	0.953	0.933	0.909
30.00	1.000	1.000	1.000	1.000	1.000	1.000	0.998	0.989	0.975	0.960	0.942
					*** BIOT= 10.0 ***						
5.00	0.700	0.631	0.567	0.507	0.455	0.400	0.357	0.321	0.290	0.263	0.240
10.00	0.894	0.850	0.803	0.756	0.709	0.653	0.603	0.555	0.512	0.474	0.441
15.00	0.965	0.942	0.913	0.882	0.849	0.810	0.768	0.726	0.685	0.646	0.610
20.00	0.994	0.982	0.965	0.946	0.924	0.901	0.873	0.842	0.808	0.773	0.738
25.00	1.000	0.999	0.991	0.980	0.966	0.952	0.935	0.914	0.890	0.864	0.836
30.00	1.000	1.000	1.000	0.997	0.989	0.981	0.969	0.956	0.942	0.924	0.903
35.00	1.000	1.000	1.000	1.000	1.000	0.996	0.991	0.982	0.972	0.960	0.947
40.00	1.000	1.000	1.000	1.000	1.000	1.000	1.000	0.997	0.992	0.984	0.974
45.00	1.000	1.000	1.000	1.000	1.000	1.000	1.000	1.000	1.000	0.998	0.993
50.00	1.000	1.000	1.000	1.000	1.000	1.000	1.000	1.000	1.000	1.000	1.000

Nondimensional Time, η	Nondimensional Length, λ 0	2	4	6	8	10	12	14	16	18	20
					*** BIOT= 15.0 ***						
5.00	0.609	0.553	0.501	0.455	0.413	0.376	0.343	0.314	0.283	0.258	0.237
10.00	0.812	0.769	0.727	0.685	0.645	0.606	0.567	0.530	0.487	0.455	0.425
15.00	0.912	0.882	0.851	0.819	0.787	0.754	0.722	0.689	0.647	0.613	0.580
20.00	0.958	0.941	0.921	0.897	0.874	0.848	0.822	0.796	0.762	0.731	0.701
25.00	0.984	0.973	0.959	0.944	0.927	0.907	0.888	0.867	0.844	0.819	0.793
30.00	0.998	0.991	0.983	0.972	0.960	0.946	0.932	0.915	0.899	0.881	0.861
35.00	1.000	1.000	0.995	0.989	0.981	0.972	0.960	0.949	0.938	0.924	0.908
40.00	1.000	1.000	1.000	0.999	0.994	0.988	0.980	0.971	0.963	0.953	0.942
45.00	1.000	1.000	1.000	1.000	1.000	0.997	0.992	0.987	0.981	0.973	0.965
50.00	1.000	1.000	1.000	1.000	1.000	1.000	1.000	0.996	0.993	0.988	0.982
					*** BIOT= 20.0 ***						
10.00	0.743	0.704	0.667	0.630	0.595	0.559	0.526	0.495	0.467	0.441	0.410
20.00	0.920	0.897	0.874	0.851	0.826	0.801	0.776	0.751	0.726	0.701	0.669
30.00	0.977	0.967	0.955	0.942	0.928	0.912	0.896	0.879	0.862	0.845	0.824
40.00	0.999	0.994	0.989	0.982	0.975	0.965	0.955	0.945	0.934	0.921	0.910
50.00	1.000	1.000	1.000	0.999	0.996	0.991	0.986	0.980	0.973	0.965	0.958
60.00	1.000	1.000	1.000	1.000	1.000	1.000	1.000	0.997	0.994	0.990	0.985
70.00	1.000	1.000	1.000	1.000	1.000	1.000	1.000	1.000	1.000	1.000	0.999
80.00	1.000	1.000	1.000	1.000	1.000	1.000	1.000	1.000	1.000	1.000	1.000
90.00	1.000	1.000	1.000	1.000	1.000	1.000	1.000	1.000	1.000	1.000	1.000
100.00	1.000	1.000	1.000	1.000	1.000	1.000	1.000	1.000	1.000	1.000	1.000
					*** BIOT= 30.0 ***						
10.00	0.641	0.610	0.579	0.549	0.521	0.495	0.471	0.447	0.425	0.405	0.385
20.00	0.835	0.812	0.789	0.767	0.744	0.722	0.700	0.678	0.657	0.635	0.615
30.00	0.927	0.912	0.896	0.880	0.864	0.848	0.831	0.814	0.797	0.780	0.763
40.00	0.967	0.958	0.949	0.940	0.929	0.917	0.905	0.893	0.880	0.868	0.855
50.00	0.989	0.984	0.978	0.972	0.965	0.957	0.949	0.941	0.931	0.921	0.911
60.00	1.000	0.997	0.994	0.990	0.986	0.981	0.976	0.970	0.963	0.956	0.949
70.00	1.000	1.000	1.000	1.000	0.997	0.995	0.991	0.988	0.984	0.979	0.974
80.00	1.000	1.000	1.000	1.000	1.000	1.000	1.000	0.998	0.995	0.992	0.990
90.00	1.000	1.000	1.000	1.000	1.000	1.000	1.000	1.000	1.000	1.000	0.999
100.00	1.000	1.000	1.000	1.000	1.000	1.000	1.000	1.000	1.000	1.000	1.000

Table 3.15 Nondimensional heat storage, hollow cylinder, $U^+ = 5.0$

Nondimensional Time, η	Nondimensional Length, λ										
	0	2	4	6	8	10	12	14	16	18	20
					*** BIOT= 2.0 ***						
3.00	0.844	0.684	0.548	0.444	0.360	0.292	0.248	0.214	0.187	0.167	0.150
6.00	0.980	0.909	0.821	0.725	0.634	0.538	0.470	0.416	0.373	0.333	0.300
9.00	1.000	0.981	0.938	0.880	0.811	0.734	0.660	0.591	0.533	0.484	0.444
12.00	1.000	1.000	0.985	0.953	0.912	0.860	0.801	0.741	0.681	0.624	0.574
15.00	1.000	1.000	1.000	0.988	0.964	0.935	0.894	0.846	0.796	0.745	0.695
18.00	1.000	1.000	1.000	1.000	0.991	0.976	0.950	0.917	0.878	0.835	0.791
21.00	1.000	1.000	1.000	1.000	1.000	0.999	0.983	0.960	0.933	0.901	0.865
24.00	1.000	1.000	1.000	1.000	1.000	1.000	1.000	0.988	0.969	0.945	0.918
27.00	1.000	1.000	1.000	1.000	1.000	1.000	1.000	1.000	0.992	0.975	0.955
30.00	1.000	1.000	1.000	1.000	1.000	1.000	1.000	1.000	1.000	0.995	0.981
					*** BIOT= 6.0 ***						
3.00	0.677	0.569	0.476	0.392	0.321	0.271	0.236	0.209	0.186	0.167	0.150
6.00	0.876	0.801	0.723	0.640	0.563	0.495	0.443	0.396	0.352	0.314	0.287
9.00	0.954	0.910	0.858	0.797	0.728	0.664	0.606	0.550	0.501	0.462	0.428
12.00	0.988	0.962	0.930	0.890	0.844	0.789	0.730	0.677	0.630	0.585	0.542
15.00	1.000	0.989	0.968	0.943	0.911	0.874	0.831	0.782	0.731	0.686	0.646
18.00	1.000	1.000	0.989	0.973	0.952	0.926	0.895	0.861	0.821	0.776	0.732
21.00	1.000	1.000	1.000	0.990	0.978	0.959	0.938	0.911	0.882	0.849	0.812
24.00	1.000	1.000	1.000	1.000	0.992	0.981	0.965	0.947	0.924	0.898	0.870
27.00	1.000	1.000	1.000	1.000	1.000	0.993	0.984	0.970	0.954	0.934	0.911
30.00	1.000	1.000	1.000	1.000	1.000	1.000	0.994	0.986	0.974	0.960	0.942
					*** BIOT= 10.0 ***						
5.00	0.733	0.658	0.587	0.522	0.464	0.410	0.362	0.320	0.287	0.262	0.240
10.00	0.910	0.868	0.821	0.771	0.720	0.666	0.616	0.568	0.524	0.485	0.452
15.00	0.974	0.951	0.925	0.895	0.862	0.823	0.781	0.738	0.696	0.658	0.622
20.00	0.998	0.987	0.972	0.953	0.933	0.909	0.883	0.854	0.821	0.786	0.749
25.00	1.000	1.000	0.994	0.984	0.970	0.956	0.939	0.919	0.898	0.874	0.848
30.00	1.000	1.000	1.000	0.999	0.991	0.982	0.971	0.958	0.943	0.926	0.908
35.00	1.000	1.000	1.000	1.000	1.000	0.996	0.989	0.982	0.972	0.960	0.947
40.00	1.000	1.000	1.000	1.000	1.000	1.000	0.999	0.994	0.988	0.981	0.972
45.00	1.000	1.000	1.000	1.000	1.000	1.000	1.000	1.000	0.988	0.983	0.979
50.00	1.000	1.000	1.000	1.000	1.000	1.000	1.000	1.000	1.000	0.989	0.985

Nondimensional Time, η	Nondimensional Length, λ										
	0	2	4	6	8	10	12	14	16	18	20
*** BIOT= 15.0 ***											
5.00	0.647	0.585	0.528	0.475	0.427	0.386	0.352	0.323	0.285	0.260	0.239
10.00	0.837	0.794	0.750	0.704	0.661	0.619	0.578	0.539	0.497	0.464	0.434
15.00	0.927	0.899	0.869	0.838	0.804	0.768	0.732	0.697	0.657	0.623	0.590
20.00	0.970	0.952	0.933	0.911	0.888	0.864	0.837	0.809	0.774	0.741	0.709
25.00	0.991	0.981	0.967	0.952	0.937	0.920	0.901	0.881	0.857	0.832	0.806
30.00	1.000	0.995	0.988	0.978	0.966	0.953	0.940	0.925	0.908	0.891	0.872
35.00	1.000	1.000	0.998	0.993	0.985	0.976	0.965	0.953	0.943	0.930	0.915
40.00	1.000	1.000	1.000	1.000	0.996	0.990	0.983	0.974	0.966	0.956	0.946
45.00	1.000	1.000	1.000	1.000	1.000	0.999	0.994	0.988	0.982	0.975	0.967
50.00	1.000	1.000	1.000	1.000	1.000	1.000	1.000	0.997	0.992	0.987	0.981
*** BIOT= 20.0 ***											
10.00	0.776	0.735	0.694	0.653	0.615	0.577	0.541	0.507	0.475	0.447	0.419
20.00	0.934	0.913	0.892	0.869	0.845	0.819	0.792	0.764	0.736	0.708	0.678
30.00	0.986	0.976	0.964	0.951	0.939	0.925	0.909	0.894	0.877	0.859	0.838
40.00	1.000	0.998	0.993	0.987	0.980	0.971	0.961	0.951	0.942	0.931	0.919
50.00	1.000	1.000	1.000	1.000	0.998	0.994	0.989	0.983	0.976	0.968	0.962
60.00	1.000	1.000	1.000	1.000	1.000	1.000	1.000	0.999	0.995	0.990	0.986
70.00	1.000	1.000	1.000	1.000	1.000	1.000	1.000	1.000	1.000	1.000	0.998
80.00	1.000	1.000	1.000	1.000	1.000	1.000	1.000	1.000	1.000	1.000	1.000
90.00	1.000	1.000	1.000	1.000	1.000	1.000	1.000	1.000	1.000	1.000	1.000
100.00	1.000	1.000	1.000	1.000	1.000	1.000	1.000	1.000	1.000	1.000	1.000
*** BIOT= 30.0 ***											
10.00	0.680	0.643	0.610	0.579	0.549	0.520	0.492	0.465	0.440	0.416	0.395
20.00	0.858	0.836	0.814	0.791	0.767	0.743	0.718	0.695	0.672	0.650	0.629
30.00	0.940	0.926	0.912	0.897	0.882	0.866	0.850	0.832	0.815	0.796	0.776
40.00	0.978	0.970	0.960	0.950	0.941	0.930	0.919	0.908	0.896	0.884	0.871
50.00	0.994	0.990	0.985	0.980	0.973	0.965	0.956	0.949	0.941	0.933	0.924
60.00	1.000	1.000	0.997	0.994	0.990	0.986	0.981	0.975	0.968	0.962	0.955
70.00	1.000	1.000	1.000	1.000	1.000	0.997	0.994	0.991	0.987	0.982	0.977
80.00	1.000	1.000	1.000	1.000	1.000	1.000	1.000	1.000	0.997	0.995	0.991
90.00	1.000	1.000	1.000	1.000	1.000	1.000	1.000	1.000	1.000	1.000	1.000
100.00	1.000	1.000	1.000	1.000	1.000	1.000	1.000	1.000	1.000	1.000	1.000

Table 3.16 Nondimensional heat storage, hollow cylinder, $U^+ = 10.0$

Nondimensional Time, η	Nondimensional Length, λ										
	0	2	4	6	8	10	12	14	16	18	20
				***	BIOT= 2.0	***					
3.00	0.843	0.681	0.542	0.433	0.354	0.289	0.245	0.212	0.187	0.167	0.150
6.00	0.976	0.905	0.815	0.717	0.626	0.535	0.469	0.416	0.372	0.333	0.300
9.00	0.999	0.976	0.931	0.872	0.802	0.727	0.653	0.588	0.531	0.484	0.444
12.00	0.999	0.998	0.980	0.945	0.903	0.852	0.794	0.733	0.674	0.619	0.571
15.00	1.000	0.998	0.998	0.983	0.955	0.925	0.884	0.837	0.788	0.737	0.687
18.00	1.000	0.999	0.998	0.999	0.985	0.966	0.939	0.906	0.868	0.826	0.783
21.00	1.000	1.000	1.000	1.000	0.999	0.990	0.972	0.948	0.922	0.890	0.855
24.00	1.000	1.000	1.000	1.000	1.000	0.998	0.992	0.976	0.956	0.933	0.907
27.00	1.000	1.000	1.000	1.000	1.000	1.000	0.998	0.994	0.980	0.962	0.942
30.00	1.000	1.000	1.000	1.000	1.000	1.000	1.000	0.998	0.995	0.983	0.967
				***	BIOT= 6.0	***					
3.00	0.685	0.574	0.478	0.389	0.323	0.276	0.238	0.208	0.184	0.165	0.149
6.00	0.879	0.804	0.726	0.646	0.565	0.492	0.434	0.386	0.348	0.318	0.292
9.00	0.953	0.910	0.858	0.799	0.735	0.672	0.611	0.550	0.497	0.454	0.417
12.00	0.988	0.960	0.927	0.890	0.844	0.793	0.739	0.686	0.637	0.588	0.541
15.00	0.998	0.987	0.964	0.940	0.911	0.874	0.832	0.787	0.741	0.696	0.654
18.00	1.000	0.998	0.987	0.969	0.948	0.925	0.896	0.861	0.822	0.783	0.742
21.00	1.000	1.000	0.997	0.988	0.973	0.955	0.935	0.911	0.882	0.849	0.814
24.00	1.000	1.000	0.999	0.997	0.989	0.976	0.960	0.942	0.923	0.899	0.870
27.00	1.000	1.000	1.000	0.999	0.998	0.991	0.979	0.964	0.948	0.931	0.911
30.00	1.000	1.000	1.000	1.000	0.999	0.998	0.991	0.981	0.968	0.953	0.937
				***	BIOT= 10.0	***					
5.00	0.741	0.667	0.596	0.531	0.473	0.410	0.362	0.324	0.293	0.266	0.243
10.00	0.914	0.871	0.824	0.775	0.726	0.672	0.620	0.571	0.527	0.485	0.447
15.00	0.974	0.950	0.925	0.895	0.862	0.825	0.784	0.742	0.702	0.662	0.623
20.00	0.997	0.986	0.970	0.951	0.931	0.911	0.885	0.856	0.823	0.788	0.753
25.00	0.999	0.998	0.993	0.982	0.968	0.954	0.938	0.922	0.902	0.878	0.851
30.00	1.000	0.999	0.999	0.996	0.989	0.981	0.969	0.956	0.942	0.929	0.913
35.00	1.000	1.000	1.000	1.000	1.000	0.995	0.988	0.980	0.969	0.958	0.945
40.00	1.000	1.000	1.000	1.000	1.000	1.000	0.998	0.994	0.987	0.979	0.970
45.00	1.000	1.000	1.000	1.000	1.000	1.000	1.000	0.998	0.997	0.992	0.986
50.00	1.000	1.000	1.000	1.000	1.000	1.000	1.000	1.000	1.000	1.000	1.000

Nondimensional Time, η	Nondimensional Length, λ										
	0	2	4	6	8	10	12	14	16	18	20
*** BIOT= 15.0 ***											
5.00	0.657	0.596	0.539	0.486	0.439	0.397	0.360	0.327	0.289	0.263	0.241
10.00	0.845	0.802	0.757	0.713	0.672	0.631	0.590	0.552	0.506	0.469	0.434
15.00	0.929	0.902	0.873	0.841	0.807	0.772	0.738	0.706	0.667	0.632	0.597
20.00	0.970	0.952	0.934	0.913	0.889	0.865	0.838	0.810	0.778	0.747	0.718
25.00	0.992	0.981	0.967	0.952	0.937	0.919	0.900	0.880	0.859	0.835	0.808
30.00	0.999	0.995	0.988	0.977	0.965	0.952	0.939	0.924	0.911	0.894	0.876
35.00	1.000	1.000	0.997	0.992	0.984	0.975	0.964	0.953	0.943	0.932	0.919
40.00	1.000	1.000	1.000	0.998	0.995	0.989	0.982	0.973	0.965	0.955	0.945
45.00	1.000	1.000	1.000	1.000	0.999	0.997	0.993	0.987	0.982	0.974	0.966
50.00	1.000	1.000	1.000	1.000	1.000	1.000	1.000	0.995	0.992	0.987	0.981
*** BIOT= 20.0 ***											
10.00	0.785	0.744	0.704	0.665	0.627	0.590	0.555	0.522	0.491	0.462	0.425
20.00	0.936	0.917	0.895	0.873	0.848	0.823	0.796	0.769	0.743	0.718	0.689
30.00	0.987	0.976	0.965	0.952	0.940	0.926	0.910	0.894	0.877	0.859	0.841
40.00	1.000	0.998	0.994	0.988	0.980	0.972	0.962	0.952	0.942	0.931	0.922
50.00	1.000	0.999	0.999	0.999	0.997	0.994	0.989	0.983	0.976	0.969	0.962
60.00	1.000	1.000	1.000	1.000	0.999	0.999	0.998	0.997	0.994	0.990	0.986
70.00	1.000	1.000	1.000	1.000	1.000	1.000	1.000	1.000	0.998	0.998	0.998
80.00	1.000	1.000	1.000	1.000	1.000	1.000	1.000	1.000	1.000	1.000	1.000
90.00	1.000	1.000	1.000	1.000	1.000	1.000	1.000	1.000	1.000	1.000	1.000
100.00	1.000	1.000	1.000	1.000	1.000	1.000	1.000	1.000	1.000	1.000	1.000
*** BIOT= 30.0 ***											
10.00	0.690	0.656	0.623	0.591	0.562	0.533	0.505	0.479	0.453	0.429	0.407
20.00	0.866	0.845	0.822	0.798	0.775	0.751	0.728	0.705	0.683	0.661	0.639
30.00	0.942	0.929	0.916	0.901	0.886	0.870	0.853	0.836	0.818	0.800	0.781
40.00	0.978	0.970	0.961	0.951	0.942	0.932	0.921	0.910	0.898	0.885	0.872
50.00	0.996	0.992	0.987	0.980	0.974	0.966	0.959	0.950	0.943	0.934	0.925
60.00	1.000	0.999	0.998	0.995	0.991	0.987	0.982	0.977	0.970	0.964	0.957
70.00	1.000	1.000	1.000	1.000	0.998	0.997	0.995	0.991	0.988	0.984	0.979
80.00	1.000	1.000	1.000	1.000	1.000	1.000	1.000	1.000	1.000	0.994	0.992
90.00	1.000	1.000	1.000	1.000	1.000	1.000	1.000	1.000	1.000	1.000	1.000
100.00	1.000	1.000	1.000	1.000	1.000	1.000	1.000	1.000	1.000	1.000	1.000

The mass rate of flow of the air is

$$\dot{m} = \rho_f v S = (1.106)(5.0)(\pi)(0.025)^2$$
$$= 1.086 \times 10^{-2}\ \text{kg/s} \quad (2.39 \times 10^{-2}\ \text{lb}_\text{m}/\text{s})$$

The nondimensional parameters are

$$\text{Bi} = \frac{h(r_o - r_i)}{k_m} = \frac{(23.4)(0.125 - 0.025)}{1.17} = 2.0$$

$$U^+ = \frac{r_w}{r_a} = \frac{0.025}{0.125} = 0.2$$

$$\lambda = \frac{(h)(2\pi r_w)(L)}{\dot{m} c_f} = \frac{(23.4)(2\pi)(0.025)(10.0)}{(1.086 \times 10^{-2})(1007.0)} = 3.36$$

$$\eta = \frac{h\tau}{(r_o - r_i)\rho_m c_m} \frac{2U^+}{1.0 - U^+} = \frac{(23.4)(3600.0)(20.0)}{(0.125 - 0.025)(2100.0)(878.0)} \frac{2.0(0.2)}{1.0 + 0.2}$$
$$= 3.046$$

The maximum amount of heat that can be stored is

$$Q_\text{max} = V\rho_m c_m (t_{fi} - t_o)$$
$$= \pi[(0.125)^2 - (0.025)^2](10.0)(2100.0)(0.878)(80.0 - 10.0)$$
$$= 6.08 \times 10^4\ \text{kJ} \quad (5.76 \times 10^4\ \text{Btu})$$

The nondimensional heat storage can be most conveniently obtained from the tabulated results and was found to be $Q^+ = 0.592$. The total amount of heat stored is

$$Q = Q^+ Q_\text{max}$$
$$= (0.592)(6.08 \times 10^4\ \text{kJ})$$
$$= 3.600 \times 10^3\ \text{kJ} \quad (3.412 \times 10^4\ \text{Btu})$$

3.4 COMPARISON OF RESULTS FOR FINITE CONDUCTIVITY MODELS OF HOLLOW CYLINDRICAL AND SLAB CONFIGURATIONS

It has previously been noted that if the finite thermal conductivity model is used, the transient response of the storage unit will depend on the unit's geometric configuration. In Figs. 3.17 and 3.20 the transient response of three geometrically different heat storage units were presented for a Biot number equal to 2.0. One unit is the flat slab, $U^+ = 1$, and the other two units have an annular cross section where the outer radius is 10 times larger than the inner radius. When $U^+ = 0.1$ the fluid is in contact with the inner surface, whereas for $U^+ = 10$ the fluid is in contact with the outer surface of the hollow cylinder. The transient response, as indicated by the nondimensional temperature of the fluid leaving and the fraction of the maximum heat storage, appears to be nearly independent of the geometry of the storage unit. As the value of λ decreases, usually a result of an increase in the heat capacity of the energy transporting fluid $\dot{m}_f c_f$ and/or decrease in the length of the unit, the differences in the nondimensional heat storage become larger. However, the differences still remain relatively small throughout the physically practical range for heat storage units. It has therefore been recommended by Schmidt and Szego [4] that if $0.1 < \text{Bi} \leqslant 2$, the effects of geometry may be neglected and the results for the flat slab presented in Sec. 3.2 for the finite conductivity model be used to predict the transient response of hollow cylindrical cross-section solid sensible heat storage units.

If the Biot number is greater than 2, a condition usually encountered when the energy transporting fluid is a liquid, curvature effects must be considered. The transient response of the different storage units for Biot numbers of 10.0 and 30.0 are presented in Figs. 3.18, 3.19, 3.21, and 3.22. The curvature effect can be detected for all values of λ; thus it is recommended [4] that the response of hollow cylindrical storage units operating under these conditions be determined by using these curves or the results tabulated in Tables 3.11 through 3.16.

3.5 ANALYSIS OF THE EFFECTS OF FINITE THERMAL CONDUCTIVITY

Three different models have been presented for the analysis of the transient response of thermal storage units: infinite heat capacity (Sec. 2.2), simplified (Sec. 2.3) and finite conductivity (Secs. 3.2 and 3.3). An increase in the sophistication of the model correspondingly results in a more complex solution. It is thus advantageous to define the regions of the independent variables where the various models can be used without the introduction of appreciable errors. An analysis of the effects of finite thermal conductivity in the single-blow heat storage unit was presented by Szego and Schmidt [5].

Two different groups of nondimensional variables have been used in the equations that govern the transient response of the storage unit. One is associated with the finite conductivity model, whereas the other is associated with the simplified model. These groups are listed and their interrelationships presented in Table 3.17 for the flat slab unit. The complete mathematical models and their solutions are presented in Table 3.18.

The finite conductivity model involves less restrictive assumptions, so it can be used to obtain the transient response of a heat storage unit throughout the entire range of variables that describe a practical unit. However, the closed-form solutions are much more convenient to use, and it is thus desirable to identify the regions where they can be used without the introduction of significant errors. The results obtained from the finite conductivity model have been used to define the boundaries of the applicable solution regions shown in Fig. 3.23.

Table 3.17 Nondimensional groups, flat slab unit, simplified and finite conductivity models [5]

Common

		Fluid	Storage material
	Temperature	$T_f = \dfrac{t_f - t_o}{t_{fi} - t_o}$	$T_m = \dfrac{t_m - t_o}{t_{fi} - t_o}$
	Heat storage	$Q^+ = \dfrac{Q}{Q_{max}} = \dfrac{\bar{t}_m - t_o}{t_{fi} - t_o}$	

Specific

Name	Simplified model	Finite conductivity model	Interrelationships
Time	$\eta = \dfrac{hA\tau}{V\rho_m c_m}$	$\text{Fo} = \dfrac{\alpha_m \tau}{w^2}$	$\eta = \text{Bi Fo}$
Coordinate			
Axial	$\xi = \dfrac{hAx}{\dot{m}_f c_f L}$	$X = \dfrac{x}{L}$	$\xi = \text{Bi} \dfrac{G^+}{V^+} \dfrac{x}{L}$
Transverse	$Y = \dfrac{y}{w}$	$Y = \dfrac{y}{w}$	–
Length of storage unit	$\lambda = \dfrac{hA}{\dot{m}_f c_f}$	$\dfrac{1}{V^+} = \dfrac{L}{w}$	$\lambda = \text{Bi} \dfrac{G^+}{V^+}$
Surface conductance	–	$\text{Bi} = \dfrac{hw}{k_m}$	–
Fluid heat capacity	–	$\dfrac{1}{G^+} = \dfrac{\dot{m}_f c_f}{P k_m}$	–

The infinite fluid heat capacity model (Sec. 2.2) is recommended for $\lambda < 0.1$. If, in addition, $\text{Bi} < 0.1$, the temperature gradients within the storage material become negligible and a lumped parameter model (Sec. 2.2.1) can be used. A comparison of the values of the nondimensional heat storage obtained from these methods at $\lambda = 0.1$ and $\text{Bi} = 0.1$, the worst possible situation in this region, is presented in Table 3.19. Close agreement is present and was the justification for establishing the boundaries shown in Fig. 3.23.

The boundary between the simplified and the finite conductivity model was drawn at $\text{Bi} = 0.1$. The values of the nondimensional heat storage and fluid outlet temperature obtained from these two models and the simplified model with the

Table 3.18 Available solutions, flat slab unit [5]

Mathematical Model:			Infinite Fluid Heat Capacity	
Conservation Equations	Simplified	Finite Conductivity	$Bi \le .1$	$Bi > .1$
Fluid	$\frac{\partial T_f}{\partial \xi} = T_m - T_f$	$\frac{dT_f}{dX} + \left(\frac{G^+}{V^+}\right)(Bi)(T_f - T_w) = 0$		
Storage Material	$\frac{\partial T_m}{\partial \eta} = T_f - T_m$	$V^{+2}\frac{\partial^2 T_m}{\partial X^2} + \frac{\partial^2 T_m}{\partial Y^2} = \frac{\partial T_m}{\partial Fo}$	$\frac{dT_m}{d\eta} = (1. - T_m)$	$\frac{\partial T_m}{\partial \eta} = \frac{1}{Bi}\frac{\partial^2 T_m}{\partial Y^2}$
Initial and Boundary Conditions	$\xi = 0 \quad T_f = 1 \quad T_m = 1 - e^{-\eta}$ $\xi = 0 \quad T_m = 0$	$Fo = 0 \quad T_m = T_f = T_o = 0$ $X = 0 \quad 0 < Y < 1 \quad T_f = 1 \quad \frac{\partial T_m}{\partial X} = 0$ $X = 1 \quad 0 < Y < 1 \quad \frac{\partial T_m}{\partial X} = 0$ $0 \le X \le 1 \quad Y = 0 \quad \frac{\partial T_m}{\partial Y} = 0$ $0 \le X \le 1 \quad Y = 1 \quad \frac{\partial T_m}{\partial Y} = Bi(T_f - T_m)$	$\eta = 0 \quad T_m = 0$	$\eta = 0 \quad T_m = 0$ $Y = 0 \quad \frac{\partial T_m}{\partial Y} = 0$ $Y = 1 \quad \frac{\partial T_m}{\partial Y} = Bi(T_{fi} - T_m)$
Solution:				
Fluid Temperature	$\eta < 2. \quad \xi < 2.$ $T_f[\eta, \xi] = 1. - e^{-(\eta+\xi)} \sum_{N=1}^{N=\infty} \left[\frac{\xi^N}{N!} \sum_{k=0}^{k=N-1} \frac{\eta^k}{k!} \right]$ $2. \le \eta < 4. \quad 2. \le \xi < 4.$ $T_f[\eta, \xi] = 1. - \frac{1}{2}\left[1 + \mathrm{erf}(\sqrt{\xi} - \sqrt{\eta})\right] - \frac{\xi^{1/4}}{\eta^{1/4} + \xi^{1/4}} e^{-(\eta+\xi)} I_o(2\sqrt{\eta\xi})$ $\eta \ge 4. \quad \xi \ge 4.$ $T_f[\eta, \xi] = 1. - \frac{1}{2}\left[1 + \mathrm{erf}(\sqrt{\xi} - \sqrt{\eta} - \frac{1}{8\sqrt{\xi}} - \frac{1}{8\sqrt{\eta}})\right]$	Fluid Outlet Temperature Only Schmidt - Szego [1]	$T_f = T_{fi} = 1$	$T_f = T_{fi} = 1$
Storage Material Temperature	$T_m[\eta, \xi] = 1. - T_f[\xi, \eta]$		$T_m = 1. - e^{-\eta}$	$T_m = 1. - \sum_{j=1}^{\infty} 2 \frac{\sin M_j}{\sin M_j \cos M_j + M_j} \exp\left[-(M_j^2)\frac{\eta}{Bi}\right] \cos(M_j Y)$ where $M_j \tan M_j = Bi$
Heat Storage	$Q^+ = \frac{1}{\lambda}\int_o^{\eta} [1 - T_{fo}]\, d\eta$	Schmidt – Szego [1]	$Q^+ = 1. - e^{-\eta}$	$Q^+ = \sum_{j=1}^{\infty} 2\left[\frac{\sin^2 M_j}{M_j \sin M_j \cos M_j + (M_j^2)}\right]\left[1. - \exp\left[-(M_j^2)\frac{\eta}{Bi}\right]\right]$

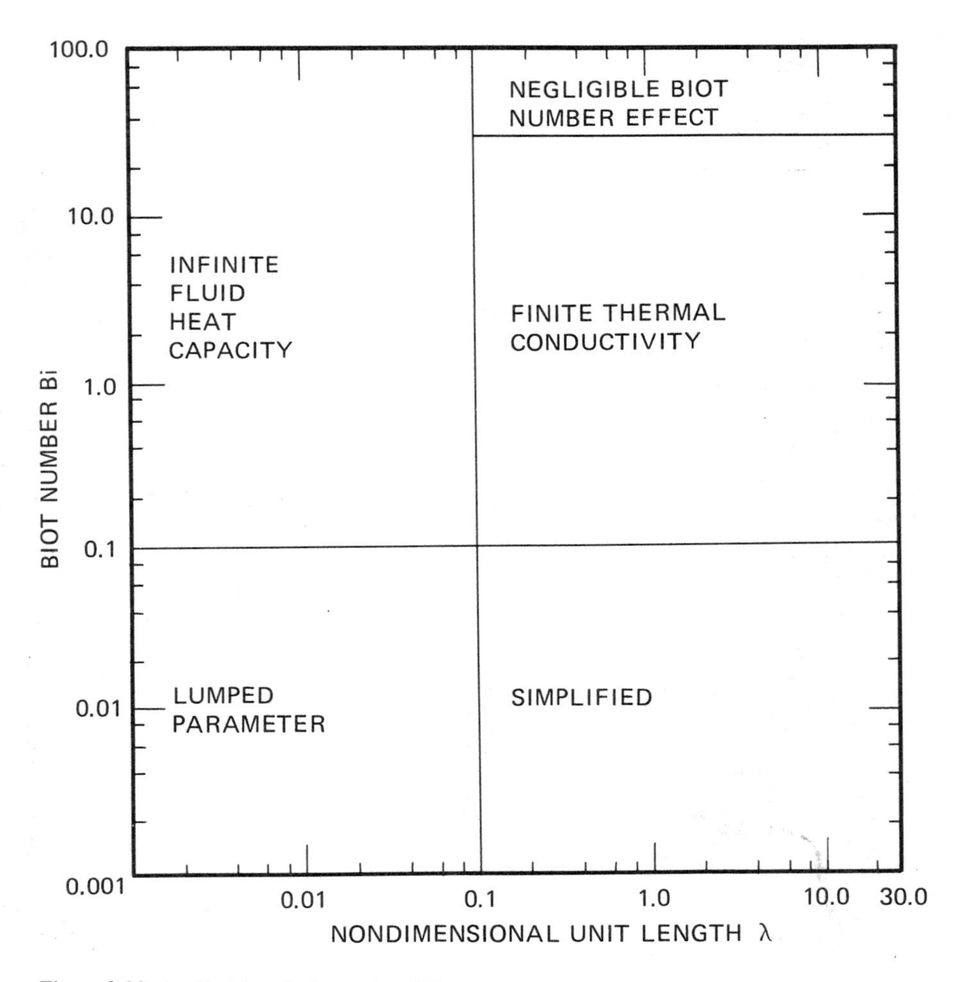

Figure 3.23 Applicable solution region [5].

modified film coefficient are compared in Table 3.20 for the flat slab configuration. At $\lambda = 5$ very good agreement was obtained between the simplified and finite conductivity models, whereas at $\lambda = 0.1$, the most severe case, the agreement is not as close, although it is still considered to be within acceptable limits. In all cases the results obtained using the simplified model with the modified film coefficient (see Sec. 2.4) were in very good agreement with the results obtained using the finite conductivity model.

When evaluating the accuracy of the results obtained using the modified heat transfer coefficient, it should be remembered that the expressions developed were based on the fact that the temperature of the storage unit, operating as a regenerator, varied linearly with time. The results obtained for the single-blow problem using the finite thermal conductivity model were analyzed in an attempt to identify the conditions for which the material temperature had a linear timewise variation. At small Biot numbers a linear variation was observed to be present throughout a wide range of variables. It was also observed that for a given Biot number the nonlinearities first appeared at small values of η and grew as the nondimensional length, λ, increased. As the Biot number increases, the deviation from a linear response becomes significant throughout the complete range of variables.

A careful comparison of the results obtained using the finite thermal conductivity model and the modified film coefficient model described in Sec. 2.4 was made in the range $0 < \lambda < 20$ and $0 < \eta < 20$. The largest differences appeared in the calculation of the nondimensional outlet fluid temperature T_{fo}. It is extremely difficult to generalize with regard to the conditions under which the modified heat transfer coefficient model will yield acceptable results. As a guide, the regions where an error of 2% or greater was introduced are shown in Figs. 3.24 and 3.25. These were obtained with Biot numbers of 1, 5, and 10. For a Biot number of 1, the percentage of error introduced in the calculation of Q^+ is slightly less than 1%. When the Biot number is greater than 10, it is recommended that the finite thermal conductivity model be used if accurate predictions of the transient response of the heat storage unit is required.

The total resistance to the transfer of heat from the fluid to the storage material is composed of two components. One is associated with the convective film coefficient and the other is related to the internal conduction process within the storage material. As the Biot number increases, the convective resistance decreases, thereby increasing the significance of the internal resistance. The heat storage at high Biot numbers and $\lambda > 0.1$ thus becomes primarily a function of G^+/V^+ and Fo and shows little dependence on the Biot number. The magnitude of these effects can be evaluated by reviewing Fig. 3.26. It was concluded that, for $\lambda > 0.1$ and $\mathrm{Bi} \geqslant 30$, the heat storage can be accurately predicted using the results for $\mathrm{Bi} = 30$ given in Sec. 3.2 for the flat slab

Table 3.19 Comparison of calculational methods, flat slab unit, Bi = 0.1 and $\lambda = 0.1$

	Heat storage, %			
η	Lumped	One-dimensional	Modified heat transfer coefficient	Finite conductivity
0.1	9.5	9.2	8.8	8.8
0.2	18.1	17.6	16.8	16.9
0.3	25.9	25.2	24.2	24.2
0.5	39.3	38.4	37.0	36.9
0.7	50.3	49.2	47.6	47.5
1.0	63.2	62.0	60.2	60.2
1.2	69.9	68.7	66.9	66.8
1.6	79.8	78.7	77.2	77.0
2.5	91.8	91.1	89.4	88.9

Table 3.20 Comparison of calculation methods, flat slab unit, Bi = 0.1

	Heat storage, %						Fluid outlet temperature					
	Finite conductivity		Modified heat transfer coefficient		Simplified		Finite conductivity		Modified heat transfer coefficient		Simplified	
η	$\lambda = 0.1$	$\lambda = 5.0$	$\lambda = 0.1$	$\lambda = 5.0$	$\lambda = 0.1$	$\lambda = 5.0$	$\lambda = 0.1$	$\lambda = 5.0$	$\lambda = 0.1$	$\lambda = 5.0$	$\lambda = 0.1$	$\lambda = 5.0$
0.5	36.9	9.9	36.95	9.81	38.6	9.8	0.941	0.034	0.942	0.033	0.941	0.030
1.0	60.2	19.6	60.23	19.29	62.3	19.4	0.963	0.072	0.963	0.070	0.963	0.065
2.5	88.9	45.2	89.42	44.87	90.1	45.1	0.992	0.241	0.992	0.237	0.991	0.231
5.0		74.3		74.71		75.1		0.568		0.564		0.563
7.5		89.4		90.07		90.4		0.800		0.801		0.804
10.0		96.1		96.56		96.8		0.918		0.923		0.926

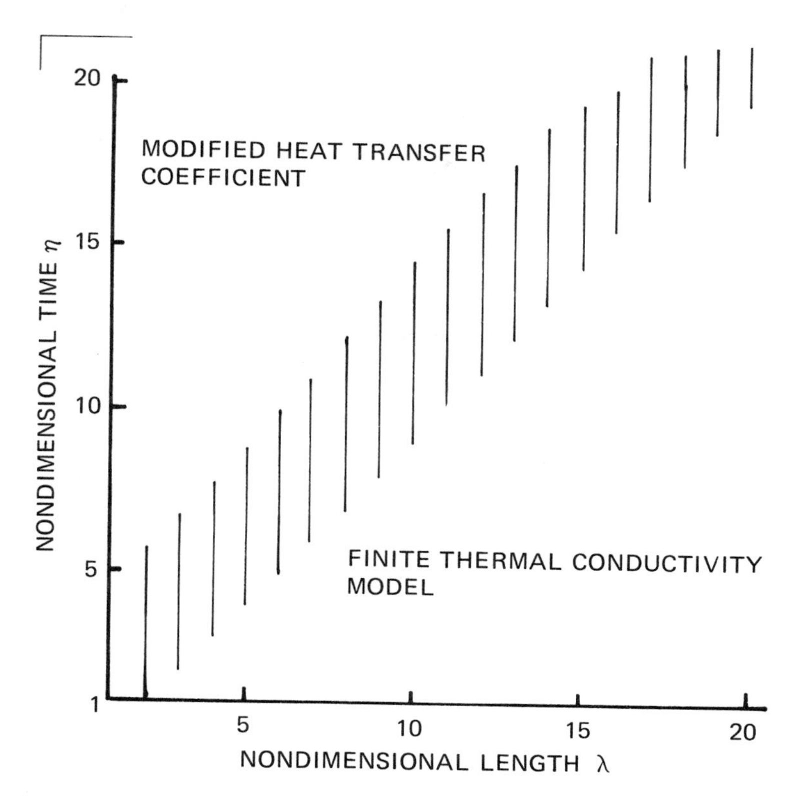

Figure 3.24 Nondimensional outlet fluid temperature.

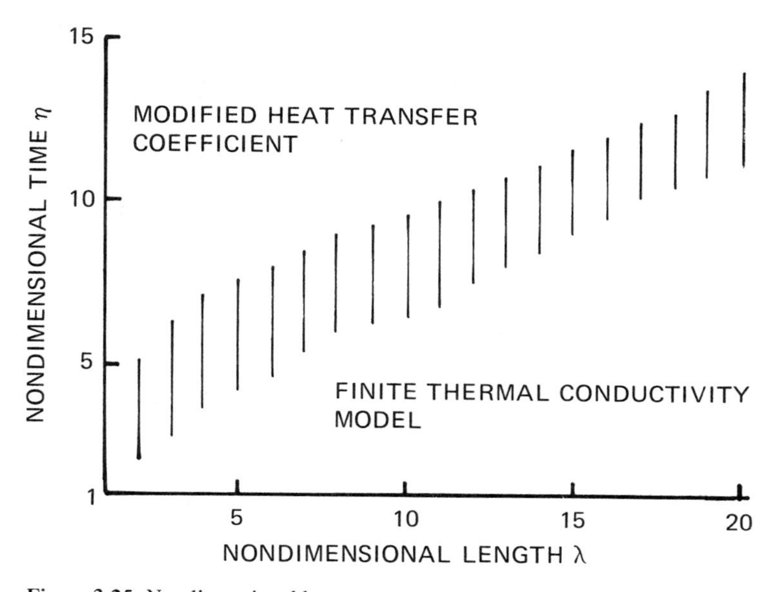

Figure 3.25 Nondimensional heat storage.

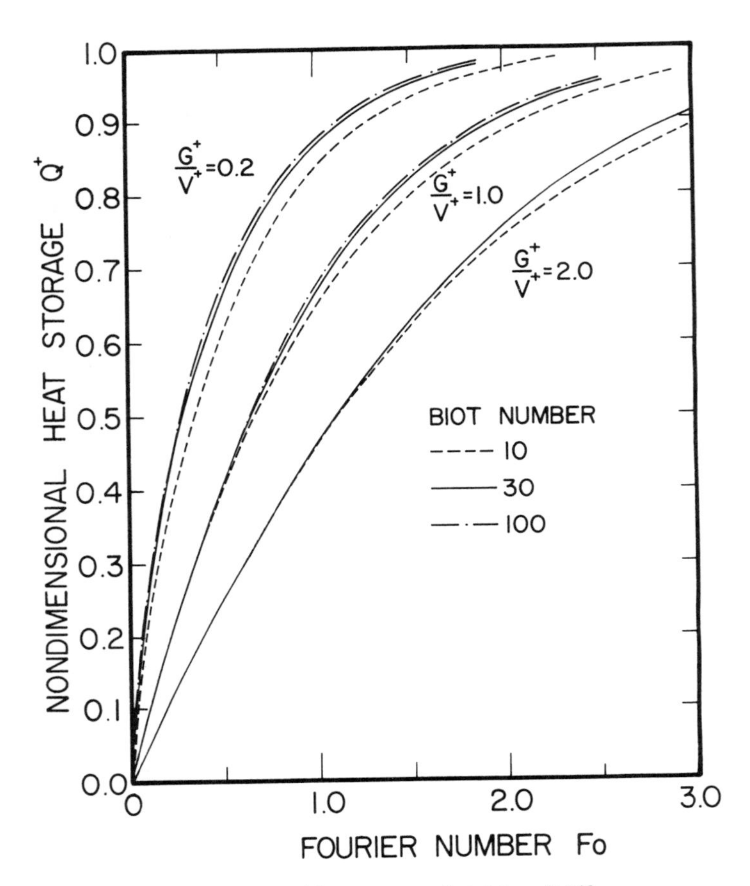

Figure 3.26 Nondimensional heat storage, flat slab unit [4].

unit. Similar conclusions can be drawn for the hollow cylindrical storage unit using the results presented in Sec. 3.3.

The effect of neglecting axial conduction is more difficult to assess. The finite conductivity model considers conduction in the axial direction but it has been shown that if axial conduction is small, the response of the storage unit can be presented by using the parameters G^+/V^+ instead of G^+ and V^+ independently. For the flat slab configuration, Handley and Heggs [6] have recommended that if $\lambda > 4$, the effect of axial conduction can be neglected where $(\lambda^2/\mathrm{Bi})(V^+)^2 \leqslant 0.1$. It must be noted, however, that the model used to obtain this criterion neglected the heat conduction normal to the flow direction. In general, the transverse conduction effects are several orders of magnitude greater than those of axial conduction in a continuous slab.

The finite conductivity model for the flat slab unit may also be expressed in terms of η and ξ.

Transfer fluid:
$$\frac{\partial T_f}{\partial \xi} = T_m - T_f \tag{3.21}$$

Storage material: $$\frac{\partial T_m}{\partial \eta} = C^* \frac{\partial^2 T_m}{\partial \xi^2} + D^* \frac{\partial^2 T_m}{\partial Y^2} \tag{3.22}$$

where $C^* = \frac{\lambda^2}{\text{Bi}} (V^+)^2$ and $D^* = \frac{1}{\text{Bi}}$

The initial and boundary conditions are

$$\eta = 0 \qquad T_m = T_f = 0$$

$$\begin{aligned}
\eta > 0 \qquad & \xi = 0 \qquad T_f = 1 \qquad \frac{\partial T_m}{\partial \xi} = 0 \\
& \xi = \lambda \qquad \frac{\partial T_m}{\partial \xi} = 0 \\
& Y = 0 \qquad \frac{\partial T_m}{\partial Y} = 0 \\
& Y = 1 \qquad \frac{\partial T_m}{\partial Y} = \text{Bi}(T_f - T_m)
\end{aligned} \tag{3.23}$$

The contribution of the axial and transverse conduction can be identified in Eq. (3.22) as

Transverse conduction: $$\frac{1}{\text{Bi}} \frac{\partial^2 T_m}{\partial Y^2}$$

Axial conduction: $$\frac{\lambda^2 (V^+)^2}{\text{Bi}} \frac{\partial^2 T_m}{\partial \xi^2}$$

If the worst condition is assumed to occur when

$$\frac{\partial^2 T_m}{\partial \xi^2} \approx \frac{\partial^2 T_m}{\partial Y^2}$$

the maximum contribution of the axial conduction term would be an order of magnitude less than that of the transverse heat conduction term if $\lambda^2 (V^+)^2 \leqslant 0.1$. Under these conditions, little error is introduced if axial heat conduction is neglected. This criterion can be written in the form $\lambda V^+ < 0.3$, which shows that for $\lambda = 30$ axial heat conduction can be neglected if $V^+ < 0.01$. Similar conclusions can again be expected for the hollow cylindrical storage unit.

REFERENCES

1. F. W. Schmidt and J. Szego, "Transient Response of Solid Sensible Heat Thermal Storage Units—Single Fluid," *J. Heat Transfer, Trans. ASME*, vol. 98, 1976, p. 471.

2. G. E. Forsythe, "Generation and Use of Orthogonal Polynomial for Data Fitting with a Digital Computer," *J. Soc. Ind. Appl. Math.*, vol. 5, 1957, p. 74.
3. D. N. De Allen and R. V. Southwell, "Relaxation Methods Applied to Determine the Motion in Two Dimensions of a Viscous Fluid Past a Fixed Cylinder," *Quart. J. Mech. Appl. Math.*, vol. 8, pt. 2, 1955, p. 129.
4. F. W. Schmidt and J. Szego, "Transient Response of a Hollow Cylindrical-Cross-Section Solid Sensible Heat Storage Unit–Single Fluid," *J. Heat Transfer, Trans. ASME*, vol. 100, 1978, p. 737.
5. J. Szego and F. W. Schmidt, "Analysis of the Effects of Finite Conductivity in the Single Blow Heat Transfer Unit," *J. Heat Transfer, Trans. ASME*, vol. 100, 1978, p. 740.
6. D. Handley and P. J. Heggs, "The Effect of Thermal Conductivity of the Packing on Transient Heat Transfer in a Fixed Bed," *Int. J. Heat Mass Transfer*, vol. 12, 1969, pp. 549–570.

CHAPTER

FOUR

PREDICTION OF TRANSIENT RESPONSE OF HEAT STORAGE UNITS WITH TIMEWISE VARIATIONS IN INLET FLUID TEMPERATURE AND MASS FLOW RATE

4.1 INTRODUCTION

In most applications involving thermal energy storage units, the inlet conditions of the energy transporting fluid experience timewise variations in temperature and/or mass flow rate. These variations are created by the transient nature of a system component, such as a solar collector or a batch-type industrial process, which supplies the energy to the fluid prior to its entry into the heat storage unit. A similar situation could be encountered when, in a closed system, the storage unit supplies heat to a fluid that passes through an energy removal device that has a timewise variation in its demand.

The transient response of heat storage units under these conditions can be obtained by utilizing numerical techniques to simulate the heat storage unit. Such a program was developed by Schmidt and Szego [1].

If the properties of the fluid and storage material are independent of temperature and if the convective heat transfer coefficient is uniform, the mathematical model describing the operation of the storage unit is linear. The transient response can thus be obtained using the superposition techniques described by Larsen [2], and the response of the unit to a single step variation in inlet temperature, single-blow operating mode, presented in Chaps. 2 and 3. These methods will be discussed in this chapter. Several examples in which superposition is used to predict the transient response of heat storage units in typical industrial applications will be presented.

4.2 SUPERPOSITION: TIMEWISE VARIATIONS IN INLET FLUID TEMPERATURE

Superposition techniques can, as previously noted, be used for the prediction of the transient response of a heat storage unit provided that the physical properties of the fluid and the storage material are independent of temperature. This also implies that the convective film coefficient is independent of temperature and is uniform. Under these conditions the differential energy equations for the fluid and storage material and the initial and boundary conditions are linear for all the models discussed in Chaps. 2 and 3. In this section the application of superposition techniques to predict the transient response of the storage unit that experiences a timewise variation in inlet fluid temperature will be presented. The initial temperature distribution in the storage unit will be uniform, and the mass flow rate of the fluid passing through the unit will remain constant. The convective film coefficient will remain uniform as a result of the invariant flow rate assumption.

The arbitrarily varying inlet fluid temperature is broken up into a series of step changes in temperature as shown in Fig. 4.1. Each step represents a subproblem that has a step change in inlet temperature, with the unit considered to be initially at a uniform temperature. For the first subproblem the initial temperature of the unit will be t_o. In all other subproblems the initial temperature will be zero. The response of the unit with a variation in fluid inlet temperature is the sum of the responses of all the subproblems.

The calculation of the response of a heat storage unit, initially at a uniform temperature, to a step change in inlet temperature was presented in Chaps. 2 and 3. The nondimensional temperature of the fluid leaving the unit and the heat storage are functions of the nondimensional length and time, $T_f(\eta, \lambda)$ and $Q^+(\eta, \lambda)$. In the definition of the nondimensional temperature the initial temperature of the unit is zero for all the subproblems except the first. The step change in inlet fluid temperature for the

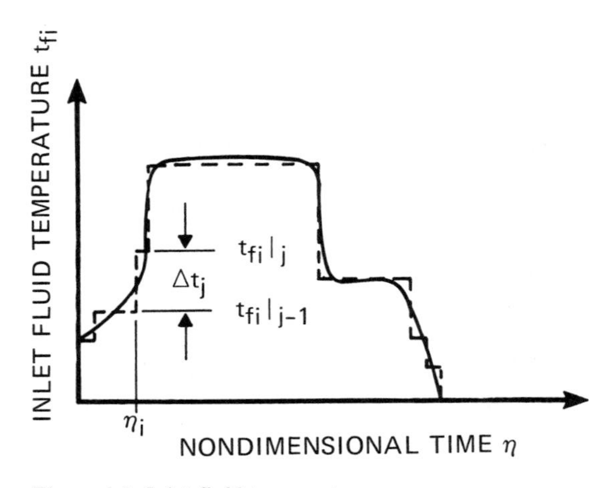

Figure 4.1 Inlet fluid temperature.

subproblem j is

$$\Delta t_j = t_{fi}|_j - t_{fi}|_{j-1}$$

Since the initial temperature is zero, the nondimensional temperature is

$$T_{fj} = \frac{t_{fi} - t_o}{t_{fi} - t_o} = \frac{t_{fj}}{\Delta t_j}$$

The contribution of the jth step in inlet fluid temperature is

$$t_{fj} = \Delta t_j(T_{fj})$$

The nondimensional time η is determined with reference to the start of the storage process. The step in inlet fluid temperature Δt_j occurs at a nondimensional time η_j. The nondimensional time used in the evaluation of t_{fj} is the elapsed time since the temperature step, $\eta - \eta_j$, so

$$t_{fj} = \Delta t_j(T_f\{\eta - \eta_j, \xi\})$$

The temperature of the fluid leaving a heat storage unit when J substeps in inlet fluid temperature have been taken is

$$t_{fo} = t_{f1} + t_{f2} + t_{f3} + \cdots + t_{fJ}$$

$$= t_o + \sum_{j=1}^{J} \Delta t_j(T_f\{\eta - \eta_j, \lambda\}) \tag{4.1}$$

A similar relationship can be developed for the total heat stored. The nondimensional heat storage is

$$Q^+ = \frac{Q}{Mc_m(t_{fi} - t_o)}$$

where M is the total mass of the heat storage unit. The heat stored for the subproblem j is

$$Q_j = Mc_m(t_{fi}|_j - t_{fi}|_{j-1})Q^+\{\eta - \eta_j, \lambda\}$$

$$= Mc_m \Delta t_j Q^+\{\eta - \eta_j, \lambda\}$$

The total heat stored when J steps in inlet temperature have occurred is

$$Q = Q_1 + Q_2 + Q_3 + \cdots + Q_J$$

$$= Mc_m \sum_{j=1}^{J} \Delta t_j(Q^+\{\eta - \eta_j, \lambda\}) \tag{4.2}$$

If the timewise variation of the inlet fluid temperature is a continuous function, Duhamel's theorem can be used to obtain the following expression for the outlet fluid temperature:

$$t_{fo} = t_o + \int_0^{\eta} \frac{d(\Delta t)}{d\beta} (T_f\{\eta - \beta, \lambda\})\, d\beta \tag{4.3}$$

A general expression for an inlet fluid temperature variation containing both continuous and step changes is

$$t_{fo} = t_o + \sum_{j=1}^{J} \Delta t_j (T_f(\eta - \eta_j, \lambda)) + \int_0^{\eta} \frac{d(\Delta t)}{d\beta} (T_f(\eta - \beta, \lambda))\, d\beta \qquad (4.4)$$

The same procedure can be used for evaluating the heat stored. The resulting general expression is

$$Q = Mc_m \left[\sum_{j=1}^{J} \Delta t_j (Q^+(\eta - \eta_j, \lambda)) + \int_0^{\eta} \frac{d(\Delta t)}{d\gamma} Q^+(\eta - \gamma, \lambda)\, d\gamma \right] \qquad (4.5)$$

The expressions for T_f and Q^+ will depend on the assumptions made in modeling the heat storage unit. A complete discussion of commonly employed models, their corresponding solutions, and regions in which they may be used without introducing appreciable errors has been presented in Chaps. 2 and 3.

Example 4.1 In a chemical processing plant it is necessary to cool a batch-type chemical reactor by using ambient air at 30°C (86°F). The mass rate of flow of the air is 20 kg/s (44.09 lb_m/s). The temperature of the air leaving the reactor varies in the timewise fashion shown in Fig. 4.2*a*. The air stream upon leaving the reactor enters a slab-type heat storage device 0.4 m (1.31 ft) long, which is initially at a uniform temperature of 20°C (68°F). The storage unit has a frontal configuration 1 m (3.28 ft) square and is composed of magnesite bricks that are 2 cm (0.787 in) thick. The distance between the bricks is 2 cm (0.787 in), and the convective film coefficient is 45 W/m² °C (7.72 Btu/ft² h °F). Determine the temperature of the air leaving the heat storage unit, the temperature of the storage material, and the amount of heat stored in the magnesite bricks, after 1.0 and 1.25 h of operation. Bricks 1 cm thick are used at the top and bottom and the unit is completely insulated.

SOLUTION The physical properties required are as follows:

Magnesite bricks @ 200°C:

$\rho_m = 3500$ kg/m³ (218.5 lb_m/ft^3)

$c_m = 0.984$ kJ/kg (0.235 Btu/lb_m °F)

$k_m = 4.62$ W/m °C (2.67 Btu/ft h °F)

Air @ 230°C:

$c_f = 1.0295$ kJ/kg (0.246 Btu/lb_m °F)

$\rho_f = 0.7048$ kg/m³ (0.044 lb_m/ft^3)

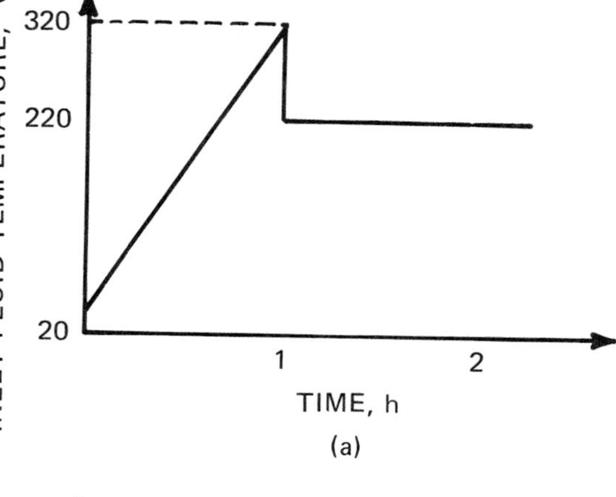

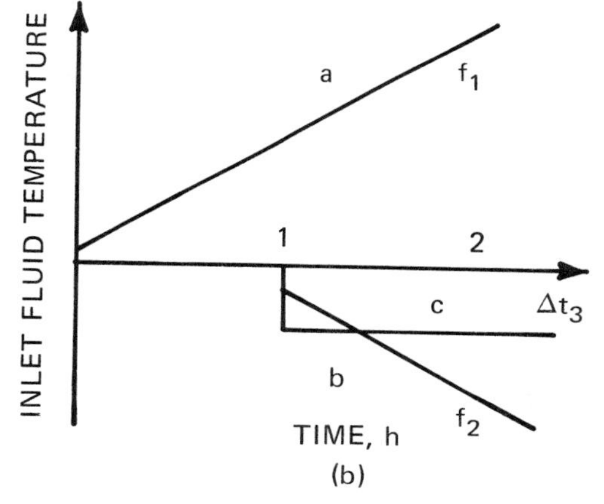

Figure 4.2 Inlet temperature variation – Example 4.1. (*a*) Inlet air temperature. (*b*) Subproblems.

The Biot number is

$$\mathrm{Bi} = \frac{hw}{k_m} = \frac{45.0(0.01)}{4.62} = 0.097$$

The surface area is

$$A = \frac{1.0(2.0)(1.0)(0.4)}{(0.04)} = 20 \text{ m}^2 \quad (215.28 \text{ ft}^2)$$

and the nondimensional length of the unit is

$$\lambda = \frac{hA}{\dot{m}_f c_f} = \frac{45.0(20.0)}{(20.0)(1029.5)} = 0.044$$

From Fig. 3.23 it can be seen that the unit is operating in the infinite heat capacity-lumped parameter region, and the results presented in Sec. 2.2.1 can be used to predict the transient response of the unit.

The volume of the storage material is $V = (0.5)(0.4) = 0.20\ \text{m}^3$ ($7.06\ \text{ft}^3$), and its mass $M = V\rho_m = 0.20(0.3500) = 700.0$ kg ($1543.0\ \text{lb}_m$). The nondimensional time is

$$\eta = \frac{hA\tau}{\rho_m c_m V} = \frac{(45.0)(20.0)\tau}{(3500)(984)(0.20)} = 1.3066 \times 10^{-3}\tau$$

or

τ	η
1 h	4.704
1.25 h	5.880

Since the heat storage unit is operating in the infinite heat capacity region, the temperature of the fluid leaving the unit is the same as its entrance temperature, $t_{fo} = t_{fi}$.

In order to determine the temperature of the storage material and the amount of heat stored, the inlet fluid temperature's timewise variation must be divided into the three subproblems as shown in Fig. 4.2*b*. The storage material temperature and the heat storage may be obtained from the following expressions:

Time: $0 \leqslant \tau \leqslant 1$ h

$$t_m = t_o + \int_0^\eta \frac{d(f_1)}{d\beta} T_m\{\eta - \beta\}\, d\beta$$

$$Q = Mc_m \int_0^\eta \frac{d(f_1)}{d\gamma} Q^+\{\eta - \gamma\}\, d\gamma$$

Time: $1\ \text{h} < \tau$

$$t_m = t_o + \int_0^\eta \frac{d(f_1)}{d\beta} T_m\{\eta - \beta\}\, d\beta + \int_{\eta_1}^\eta \frac{d(f_2)}{d\beta} T_m\{\eta - \beta\}\, d\beta + \Delta t_3 T_m\{\eta - \eta_1\}$$

$$Q = Mc_m \left(\int_0^\eta \frac{d(f_1)}{d\gamma} Q^+\{\eta - \gamma\}\, d\gamma + \int_{\eta_1}^\eta \frac{d(f_2)}{d\gamma} Q^+\{\eta - \gamma\}\, d\gamma + \Delta t_3 Q^+\{\eta - \eta_1\} \right)$$

The continuous temperature functions are

$$f_1 = t_{fi1} = 20.0 + 0.0833\tau \qquad \text{or} \qquad f_1 = \left[20.0 + 0.0833 \left(\frac{Mc_m}{hA}\right) \eta\right]$$

$$f_2 = t_{fi2} = -0.0833(\tau - \tau_1) \qquad \qquad f_2 = -0.0833 \left(\frac{Mc_m}{hA}\right)(\eta - \eta_1)$$

Thus,

$$\frac{d(f_1)}{d\beta} = \frac{d(f_1)}{d\gamma} = 0.0833\left(\frac{Mc_m}{hA}\right) = 0.0833\left[\frac{700.0(984)}{45.0(20.0)}\right] = 63.75°\text{C}$$

$$\frac{d(f_2)}{d\beta} = \frac{d(f_2)}{d\gamma} = -0.0833\left(\frac{Mc_m}{hA}\right) = -63.75°\text{C}$$

whereas $\Delta t_3 = 220 - 320 = -100°\text{C}$.

The temperature of the storage material and the nondimensional heat flux may be obtained from Eqs. (2.5) and (2.8). These expressions are identical, so $T_m(\eta) = Q^+(\eta) = 1.0 - e^{-\eta}$. At 1.0 h,

$$t_m = 20.0 + (63.75)\int_0^{4.704} [1.0 - e^{-(4.704-\beta)}]\, d\beta$$

$$= 20.0 + (63.75)[\beta - e^{-(4.704-\beta)}]_0^{4.704}$$

$$= 20.0 + (63.75)(3.713) = 256.70°\text{C} \quad (494.06°\text{F})$$

$$Q^+ = (700.0)(984.0)(63.75)\int_0^{4.704} [1.0 - e^{-\gamma}]\, d\gamma$$

$$= (700.0)(984.0)(63.75)(3.713)$$

$$= 1.630 \times 10^5 \text{ kJ} \quad (1.545 \times 10^5 \text{ Btu})$$

and at 1.25 h of operation,

$$t_m = 20.0 + (63.75)\left\{\int_0^{5.880} [1.0 - e^{-(5.88-\beta)}]\, d\beta - \int_{4.704}^{5.880} [1.0 - e^{-(5.88-\beta)}]\, d\beta\right\} - 100.0T_m(5.880 - 4.704)$$

$$= 20.0 + 63.75(4.883 - 0.485) - 100.0(0.691)$$

$$= 20.0 + 280.37 - 69.1$$

$$= 231.27°\text{C} \quad (448.29°\text{F})$$

$$Q = (700.0)(984.0)[(63.75)(4.883 - 0.485) - 100.0(0.691)]$$

$$= (700.0)(984.0)(211.27)$$

$$= 1.455 \times 10^5 \text{ kJ} \quad (1.379 \times 10^5 \text{ Btu})$$

Example 4.2 The inlet fluid temperature of the heat storage unit described in Example 2.1 varies in the timewise fashion shown in Fig. 4.3. Determine the transient response of the unit during the 20-h period of operation. The initial temperature of the unit is 10°C (50°F).

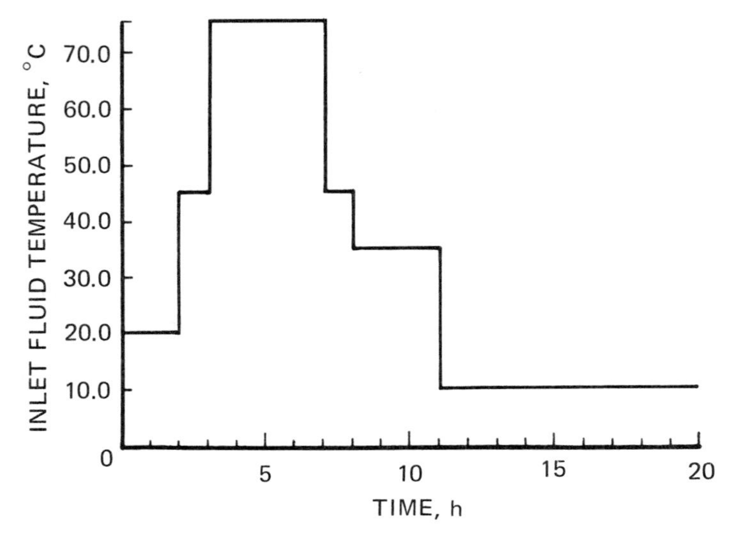

Figure 4.3 Inlet fluid temperature.

SOLUTION The nondimensional length of the storage unit is 1.847, the total mass of the heat storage material is 1.086×10^4 kg (2.394×10^4 lb_m), and the nondimensional time is $\eta = 3.5 \times 10^{-4}\tau$. The inlet fluid temperature is subdivided into six steps. The thermal conductivity of the Feolite storage material is 2.1 W/cm °C (1.370 Btu/ft^2 h °F). The Biot number is

$$\text{Bi} = \frac{hw}{k_m} = \frac{(50.23)(0.04)}{2.1} = 0.957$$

which necessitates the use of the finite thermal conductivity model or a modified heat transfer coefficient.

The temperature of the fluid leaving the storage unit and the heat storage are obtained from Eqs. (4.4) and (4.5).

$$t_{fo} = t_o + \sum_{j=1}^{6} \Delta t_j (T_f(\eta - \eta_j, \lambda))$$

and

$$Q = Mc_m \sum_{j=1}^{6} \Delta t_j (Q^+(\eta - \eta_j, \lambda))$$

The information needed for the solution is as follows:

	Step Number					
	1	2	3	4	5	6
Time, h	0	2.0	3.0	7.0	8.0	11.0
Δt_j, °C	10.0	25.0	30.0	–30.0	–10.0	–25.0
η_j	0.0	2.52	3.78	8.82	10.08	13.86

The finite conductivity model is selected and the expressions for T_f and Q^+ are obtained by using the tables presented in Chap. 3 or the interpolation program HSSF [3].

After 10 h the temperature of the fluid leaving the unit is

$$\begin{aligned} t_{fo} &= 10.0 + (10.0)(1.00) + (25.0)(0.993) + (30.0)(0.982) \\ &\quad + (-30.0)(0.836) + (-10.0)(0.719) \\ &= 42.02°\text{C} \quad (107.64°\text{F}) \end{aligned}$$

The heat storage is

$$\begin{aligned} Q &= 1.086 \times 10^4(0.920)[(10.0)(1.000) + (25.0)(0.998) + (30.0)(0.990) \\ &\quad + (-30.0)(0.811) + (-10.0)(0.663)] \\ &= 3.368 \times 10^5 \text{ kJ} \quad (3.192 \times 10^5 \text{ Btu}) \end{aligned}$$

The temperature of the fluid leaving the storage unit and the amount of heat stored within the unit have been determined at each hour. These results as well as those obtained from the simplified and the modified heat transfer coefficient models are presented in Table 4.1. A comparison of the results shows that the simplified model overpredicts the heat storage or removal. This is an expected result, since the resistance to the transfer of heat within the storage material present in the finite thermal conductivity model will decrease the flow of heat into or out of the storage material. The results obtained using a modified heat transfer coefficient are in very good agreement with those obtained using the finite thermal conductivity model.

4.3 SUPERPOSITION: ARBITRARY INITIAL TEMPERATURE DISTRIBUTION IN THE STORAGE MATERIAL

The prediction of the transient response of a heat storage unit with a nonuniform initial temperature will be restricted to the units with fluids that have finite heat capacity, $\lambda > 0.1$. If $\text{Bi} \leqslant 0.1$ the simplified model can be used, but for $\text{Bi} > 0.1$ either the finite thermal conductivity or the modified heat transfer coefficient model is required.

The basic conservation equations describing the heat storage unit are the same as those discussed previously. The initial conditions are no longer uniform, but the linearity of the mathematical model is preserved and superposition can be used in obtaining solutions. The initial temperature distribution is subdivided into a series of constant temperature sections. Each isothermal section will represent a subproblem in which the initial temperature of the storage unit is equal to the step in temperature that occurred as one moves in the direction of flow from the last subproblem. The first subproblem at the inlet of the storage unit uses the actual inlet fluid temperature. All other subproblems assume the inlet temperature of the fluid to be equal to zero.

Table 4.1 Transient response of storage unit under time-varying inlet fluid conditions (Example 4.2)

	T_{fo}, °C			Q, 10^5 kJ		
Time, h	Simplified model	Modified heat transfer coefficient model	Finite conductivity model	Simplified model	Modified heat transfer coefficient model	Finite conductivity model
1	14.88	15.28	15.33	0.457	0.412	0.412
2	17.17	17.18	17.19	0.721	0.665	0.662
3	30.75	31.57	31.69	2.004	1.844	1.840
4	51.86	52.86	53.07	4.107	3.797	3.785
5	62.52	61.96	61.99	5.291	4.973	4.954
6	68.65	67.56	67.47	5.912	5.655	5.638
7	71.91	70.86	70.89	6.221	6.039	6.030
8	58.90	56.91	56.63	5.000	5.014	5.019
9	47.94	46.99	46.61	3.818	3.959	3.989
10	41.92	42.07	42.02	3.159	3.322	3.368
11	38.50	39.04	39.15	2.816	2.952	2.995
12	24.48	24.05	24.04	1.504	1.712	1.742
13	17.86	18.28	18.53	0.762	0.966	0.982
14	14.02	14.73	14.99	0.370	0.532	0.530
15	11.96	12.63	12.66	0.173	0.288	0.276
16	10.93	11.43	11.28	0.079	0.153	0.143
17	10.42	10.77	10.87	0.035	0.080	0.069
18	10.19	10.41	10.45	0.015	0.042	0.025
19	10.08	10.21	10.17	0.007	0.021	0.043
20	10.04	10.11	10.00	0.003	0.011	0.000

The fluid outlet temperature for the first subproblem is

$$t_{f1} = t_{o1} + (t_{fi} - t_{o1})T_f(\eta, \lambda)$$

whereas the other fluid outlet temperatures are

$$t_{f2} = (t_{o2} - t_{o1})(1.0 - T_f(\eta, \lambda - \xi_2))$$

and $$t_{fj} = (t_{oj} - t_{oj-1})(1.0 - T_f(\eta, \lambda - \xi_j))$$

where the t_{oj}'s are the temperatures of the uniform subsections and ξ_j is the nondimensional length at which the step in temperature occurred. The nondimensional length, from the location where the temperature step occurred to the end of the heat storage unit, is $\lambda - \xi_j$. The subdivision of a heat storage unit initially at a nonuniform temperature and the resulting subproblems are given in Fig. 4.4. The temperature of

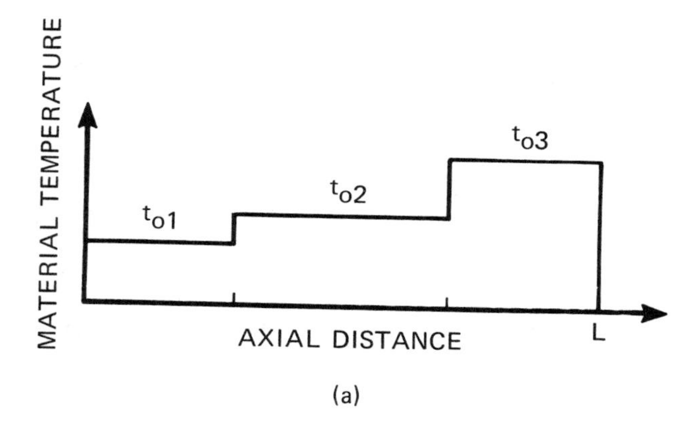

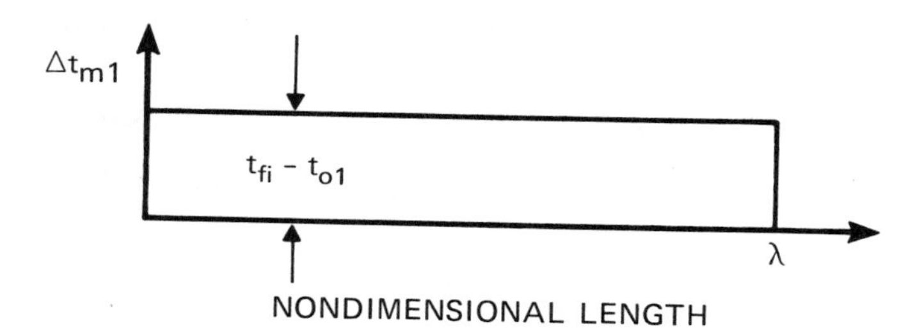

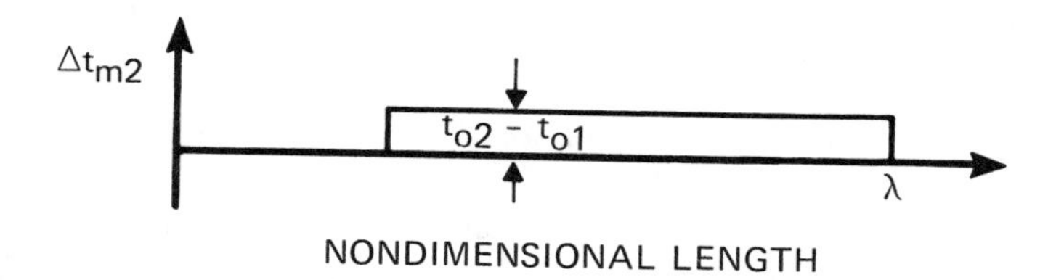

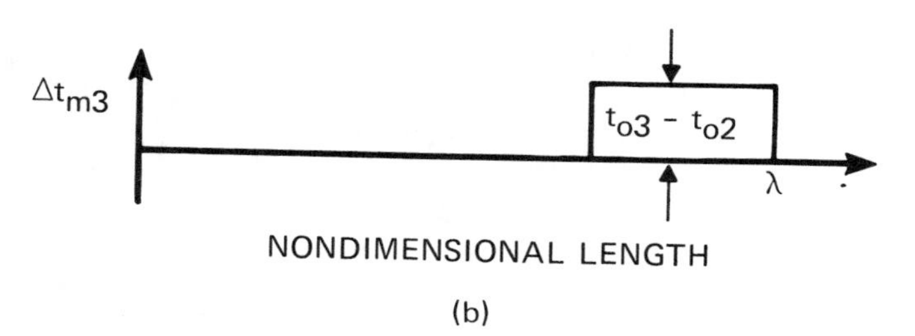

Figure 4.4 (*a*) Initial temperature distribution. (*b*) Nonuniform initial temperature superposition subproblems.

the fluid leaving the storage unit is

$$t_{fo} = t_{o1} + (t_{fi} - t_{o1})T_f\{\eta, \lambda\} + \sum_{j=2}^{J} (t_{oj} - t_{oj-1})(1.0 - T_f\{\eta, \lambda - \xi_j\}) \tag{4.6}$$

If the heat storage unit has continuous as well as discrete steps in the initial temperature, the expression for the outlet fluid temperature is

$$t_{fo} = t_{o1} + (t_{fi} - t_{o1})T_f\{\eta, \lambda\} + \sum_{j=2}^{J} (t_{oj} - t_{oj-1})(1.0 - T_f\{\eta, \lambda - \xi_j\}) + \int_0^\lambda \frac{d(\Delta t_o)}{d\alpha}(1.0 - T_f\{\eta, \lambda - \alpha\})\, d\alpha \tag{4.7}$$

A similar relationship can be developed for the heat storage. The heat stored for subproblem j is

$$Q_j = Q^+\{\eta, \lambda - \xi_j\}[S_m \rho_m (L - x_j) c_m (t_{oj} - t_{oj-1})] \tag{4.8}$$

where S_m is the cross-sectional area of the storage material. The general expression is

$$Q = S_m \rho_m c_m \left[Q^+\{\eta, \lambda\}(t_{fi} - t_{o1})L - \sum_{j=2}^{J} Q^+\{\eta, \lambda - \xi_j\}(t_{oj} - t_{oj-1})(L - x_j) - \frac{\dot{m}_f c_f L}{hA} \int_0^\lambda Q^+\{\eta, \lambda - \kappa\} \frac{d(\Delta t_o)}{d\kappa}(\lambda - \kappa)\, d\kappa \right] \tag{4.9}$$

Example 4.3 The initial temperature distribution in the heat storage unit described in Example 2.1 is shown in Fig. 4.5. If the inlet fluid temperature is 80°C (176.0°F), determine the temperature of the fluid leaving and the amount of heat stored at the end of 3 h of operation.

SOLUTION The nondimensional time is $\eta = 3.78$, the Biot number $= 0.957$, and the nondimensional length of the unit is $\lambda = 1.847$. The following table will be useful in organizing the information required to solve this problem:

	Subproblems			
	1	2	3	4
t_{fi}, °C	80.0	0.0	0.0	0.0
t_m, °C	10.0	30.0	50.0	70.0
$t_{oj} - t_{oj-1}$, °C	—	20.0	20.0	20.0
$L - x_j$, m	5.8	4.3	2.8	1.3
$\lambda - \xi_j$	1.847	1.369	0.892	0.414
$T\{\eta, \lambda - \xi_j\}$	0.836	0.891	0.939	0.976
$Q^+\{\eta, \lambda - \xi_j\}$	0.811	0.846	0.878	0.913

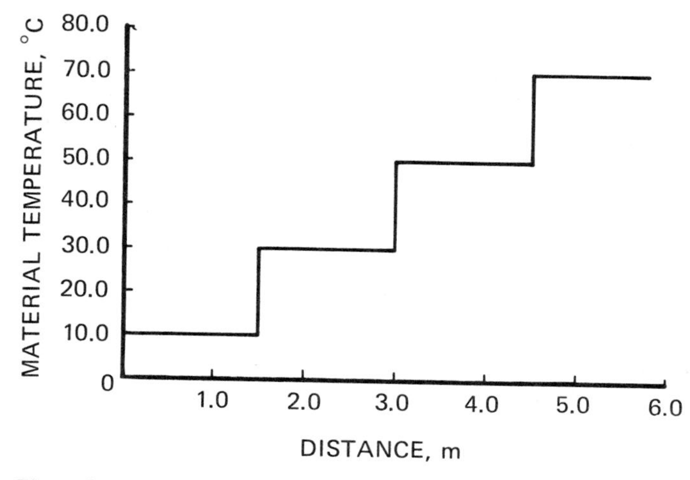

Figure 4.5 Initial temperature distribution in heat storage unit.

The temperature of the fluid leaving the unit is obtained by using Eq. (4.6):

$$t_{fo} = 10.0 + (80.0 - 10.0)(0.836) + (20.0)(1.0 - 0.891) + (20.0)(1.0 - 0.939) + (20.0)(1.0 - 0.976) = 72.4°\text{C} \quad (162.32°\text{F})$$

The total amount of heat stored per channel is calculated from Eq. (4.9):

$$Q = (0.04)(3900.0)(0.920)[(70.0)(5.8)(0.811) - (20.0)(4.3)(0.846) - (20.0)(2.8)(0.878) - (20.0)(1.3)(0.913)] = 2.635 \times 10^4 \text{ kJ per channel} \quad (2.626 \times 10^4 \text{ Btu per channel})$$

4.4 SUPERPOSITION: ARBITRARY VARIATION IN FLUID MASS FLOW RATE

When the flow rate of the fluid varies in a timewise fashion, there are two possible modes of operation for the heat storage unit. If the number of heat storage channels or elements is fixed, the mass rate of flow of the fluid through the unit will vary and the unit's transient response can be obtained by superposition. A second possibility would be to divert a portion of the fluid stream so that the mass rate of flow per unit heat storage channel would remain constant. Thus, if the flow rate were increased, new storage channels would be used; whereas if it were reduced, the flow through some of the channels would be terminated. The opening or closing of the channels can be accomplished by using dampers. The transient response for this latter case is obtained by the determination of the transient response of each storage channel and summing the results.

The superposition procedure to be followed when the fluid flow rate varies in an arbitrary manner requires that the temperature distribution in the storage material be known at the instant that the flow rate changes. It is thus impractical to use superposition when working with the finite thermal conductivity model, since the required temperature distribution can be obtained only with the computer program. Under these conditions one should use the program to obtain the desired solution directly.

The symmetry of the mathematical model used in the simplified model makes it possible to calculate the temperature distribution in the storage material from Eq. (2.33). The computational steps to be followed when the flow rate varies in the manner shown in Fig. 4.6 will be presented. The initial temperature of the storage unit is uniform, and the inlet fluid temperature is independent of the mass rate of flow of the fluid and remains constant during the entire storage process. Since there are three steps in mass flow rates, there will be three subproblems:

$0 \leqslant \tau \leqslant \tau_1$

1. The response of the unit is calculated using the basic solution.

$\tau_1 \leqslant \tau \leqslant \tau_2$

2. The temperature distribution in the storage unit is obtained at τ_1. The convective film coefficient is recalculated for the new mass flow rate and the nondimensional length of the storage unit under these conditions is found. The transient response of the unit is obtained using the procedure described in Sec. 4.3 for a storage unit with nonuniform temperature distribution.

$\tau_2 < \tau$

3. The temperature distribution in the storage unit is determined at τ_2. The procedure used in step 2 is repeated.

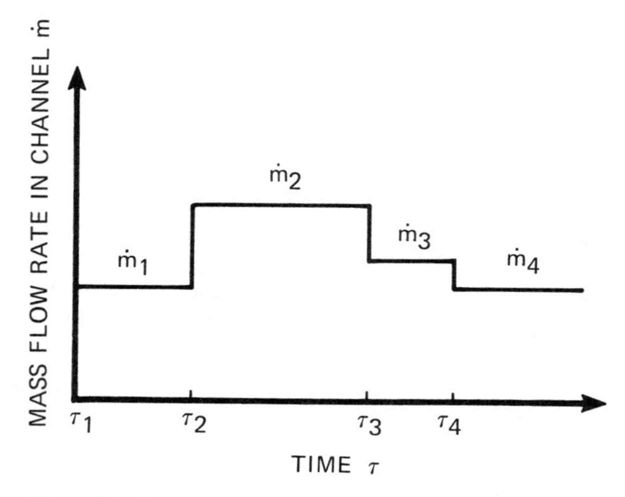

Figure 4.6 Variation in mass flow rate.

Two example problems involving step changes in the mass rate of flow will be presented. In the first, the mass flow rate per channel varies. In the second, the number of storage channels used will be selected in order to maintain the flow rate per channel at a constant value.

Example 4.4 A thermal storage unit of the flat slab configuration is composed of 12 channels separated by low-carbon-steel plates. The unit is 5.8 m (19.03 ft) long, 0.5 m (1.64 ft) wide, and the plates are 2 cm (0.79 in) thick. Plates 1 cm (0.395 in) thick are installed at the top and the bottom of the unit. The unit is completely insulated. The temperature of the hot gas entering the storage unit is 80°C (176°F). The initial temperature of the unit is 10°C (50°F). The total mass rate of flow through the unit varies in the following fashion:

Time, min	$\dot{m}$, kg/s	h, W/m² °C
$0 \leqslant \tau < 15.0$	1.872	50.03
$15.0 \leqslant \tau < 30$	2.496	63.23
$30.0 \leqslant \tau$	0.936	28.84

Determine the transient response of the unit during a 1-h operating period.

Solution The physical properties of the carbon steel and the air are

Carbon steel @ 45°C (113°F):

$\rho_m = 7833.0$ kg/m³ (489.0 lb$_m$/ft³)

$c_m = 0.465$ kJ/kg °C (0.111 Btu/lb$_m$ °F)

$k_m = 53.5$ W/m °C (30.91 Btu/h ft °F)

Air @ 50°C (122°F):

$\rho_f = 1.095$ kg/m³ (0.068 lb$_m$/ft³)

$c_f = 1.011$ kJ/kg °C (0.24 Btu/lb$_m$ °F)

The Biot number for the highest mass flow rate is

$$\text{Bi} = \frac{hw}{k_m} = \frac{(63.23)(0.01)}{53.5} = 0.012$$

The transient response of the unit can thus be obtained with a high degree of accuracy using the simplified model.

The nondimensional parameters are

$$\xi = \frac{hAx}{\dot{m}_f c_f L} = \frac{h(5.8)x}{\dot{m}_f(1011.0)(5.8)} = (9.89 \times 10^{-4})\frac{hx}{\dot{m}_f}$$

$$\lambda = \frac{hA}{\dot{m}_f c_f} = \frac{h(5.8)}{\dot{m}_f(1011.0)} = (5.74 \times 10^{-3})\frac{h}{\dot{m}_f}$$

$$\eta = \frac{hA\tau}{S_m L \rho_m c_m} = \frac{h(5.8)\tau}{(0.01)(5.8)(7833.0)(465.0)} = (2.745 \times 10^{-5})h\tau$$

and

$$Q_{\max} = S_m \rho_m c_m (L - x_j)(t_{fi} - t_o) = (0.01)(7833.0)(0.465)(5.8 - x_j)(t_{fi} - t_o)$$
$$= 36.42(5.8 - x_j)(t_{fi} - t_o) \text{ kJ per channel}$$

The organization of the calculation as shown in the following outline will help to make the required computations where

$0 \leqslant \tau < 15$ min $\quad \dot{m}_f = 0.156$ kg/s per channel $\quad h = 50.03$ W/m^2 °C

$\lambda_1 = 1.841 \quad \eta_1 = 1.373 \times 10^{-3}\tau$

$$t_{fo} = t_o + (t_{fi} - t_o)T_f\{\eta_1, \lambda_1\}$$
$$Q = Q_{\max} Q^+\{\eta_1, \lambda_1\}$$

At 15 min,

$$t_{fo} = 10.0 + (80.0 - 10.0)(0.484)$$
$$= 43.88\text{°C} \quad (110.99\text{°F})$$
$$Q = (36.42)(5.8)(80.0 - 10.0)(0.451)$$
$$= 6.67 \times 10^3 \text{ kJ} \quad (6.32 \times 10^3 \text{ Btu})$$

The temperature distribution in the storage unit at 15 min is obtained by

$\eta = 1.236 \qquad \xi = 0.3172x$

$$T_m\{\eta, \xi\} = 1.0 - T_f\{\xi, \eta\}$$
$$t_m = t_o + (t_{fi} - t_o)T_m\{\eta, \xi\}$$

A general expression for the determination of the temperature distribution in the storage unit with arbitrary initial temperature distribution is

$$t_m = t_{o1} + (t_{fi} - t_{o1})T_m\{\eta, \xi - \xi_1\} + \sum_{j=2}^{J} (t_{oj} - t_{oj-1})(1.0 - T_m\{\eta, \xi - \xi_j\})$$

The number of locations in the storage unit where the temperature is calculated will depend to a certain degree on the axial temperature gradient, since it will be necessary to subdivide the heat storage unit into sections that are approximately isothermal. This is necessary for the subsequent calculations of the transient response of the heat storage unit. Four subdivisions will be used in this example, with each section considered to be at the temperature of the center of the section.

	Location			
x, m	0.0	1.93	3.87	5.8
t_{o_1}, °C	59.67	46.04	35.61	27.95

15 min $\leqslant \tau \leqslant$ 30 min:

$$\dot{m}_f = 0.208 \text{ kg/s per channel} \qquad h = 63.23 \text{ W/m}^2\,°\text{C}$$

$$\lambda_2 = 1.745 \qquad \eta_2 = 1.736 \times 10^{-3}\,[\tau - (15.0)(60.0)]$$

$$\xi_j = 0.3006(x_j)$$

$$\lambda_2 - \xi_j = 0.3006(5.8 - x_j)$$

The temperature of the fluid leaving is obtained by using Eq. (4.6).

$$t_{fo} = t_{o1} + (t_{fi} - t_{o1})T_f\{\eta_2, \lambda_2\}$$

$$+ \sum_{j=2}^{J} (t_{oj} - t_{oj-1})(1.0 - T_f\{\eta_2, \lambda_2 - \xi_j\})$$

Four subproblems will be used.

	Subproblems			
	1	2	3	4
t_{fi}, °C	80.0	0.0	0.0	0.0
t_m, °C	59.67	46.04	35.61	27.95
$t_{oj} - t_{oj-1}$, °C		−13.59	−10.43	−7.66
$L - x_j$, m	5.8	4.835	2.90	0.965
$\lambda_2 - \xi_j$	1.745	1.453	0.872	0.290
$1.0 - T_f\{\eta_2, \lambda_2 - \xi_j\}$	0.424	0.355	0.210	0.065
$Q^+\{\eta_2, \lambda_2 - \xi_j\}$	0.547	0.581	0.658	0.744

At $\tau = 30$ min, $\eta_2 = 1.562$,

$$t_{fo} = 59.67 + (80.0 - 59.67)(0.424) - (13.59)(0.355)$$

$$- (10.43)(0.210) - 7.66(0.065)$$

$$= 60.78°\text{C} \quad (141.40°\text{F})$$

The heat storage is obtained by the use of Eq. (4.9). The amount of heat stored during the second 15 min of operation is

$$Q = 36.42[(5.8)(20.33)(0.547) + (4.835)(13.59)(0.581)$$

$$+ (2.90)(10.43)(0.658) + (0.965)(7.66)(0.744)]$$

$$= 4.67 \times 10^3 \text{ kJ per channel} \quad (4.42 \times 10^3 \text{ Btu per channel})$$

The total heat stored after 30 min of storage is

$$Q = 6.67 \times 10^3 + 4.67 \times 10^3 = 1.13 \times 10^4 \text{ kJ per channel}$$

$$(1.07 \times 10^4 \text{ Btu per channel})$$

The temperature distributed in the storage unit at τ equal to 30 min is obtained from the general expression given previously.

	Location			
x, m	0.0	1.93	3.87	5.8
t_o, °C	75.74	68.05	59.68	51.47

30 min $\leqslant \tau$:

$$\dot{m}_f = 0.078 \text{ kg/s} \qquad h = 28.84 \text{ W/m}^2 \text{ s}$$

$$\lambda_3 = 2.122 \qquad \eta_3 = 7.917 \times 10^{-4} [\tau - (30.0)(60.0)]$$

$$\xi_j = 0.366(5.8 - x_j)$$

The tabulated quantities needed in the solution are

	Subproblems			
	1	2	3	4
t_{fi}, °C	80.0	0.0	0.0	0.0
t_m, °C	75.74	68.05	59.68	51.47
$t_{oj} - t_{oj-1}$		−7.67	−8.37	−8.21
$L - x_j$, m	5.8	4.835	2.900	0.965
$\lambda_3 - \xi_i$	2.122	1.770	1.061	0.353
$1.0 - T_f\{\eta_3, \lambda_3 - \xi_j\}$	0.536	0.458	0.281	0.090
$Q^+\{\eta_3, \lambda_3 - \xi_j\}$	0.472	0.510	0.597	0.701

After 1 h of operation, $\eta = 1.425$,

$$\begin{aligned} t_{fo} &= 75.74 + (4.26)(0.536) - (7.67)(0.458) - (8.37)(0.281) \\ &\quad - (8.21)(0.090) \\ &= 71.42°\text{C} \quad (160.56°\text{F}) \end{aligned}$$

The heat storage is obtained by Eq. (4.8). The amount of heat stored during the last 30 min is

$$\begin{aligned} Q &= (36.42)[(5.8)(0.472)(4.26) + (4.835)(0.510)(7.67) \\ &\quad + (2.900)(0.597)(8.37) + (0.965)(0.701)(8.21)] \\ &= 1.84 \times 10^3 \text{ kJ per channel} \quad (1.74 \times 10^3 \text{ Btu per channel}) \end{aligned}$$

The total heat stored during the first hour of operation of the unit is

$$Q = 1.317 \times 10^4 \text{ kJ per channel} \quad (1.248 \times 10^4 \text{ Btu per channel})$$

Example 4.5 The mass rate of flow to the storage unit is as described in Example 4.4. It is desired to control the flow so that the flow rate per channel remains constant at 0.156 kg/s. Determine the average temperature of the fluid leaving the storage unit and the amount of heat stored during a 1-h operating period.

SOLUTION The number of fluid channels used during the heat storage is

$0 \leqslant \tau < 15$ min: 12 channels (numbers 1, 2, . . ., 12)

15 min $\leqslant \tau < 30$ min: 16 channels (numbers 1, 2, 3, . . ., 16)

30 min $< \tau$: 6 channels (numbers 1, 2, . . ., 6)

$\dot{m}_f = 0.156$ kg/s per channel $\qquad h = 50.03\ \text{W/m}^2\,^\circ\text{C}$

$\lambda = 1.841 \qquad \eta = 1.373 \times 10^{-3}\,\tau$

The temperature of the fluid leaving the heat storage unit and the heat storage is

$0 \leqslant \tau < 15$ min, channels 1 → 12 active:

$$t_{fo} = t_o + (t_{fi} - t_o)T_f\{\eta, \lambda\}$$
$$= 10.0 + (80.0 - 10.0)T_f\{\eta, 1.841\}$$

At 15 minutes,

$$t_{fo} = 10.0 + (80.0 - 10.0)T_f\{1.236, 1.841\}$$
$$= 10.0 + (70.0)(0.484)$$
$$= 43.88^\circ\text{C} \quad (110.99^\circ\text{F})$$

The heat storage is

$$Q = (0.01)(7833.0)(0.465)(5.8)(70.0)Q^+\{1.236, 1.841\}$$
$$= 36.42(5.8)(70.0)(0.451)$$
$$= 6.67 \times 10^3 \text{ kJ per channel} \quad (6.321 \times 10^3 \text{ Btu per channel})$$

or $Q_{\text{total}} = 8.004 \times 10^4$ kJ (7.585×10^4 Btu).

15 min $\leqslant \tau < 30$ min, channels 1 → 16 active:

Channels 1 → 12

$$t_{fo}^{1-12} = t_o + (t_{fi} - t_o)T_f\{\eta, \lambda\}$$

Channels 13 → 16

$$t_{fo}^{13-16} = t_o + (t_{fi} - t_o)T_f\{\eta', \lambda\}$$

where $\eta' = 1.373 \times 10^{-3}\,[\tau - (60.0)(15.0)]$.

At 30 min, $t_{fo}^{1-12} = 59.81^\circ$C and $t_{fo}^{13-16} = 43.88^\circ$C. The average temperature of the mixed outlet streams is

$$\bar{t}_{fo} = \frac{12t_{fo}^{1-12} + 4t_{fo}^{13-16}}{16} = 55.83^\circ\text{C} \quad (132.49^\circ\text{F})$$

The total heat stored in all the channels at 30 min is

$$
\begin{aligned}
Q &= S_m \rho_m c_m L(t_{fi} - t_{fo})(12Q^+\{\eta, \lambda\} + 4\dot{Q}^+\{\eta', \lambda\}) \\
&= (36.42)(5.8)(70.0)[12(0.715) + 4(0.451)] \\
&= 1.535 \times 10^5 \text{ kJ} \qquad (1.455 \times 10^5 \text{ Btu})
\end{aligned}
$$

30 min $\leqslant \tau$, channels 1 → 6 active:

$$
t_{fo}^{1-6} = t_o + (t_{fi} - t_o)T_f\{\eta, \lambda\}
$$

At 1 h,

$$
\begin{aligned}
t_{fo}^{1-6} &= 10.0 + 70.0(0.9242) \\
t_{fo}^{1-6} &= 74.69^\circ\text{C} \qquad (166.4^\circ\text{F})
\end{aligned}
$$

The total heat stored at 1 h is

$$
\begin{aligned}
Q &= S_m \rho_m c_m L(t_{fi} - t_o)(6Q^+\{4.943, 1{,}841\} + 6Q^+\{2.4714, 1.841\} \\
&\quad + 4Q^+\{1.236, 1.841\}) \\
&= (36.42)(5.8)(70.0)[6(0.932) + 6(0.715) + 4(0.451)] \\
&= 1.728 \times 10^5 \text{ kJ} \qquad (1.63 \times 10^5 \text{ Btu})
\end{aligned}
$$

REFERENCES

1. F. W. Schmidt and J. Szego, "Computer Program for the Prediction of the Transient Response of Solid Sensible Heat Storage Units," TESR-4, Mechanical Engineering Department, Pennsylvania State University, University Park, Pa., 1978.
2. F. W. Larsen, "Rapid Calculation of Temperature in a Regenerative Heat Exchanger Having Arbitrary Initial Solid and Entering Fluid Temperatures," *Int. J. Heat Mass Transfer*, vol. 10, 1967, p. 149.
3. F. W. Schmidt and J. Szego, "Transient Response of Solid Sensible Heat Thermal Storage Units–Single Fluid," *J. Heat Transfer, Trans. ASME*, vol. 98, 1976, p. 471.

CHAPTER

FIVE

BASIC CONCEPTS IN COUNTERFLOW REGENERATORS

5.1 INTRODUCTION

In previous chapters the single-blow problem has been considered in terms of the spatial and time variation of the temperature of the gas passing over the available heating surface area of a heat storing solid. The corresponding variations of solid temperature were considered together with a heat storage factor representing the chronological variation of the proportion of the heat held in the solid relative to the thermodynamically maximum heat that could be accumulated. The recovery of heat from the storage medium was considered as a single-blow problem and treated in an analogous way.

Thermal regenerators can be viewed from two complementary viewpoints. On the one hand, they can be viewed as regenerative heat exchangers used to effect the transfer of heat between two fluids (usually gases) by temporarily storing heat from the hotter gas in a permeable packing and subsequently passing this thermal energy to the cooler gas. All the surface area of the regenerator packing or "checkerwork" is washed by the hot gas for the duration of the hot period during which sensible heat is stored in the packing. At the end of this period, a reversal takes place at which the hot gas is switched off and the cooler gas starts to flow through the same channels in the packing over the whole surface area. The thermal energy is recovered by the cold gas during the cold period at the end of which occurs another reversal.

On the other hand, the surface of the regenerator packing can be considered as being submitted to a succession of linked single blows, during which heat is alternately accumulated for the length of the hot period and then recovered from the packing for the duration of the cold period.

A view of regenerator performance emerges rather different from that of a sensible heat store submitted to a single blow. It will be seen that the ability of the packing to store heat is matched by a cycle time (hot period plus cold period) appropriately chosen in order to minimize the heat not recovered from the hot gas while meeting a specified thermal demand in the cold period, or in some cases to adjust the thermal performance of the regenerator to meet required operating conditions, which may be less than optimal. Regenerators appear in two distinct forms, the fixed bed regenerator and the rotary regenerator.

The reversals in the fixed bed system are effected by the closing and opening of the relevant valves, shutting off the hot/cold gas and allowing the start of the flow of the cold/hot gas through the packing. In order to provide an almost continuous exchange of heat between the two gases, at least two fixed bed regenerators must be used, one of which at any instant will be preheating the cold gas (Fig. 5.1).

The Ljungstrom or rotary regenerator (Fig. 5.2) consists of a cylindrical packing through which the hot and cold gases pass simultaneously. Heat temporarily stored in this packing from the hot gas is physically moved into the cold gas stream by steadily rotating the cylindrical body of the checkerwork, whose axis of rotation is parallel to the gas flow. Considerable attention must be paid to providing gas seals that minimize the leakage between hot and cold gas at the gas entrance/exit faces of the packing. As a section of the packing rotates between the gas seals from one gas stream to the other, it experiences what is the equivalent of a reversal in the fixed bed system. The time taken for this section to pass through the hot/cold gas stream corresponds to the hot/cold period. Regenerators are commonly operated in counterflow mode: that is, the hot gas passes through the checkerwork in the opposite direction to the cold gas. If the gases pass through the packing in the same direction, the regenerator is said to operate in parallel-flow mode.

Regenerators of the fixed bed type are employed under conditions of high temperature and/or significant pressure differences between the hot and cold gas streams. Cowper stoves (Fig. 5.3) are used to preheat the blast for the iron making furnace and are required to deliver preheated air at a temperature of 800–1200°C. Regenerators are constructed as part and parcel of a glass making furnace and must be able to withstand entrance gas temperatures of the order of 1600°C. In these applications, the checkerwork is made of ceramic brick material; any design methodology has to balance desirable heat transfer characteristics of the checkerwork with the ability of that checkerwork to withstand possibly harsh operating conditions.

Under less severe conditions, what proves to be the very compact rotary regenerator is employed as a heat exchanger. It is commonly used to recover the waste heat from boilers in power stations and aboard ships, from gas turbines, and from process drying plants. More recently, the rotary regenerator has been exploited to recover waste heat (in cold climates) or waste cold (in hot climates) in the air conditioning and ventilating systems of buildings.

In this introduction, the thermal regenerator has been viewed on the whole as a heat exchanger. The necessary switching of the bed from the hot to the cold gas stream—by the opening and closing of valves for a two or more unit fixed bed system or by rotating the checkerwork—has often been regarded as a nuisance. Indeed, Hausen

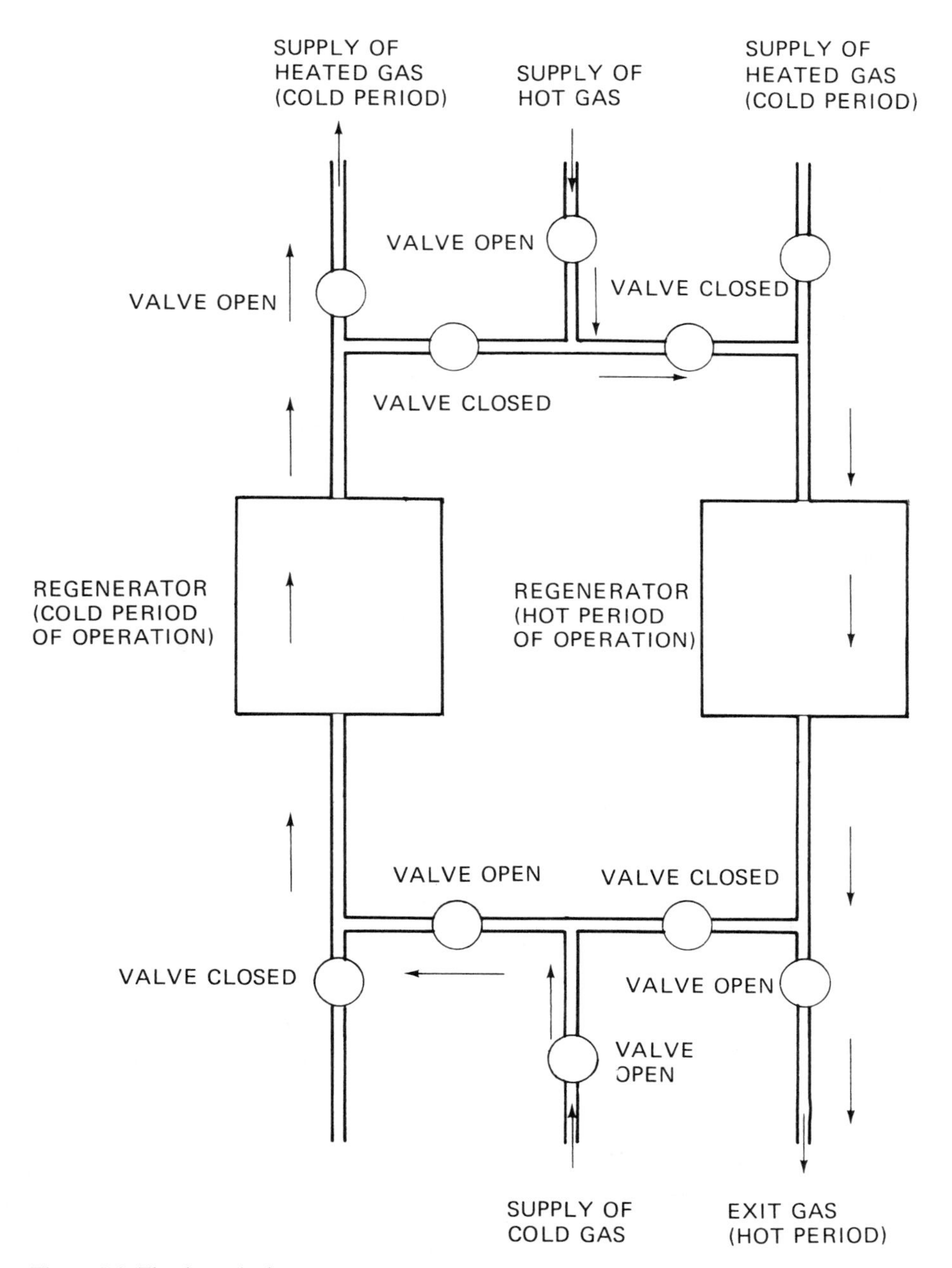

Figure 5.1 Fixed two-bed regenerator system.

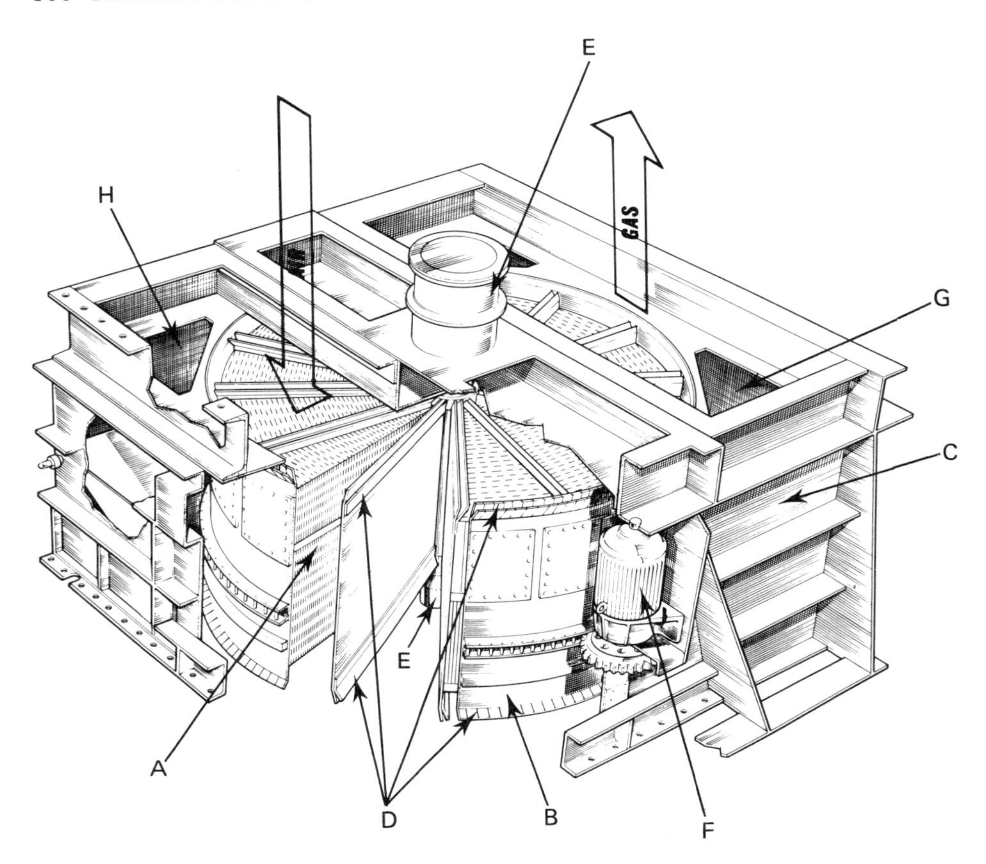

Figure 5.2 Rotary regenerator. A, Heating surface elements. B, Rotor in which the elements are packed. C, Housing in which the rotor rotates. D, Seals and sealing surfaces. E, Support and guide bearing assemblies. F, Drive mechanism. G, Gas by-pass. H, Air by-pass. (*Courtesy of James Howden & Co., Glasgow, Scotland.*)

[1] suggests that a regenerator ideally should be used only where the cyclic operation of the system can be coupled directly to the necessary cyclic operation of the regenerator. Hausen goes on to propose that the walls of the throat and nasal passages act regeneratively by transferring heat from expelled breath to the cold air subsequently breathed into the lungs. Nevertheless, advantage is taken of the alternate passing of the hot and cold gases through the same passages in the checkerwork; dirt or frozen fluid (in low-temperature applications) deposited by one stream can be relied upon to be purged from the passages by the other stream subsequently, thereby avoiding the blocking of the channels possible in the recuperative type of heat exchanger.

Only comparatively recently have regenerators begun to become important as sensible heat storage units. The cyclic operation of the regenerator is not seen as a nuisance, but rather as an ideal mechanism to transfer thermal energy in circumstances where the availability of the heat or cold does not coincide chronologically with

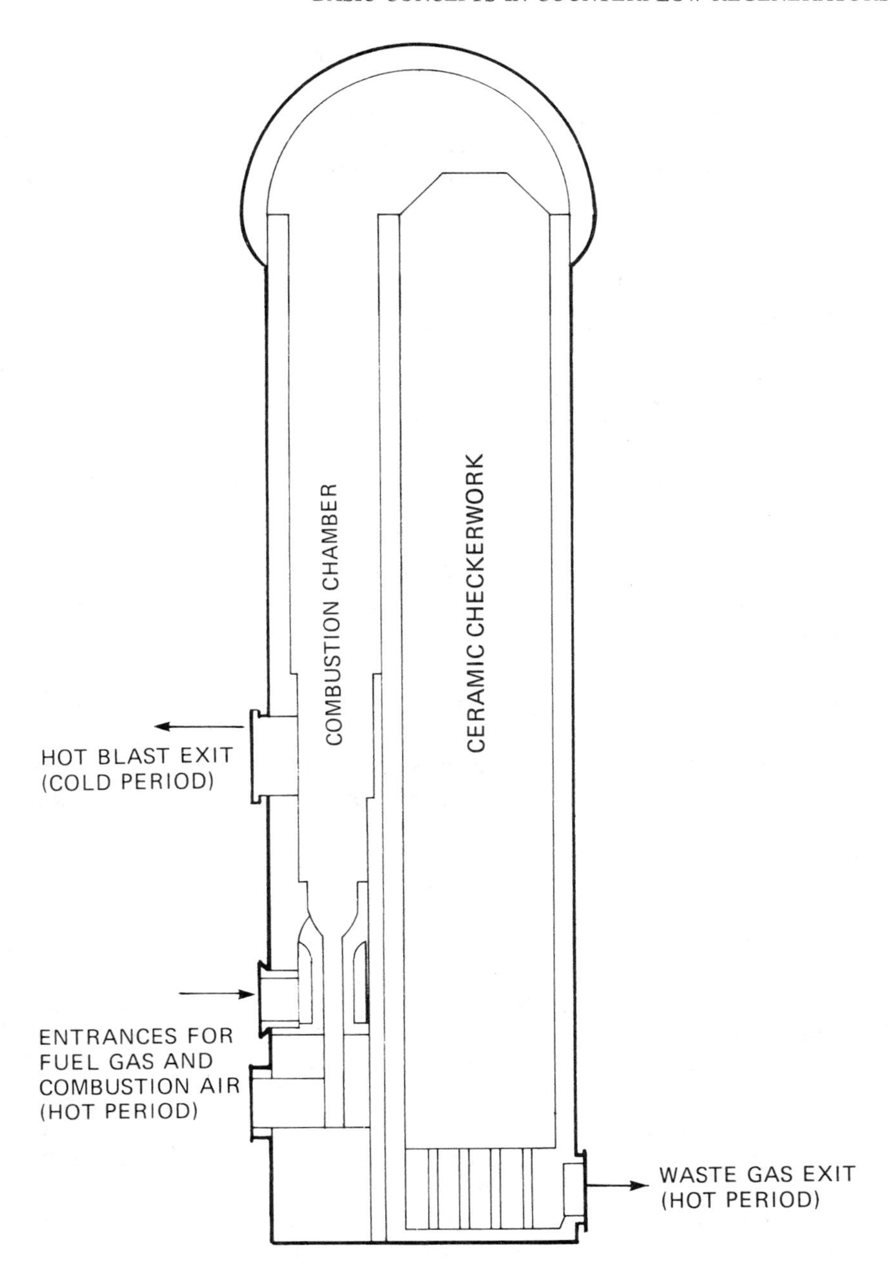

Figure 5.3 Cowper stove regenerator system.

demand. Here the thermal storage mechanism, which is part of the regenerative cycle, is specifically exploited.

Fortunately, rotary and fixed bed regenerators used either as heat exchangers or as heat storage units can be idealized into a common mathematical model, which can be employed to predict the likely performance of existing systems or of regenerators under design for possibly new and novel applications.

5.2 MATHEMATICAL MODEL

The descriptive differential equations in one period of regenerator operation are similar to those developed for the single-blow unit. They may be developed by consideration of the exchange of heat between the gas flowing through a single channel of the checkerwork and the wall of solid material through which the channel is driven. The overall behavior of the regenerator is then obtained by regarding the checkerwork as a bundle of channels, identical in configuration and operation.

These differential equations represent the relation between the rate of heat transferred between gas and solid and the rate of heat given up/absorbed by the gas flowing through the regenerator:

$$\bar{h}A(t_m - t_f) = \dot{m}_f c_f L \frac{\partial t_f}{\partial x} + m_f c_f \frac{\partial t_f}{\partial \tau} \tag{5.1}$$

where $m_f = \rho_f S_f L$ and the relation between the same rate of exchange of heat across the surface of solid packing and the heat stored in/recovered from the checkerwork is

$$\bar{h}A(t_f - t_m) = M_m c_m \frac{\partial t_m}{\partial \tau} \tag{5.2}$$

These equations are identical in form to those employed for the single-blow problem. They are applied separately to the hot and cold periods of operation. Whereas the initial temperature distribution $t_m(x, 0)$ in the solid is specified arbitrarily for the single-blow problem—sometimes, the solid is treated to be initially isothermal—for regenerators, at the reversals, the temperature distribution $t_m(x, 0)$ at the start of a period is set equal to that at the end of the previous period of operation.

To represent this mathematically, two nomenclature problems must be resolved. First, it is necessary to distinguish between the hot and cold periods: we use a single prime to signify the hot period, and double primes the cold period. In this way, different gas flow rates in the hot and cold periods, for example, can be represented. Second, it is important to embody in the model the counterflow operation of the regenerator; by specifying that the direction x is always measured from the gas entrance to the checkerwork in the direction of gas flow, Eqs. (5.1) and (5.2) (with single/double primes as appropriate) can be applied to both periods.

The reversal condition can now be specified. At the start of the hot period,

$$t'_m(x, 0) = t''_m(L - x, P'') \tag{5.3}$$

and at the start of the cold period,

$$t''_m(x, 0) = t'_m(L - x, P') \tag{5.4}$$

The mathematical model embodied in Eqs. (5.1)–(5.4) is based on an idealization of a physical regenerator system. Such an idealization seeks on the one hand to provide a generally applicable model; thus the duration of hot or cold period, P' or P'', corresponds to the time a channel of the packing is moving through the hot or cold gas stream in the case of the rotary regenerator, and the length of time the hot or cold gas is allowed to pass through the packing of the fixed bed system. On the other hand, a useful idealization will be mathematically and computationally tractable and must embody assumptions about the physical system. Some of the assumptions add to the generality of the model, others provide what prove to be acceptable simplifications in the model in the sense that they do not inhibit the ability of the model to represent the physical system to an acceptable degree of accuracy.

In particular, this model assumes that:

1. There is a uniform flow of gas across the whole cross section of the packing for both hot and cold gas streams; the gas flow rates do not vary with time.
2. There is no longitudinal conduction of heat in the direction of gas flow, either in the solid or in the gas.
3. The heat transfer coefficients and the thermal properties of both gas and solid do not vary spatially or with time in either the hot or cold period, although the parameter values applicable to the hot period may be different from those for the cold period.

In many papers it is assumed also that the inlet gas temperature to the regenerator, in either period of operation, does not vary with time. This is not a necessary assumption for the model to retain the linearizing effect of assumption 3 above; it is a sufficient boundary condition for the differential Eqs. (5.1) and (5.2) to specify that the inlet gas temperatures vary with time in a prescribed manner, or in a way determined by an external process that is part of the whole system in which the regenerator may be embodied. At this stage it is useful to introduce the dimensionless parameters

$$\xi = \frac{\bar{h}Ax}{\dot{m}_f c_f L} \tag{5.5}$$

$$\eta = \frac{\bar{h}A}{M_m c_m}\left(\tau - \frac{m_f x}{\dot{m}_f L}\right) \tag{5.6}$$

These allow Eqs. (5.1) and (5.2) to be simplified into the nondimensional form developed in Chap. 2 for single-blow units,

$$\text{Fluid:} \qquad \frac{\partial t_f}{\partial \xi} = t_m - t_f \tag{5.7}$$

$$\text{Storage material:} \qquad \frac{\partial t_m}{\partial \eta} = t_f - t_m \tag{5.8}$$

The term $m_f c_f(\partial t_f/\partial\tau)$ in Eq. (5.1) represents the effect of the heat capacity of the gas resident in the channels of the regenerator. The corresponding term $m_f x/\dot{m}_f L$ represents the delay in the effect of any gas entering the regenerator becoming manifest at a distance from the gas entrance which is a proportion x/L of the total length L of the regenerator. At the start of a period, the gas residing in the regenerator from the previous period must be driven out by the incoming new gas; assuming no mixing between the old and new gases, $m_f/\dot{m}_f$ is the time taken for the residual gas to be expelled from the regenerator.

In most models it is assumed that the filling of the regenerator channels with the new gas, called the *reversal effect*, has no influence on regenerator performance. At any position x in the packing, the period is considered only to start at time $\tau = m_f x/\dot{m}_f L$, that is, at $\eta = 0$.

The parameters ξ and η allow a natural set of design parameters to be evolved; these correspond to the values of ξ and η for $x = L$ and $\tau = P$. The design parameters have been called by Hausen [1] *reduced length* Λ and *reduced period* Π and are as follows:

Hot period	Cold period
$\Lambda' = \bar{h}' A/\dot{m}'_f c'_f$	$\Lambda'' = \bar{h}'' A/\dot{m}''_f c''_f$
$\Pi' = \bar{h}' A(P' - m'_f/\dot{m}'_f)/M_m c_m$	$\Pi'' = \bar{h}'' A(P'' - m''_f/\dot{m}''_f)/M_m c_m$

The reversal conditions for counterflow operation can be now written

$$t'_m\{\xi', 0\} = t''_m\{\Lambda''(\Lambda' - \xi')/\Lambda', \Pi''\} \tag{5.9}$$

and

$$t''_m\{\xi'', 0\} = t'_m\{\Lambda'(\Lambda'' - \xi'')/\Lambda'', \Pi'\} \tag{5.10}$$

If t'_{fi} and t''_{fi} are constant inlet gas temperatures in the hot and cold periods, respectively, a nondimensional [0, 1] temperature scale can be used, similar in form to that used for the single-blow problem. The nondimensional temperatures T_f and T_m for regenerators are defined by

$$T_f = \frac{t_f - t''_{fi}}{t'_{fi} - t''_{fi}} \tag{5.11}$$

$$T_m = \frac{t_m - t''_{fi}}{t'_{fi} - t''_{fi}} \tag{5.12}$$

Equations (5.7) and (5.8) then become simply

$$\frac{\partial T_f}{\partial \xi} = T_m - T_f \tag{5.13}$$

$$\frac{\partial T_m}{\partial \eta} = T_f - T_m \tag{5.14}$$

with hot inlet gas temperature equal to unity, cold inlet gas temperature equal to zero.

5.3 DISCUSSION OF DESIGN PARAMETERS

The reduced lengths Λ' and Λ'' and the reduced periods Π' and Π'' form a natural set of parameters in two senses. First, they are evolved directly from the descriptive differential equations; and second, consequent upon this, they separate conveniently the distinct mechanisms of the transfer of heat to the gases passing through the packing and of the storage of heat in the packing.

The ability of the regenerator to transfer heat to the gas is proportional to the product $\bar{h}'A$ in the hot period and $\bar{h}''A$ in the cold period. The corresponding loads imposed on the regenerator are proportional to the thermal flow capacities $\dot{m}_f'c_f'$ and $\dot{m}_f''c_f''$; these are combined with the regenerator's ability to transfer heat to yield the reduced lengths Λ' and Λ''. The cold period Λ'' can be regarded as the "thermal size" of the regenerator relative to the load imposed by the gas of specific heat c_f'' flowing through the regenerator at rate $\dot{m}_f''$.

The effectiveness of regenerator behavior is measured in terms of the thermal ratio η_{REG}. This is defined to be the ratio of the actual heat transfer rate to the thermodynamically limited maximum obtainable heat transfer rate in a counterflow regenerator of infinite heat transfer area. Could this maximum rate be achieved, the temperature of the gas leaving the regenerator in hot/cold period would be equal to the entrance temperature t_{fi}'' (cold period)/t_{fi}' (hot period).

In the fixed bed regenerator the exit gas temperatures vary with time, whereas in the rotary regenerator the local exit gas temperature varies across the exit face of the revolving packing. In the former case a chronological average exit temperature can be computed, and this corresponds in the latter case to the temperature of the exit gas, which consists of a mixture of gas coming from the several sectors of the rotating checkerwork. If these average exit temperatures are denoted by t_{fo}' and t_{fo}'' (T_{fo}' and T_{fo}'' on the [0, 1] scale), then the thermal ratios may be specified:

$$\eta_{REG}' = \frac{t_{fi}' - t_{fo}'}{t_{fi}' - t_{fi}''} = 1 - T_{fo}' \tag{5.15}$$

$$\eta_{REG}'' = \frac{t_{fo}'' - t_{fi}''}{t_{fi}' - t_{fi}''} = T_{fo}'' \tag{5.16}$$

As one might expect, the larger the reduced lengths of the regenerator, the greater the thermal ratio. Indeed, for the symmetric regenerator ($\Lambda = \Lambda' = \Lambda''$ and $\Pi = \Pi' = \Pi''$, where $\eta_{REG} = \eta_{REG}' = \eta_{REG}''$), an estimate of thermal ratio is provided by

$$\eta_{REG} = \frac{\Lambda}{\Lambda + 2} \tag{5.17}$$

This estimate becomes exact for infinitely small reduced period and is called the "ideal" thermal ratio. Displayed in Fig. 5.4 is a graph displaying thermal effectiveness as a function of Λ and Π for the symmetric case. The curve for $\Pi = 0$ is the ideal computed from Eq. (5.17). The remaining curves for $\Pi = 10, 20, 30, 40$, and 50 were

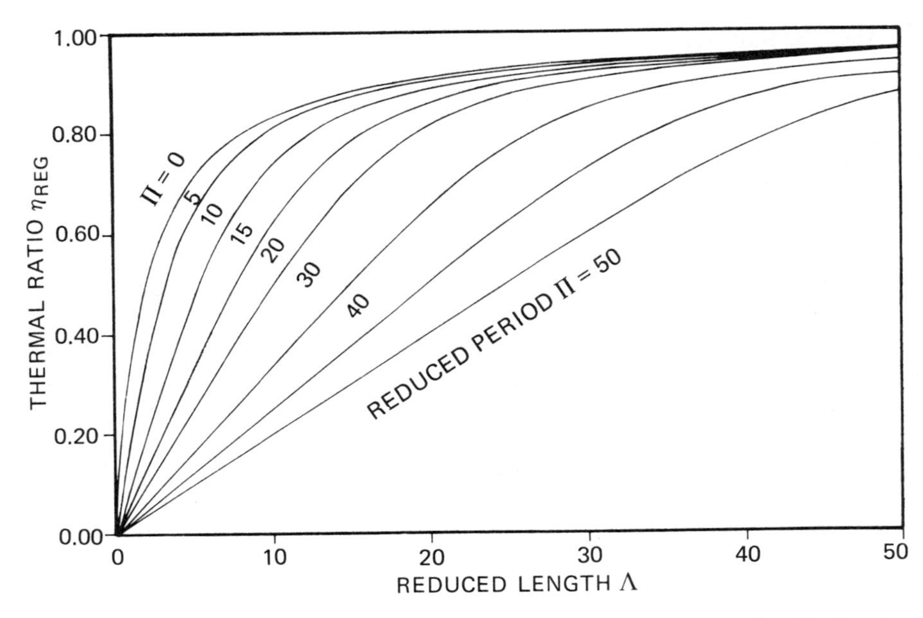

Figure 5.4 Graph of thermal effectiveness (ratio) in relation to reduced length and reduced period for symmetric counterflow regenerators.

computed initially by Hausen [1] by solving numerically Eqs. (5.13) and (5.14) by a method similar to that of Illife [2], a method discussed in Sec. 5.6.

These curves have been recomputed using Nahavandi and Weinstein's method [3] for some cases and by Willmott's method [4] for the remainder. The thermal ratios are presented in Table 5.1 and in graphical form in Fig. 5.4.

Table 5.1 Thermal effectiveness (ratio) (η_{REG}) in relation to reduced length and reduced period for symmetric counterflow regenerators

Reduced length, Λ	Reduced period, Π					
	0	10	20	30	40	50
5	0.714	0.469	0.250	0.167	0.125	0.100
10	0.833	0.738	0.494	0.333	0.250	0.200
15	0.882	0.840	0.693	0.498	0.375	0.300
20	0.909	0.886	0.811	0.651	0.400	0.400
25	0.926	0.911	0.871	0.770	0.620	0.500
30	0.936	0.927	0.903	0.845	0.727	0.598
35	0.946	0.939	0.922	0.888	0.810	0.692
40	0.952	0.947	0.935	0.912	0.865	0.773
45	0.957	0.953	0.944	0.928	0.898	0.835
50	0.962	0.958	0.951	0.939	0.909	0.875

Example 5.1 In Chap. 2 a Feolite thermal storage unit was considered composed of 12 channels 5.8 m long and 0.5 m wide. The thickness of each channel was 1.9 cm. By considering the performance of a single channel, the whole unit can be treated as a bundle of identical channels.

SOLUTION The available heating surface area A for one such channel is 5.8 m^2 and the corresponding heat storing mass M_m is 904.8 kg. The mass flow rate of the gas $\dot{m}_f$ through each channel is 0.156 kg/s; the gas specific heat c_f is 1011.0 J/kg °C. The specific heat of the packing c_m is 0.92 kJ/kg °C. The convective heat transfer coefficient is 50.23 W/m^2 °C.

The storage unit is now considered to be operated as a regenerator with hot inlet temperature $t'_{fi} = 80°C$ and cold inlet temperature $t''_{fi} = 10°C$ with the gas flow rates in both hot and cold period equal to 0.156 kg/s, the same flow rate used for the single-blow problem.

This is a symmetric regenerator with $\Lambda' = \Lambda'' = \Lambda$, where

$$\Lambda = \frac{\bar{h}A}{\dot{m}_f c_f} = \frac{(50.23)(5.8)}{(0.156)(1011.0)} = 1.847$$

The ideal thermal ratio for very short cycles is

$$\eta_{\text{REG}} = \frac{\Lambda}{2 + \Lambda} = \frac{1.847}{2 + 1.847} = 0.4801$$

The corresponding exit temperatures are computed using

Hot period:

$$\eta'_{\text{REG}} = \eta_{\text{REG}} = 0.4810 = \frac{t'_{fi} - t'_{fo}}{t'_{fi} - t''_{fi}}$$

$$= \frac{80.0 - t'_{fx}}{80.0 - 10.0}$$

$$t'_{fo} = 80.0 - (70.0)(0.4801)$$

$$= 46.4°\text{C}$$

Cold period:

$$\eta''_{\text{REG}} = \eta_{\text{REG}} = 0.4801 = \frac{t''_{fo} - t'_{fi}}{t'_{fi} - t''_{fi}}$$

$$= \frac{t''_{fo} - 10.0}{80.0 - 10.0}$$

$$t''_{fo} = (70.0)(0.4801) + 10.0$$

$$= 43.6°\text{C}$$

Operation of the regenerator with other than very short cycles would yield poorer thermal ratios.

Example 5.2 How large should such a Feolite thermal regenerator be to yield an ideal effectiveness of 0.85 for the same gas flow rates and heat transfer coefficients as in Example 5.1?

SOLUTION

$$\eta_{\mathrm{REG}} = 0.85 = \frac{\Lambda}{\Lambda + 2}$$

The required reduced length $\Lambda = \Lambda' = \Lambda''$ is

$$\Lambda = \frac{(2.0)(0.85)}{1.0 - 0.85} = \frac{1.7}{0.15} = 11.33$$

This reduced length is computed:

$$\Lambda = \frac{\bar{h}A}{\dot{m}_f c_f} = \frac{(50.23)(A)}{(0.156)(1011.0)} = 11.33$$

and the required heating surface area is

$$A = \frac{(11.33)(0.156)(1011.0)}{50.23}$$

$$= 35.56 \text{ m}^2$$

The required length L of the regenerator is in this case 35.56 m; the width remains 0.5 m. Alternatively, the width of the regenerator can be increased. In this case the heat transfer coefficient $\bar{h}$ will drop as the gas velocity in the channels decreases. The mass flow rate remains unaltered. To maintain a reduced length Λ of 11.33, the *product* of $\bar{h}$ and A must not change; as $\bar{h}$ decreases, so A, the necessary heating surface area, must be increased in the same proportion.

5.4 EFFECT OF CYCLE TIME ON REGENERATOR PERFORMANCE

The ideal thermal ratio is a function of reduced length only and does not involve the heat capacity of the checkerwork. This ideal is estimated using an infinitely short cycle time, $\Pi = 0$.

For longer periods, the storage capability of the regenerator becomes important. The heat capacity of the packing per period is $M_m c'_m/P'$ (hot period) and $M_m c''_m/P''$ (cold period). The larger the effective interface between the gas flowing through the regenerator and thermal storage medium, namely, $\bar{h}'A$ and $\bar{h}''A$ (hot and cold periods), the greater must be the heat capacity of the packing per period if over heating/cooling of the checkerwork is not to occur, thereby reducing the corresponding thermal ratio. As period length increases, thermal effectiveness becomes smaller than the ideal; this is shown in Fig. 5.4.

Translated into practical terms, this means that for a given packing, operating under specified flow rate conditions, the ratios $\bar{h}'A/M_m c'_m$ and $\bar{h}''A/M_m c''_m$ can be computed. The ratios $\bar{h}'A/M_m c'_m$ and $\bar{h}''A/M_m c''_m$ are therefore matched by period lengths P' and P'' to yield as large a pair of thermal ratios as practicable, close to the ideal thermal ratio if possible.

In rotary regenerators, the packing often consists of a plated wire meshwork or sheets of metal or synthetic material; in both cases, a very large heating surface is provided for the exchange of heat between gas and solid. In this way, for a desired reduced length Λ' (hot period), Λ'' (cold period), a very compact heat exchanger can be designed. On the other hand, the large heating surface area is not supported by a corresponding large heat capacity; the area-to-checkerwork thermal capacity ratio $A/M_m c_m$ is large. Small period lengths compensate for this, and it is common for the checkerwork of a small but compact regenerator to rotate at 2, 3, or even more revolutions per minute. The very large rotary regenerators in power stations have rotational speeds of the order of 1 rpm.

In contrast, the harsh operating conditions that occur in high-temperature regenerators require the checkerwork to consist of ceramic bricks, sometimes several centimeters thick. Here, the period lengths P' and P'' are very much longer. Indeed, for Cowper stoves it is common for the periods to be $P' = 2$ h, $P'' = 1$ h.

The rotary regenerator provides one extreme: the heat capacity of the packing is relatively small, whereas the period lengths are, correspondingly, so small as to make the heat storage component of the cycle comparatively unimportant. The other extreme is a heat storage unit in which the specific aim is to store heat between the times when this thermal energy is available and when it is required. Here, once the average time between the availability and the demand for heat is known (say, 10–12 h for solar energy to be used to provide home heating in the evening), this can be matched by a packing of sufficient size to generate a large enough value of reduced length (say, $\Lambda = 10$) for about 80% effectiveness and an area/heat capacity of packing ratio $A/M_m c_m$ to minimise the degrading effect of too large a reduced period (this means keeping $\Pi < 1$ or at most 2).

Example 5.3 We now consider possible cycle times for a regenerator of the configuration developed in Example 5.2. This is illuminating because it illustrates the significance of the dimensionless parameter Π.

SOLUTION Each channel is considered to have the following dimensions:

$$A = 35.56 \text{ m}^2$$

$$M_m = 5547.4 \text{ kg}$$

$$c_m = 920.0 \text{ J/kg °C}$$

$$\bar{h} = 50.23 \text{ W/m}^2 \text{ °C}$$

$$\Lambda = 11.33$$

Note that the development of the reduced length $\Lambda = 11.33$ in Example 5.2 did not involve any consideration of the packing material of the regenerator. Those characteristics are embodied separately in the reduced period Π, which we now consider. Again, this is a symmetric regenerator with $\Pi' = \Pi'' = \Pi$, where

$$\Pi = \frac{\bar{h}AP}{M_m c_m} = \frac{(50.23)(35.56)(P)}{(5547.4)(920.0)}$$

$$= 3.5 \times 10^{-4} P$$

For $P = 1$ h $= 3600$ s, $\Pi = 1.26$; from Fig. 5.4 it can be seen that the thermal ratio is almost as large as the ideal of 0.85 for very short cycle times. For $\Lambda = 11.26$, $\Pi = 1.26$, $0.8 < \eta_{REG} < 0.85$. In fact, $\eta_{REG} = 0.848$.

Example 5.4 Can this regenerator configuration support heat storage for a 1-day cycle time, assuming that $P' = P'' = P = 12$ h?

SOLUTION The reduced period Π $(= \Pi' = \Pi'')$ is equal to $(12.0)(1.26) = 15.12$. Notice that Λ/Π is now less than unity, which usually signifies a regenerator of low efficiency. From Fig. 5.4 it will be seen that η_{REG} is approximately 0.65 for $\Lambda = 11.26$ and $\Pi = 15.12$; precisely, $\eta_{REG} = 0.669$.

If the unit is made bigger by extending its surface area, say, by increasing its height, the heat transfer coefficient may be assumed not to be changed; the area/heat capacity ratio of the packing is not altered, and thus the reduced period does not change. To maintain an effectiveness $\eta_{REG} > 0.8$, from Fig. 5.4 it will be seen that for $\Pi = 15.12$, $\Lambda > 17.0$, almost double that of the regenerator considered in Example 5.3. If this increase was obtained by increasing the height of the regenerator, this would yield a unit almost 60 m high, as tall as a big chimney!

Notice that another alternative of leaving the reduced length $\Lambda = 11.26$ unaltered but reducing the dimensionless period Π to less than 1 would involve increasing the mass of the packing to something *ten* times larger than the original!

5.5 IMBALANCE IN REGENERATOR PERFORMANCE

Earlier in this chapter, a symmetric regenerator ($\Lambda = \Lambda' = \Lambda''$ and $\Pi = \Pi' = \Pi''$) was mentioned for which it was specified that $\eta'_{REG} = \eta''_{REG}$. For equilibrium or steady operation of a regenerator, the heat input into the system must equal the heat recovered, assuming no heat losses. This can be represented by

$$\dot{m}'_f c'_f P'(t'_{fi} - t'_{fo}) = \dot{m}''_f c''_f P''(t''_{fo} - t''_{fi}) \tag{5.18}$$

Dividing Eq. (5.18) throughout by $(t'_{fi} - t''_{fi})$ yields

$$\dot{m}'_f c'_f P' \eta'_{REG} = \dot{m}''_f c''_f P'' \eta''_{REG} \tag{5.19}$$

If the reversal effects are ignored,

$$\frac{\dot{m}'_f c'_f P'}{M_m c'_m} = \frac{\Pi'}{\Lambda'} \quad \text{and} \quad \frac{\dot{m}''_f c''_f P''}{M_m c''_m} = \frac{\Pi''}{\Lambda''}$$

Assuming that $M_m c'_m = M_m c''_m$, Eq. (5.19) becomes

$$\frac{\Pi'}{\Lambda'} \eta'_{\text{REG}} = \frac{\Pi''}{\Lambda''} \eta''_{\text{REG}} \tag{5.20}$$

A regenerator is said to be balanced if, for steady-state operation, $\eta'_{\text{REG}} = \eta''_{\text{REG}}$. Hence for a balanced regenerator, Π'/Λ' is equal to Π''/Λ''. It will be seen that for a regenerator to be balanced, it is necessary that

$$\frac{\Pi'}{\Pi''} = \frac{\Lambda'}{\Lambda''} = k \tag{5.21}$$

where k is a constant. The symmetric regenerator is the special case where $k = 1$.

It is far more convenient to deal with symmetric regenerators, and the question arises whether an unsymmetric ($k \neq 1$) but balanced regenerator can be approximated by an equivalent symmetric regenerator. Hausen [1] suggested that the two reduced lengths and the two reduced periods be combined using *harmonic means*:

$$\frac{1}{\Lambda_H} = \frac{1}{2}\left(\frac{1}{\Lambda'} + \frac{1}{\Lambda''}\right) \tag{5.22}$$

$$\frac{1}{\Pi_H} = \frac{1}{2}\left(\frac{1}{\Pi'} + \frac{1}{\Pi''}\right) \tag{5.23}$$

Subsequently, Iliffe [2] verified the acceptability of this proposal for $k = 2$ and $k = 3$ for $3 < \Lambda'' < 18$, $3 < \Pi'' < 18$.

Hausen suggests that this use of harmonic means can be extended to the *unbalanced case* where $\Pi'/\Lambda' \neq \Pi''/\Lambda''$. However, for the balanced cases, η_{REG} is the same for the hot and cold periods and the same for the equivalent harmonic mean symmetric regenerator. This is not the case for the unbalanced regenerator, but if η^H_{REG} is used to denote the thermal ratio of the equivalent symmetric regenerator and if it is assumed that the heat exchanged in the unbalanced regenerator is equal to that exchanged in its symmetric equivalent, the following equation can be used to obtain estimates of the required thermal effectiveness:

$$\frac{\Pi'}{\Lambda'} \eta'^{*}_{\text{REG}} = \frac{\Pi''}{\Lambda''} \eta''^{*}_{\text{REG}} = \frac{\Pi_H}{\Lambda_H} \eta^H_{\text{REG}} \tag{5.24}$$

In fact, Eq. (5.24) is not very precise; the assumption that the heat exchanged in the equivalent harmonic mean symmetric regenerator is equal to that exchanged in the original unbalanced one is not universally valid.

A measure of imbalance is β, where

$$\beta = \frac{\min(\eta'_{\text{REG}}, \eta''_{\text{REG}})}{\max(\eta'_{\text{REG}}, \eta''_{\text{REG}})}$$

or
$$\beta = \frac{\min(\Pi'/\Lambda', \Pi''/\Lambda'')}{\max(\Pi'/\Lambda', \Pi''/\Lambda'')}$$

For balanced regenerators, $\beta = 1$. Equation (5.24) appears to be applicable for $\beta > 0.8$.

Example 5.5 Another thermal regenerator is examined. The descriptive data are set out for each channel:

Available heating surface area $A = 31.40\ \text{m}^2$
Corresponding heat storing mass $M_m = 4898.4$ kg
Specific heat of packing $c_m = 0.92$ kJ/kg °C
Specific heat of gas (both periods) $c_f = 1.011$ kJ/kg °C

Operation is as follows:

Gas mass flow per channel (hot period) $\dot{m}'_f = 0.156$ kg/s
Gas mass flow per channel (cold period) $\dot{m}''_f = 0.078$ kg/s
Heat transfer coefficient (hot period) $\bar{h}' = 50.23\ \text{W/m}^2\ °\text{C}$
Heat transfer coefficient (cold period) $\bar{h}'' = 25.11\ \text{W/m}^2\ °\text{C}$
Length of hot period $P' = 3600$ s
Length of cold period $P'' = 10{,}800$ s

The dimensionless parameters are

Hot period:

Reduced length $$\Lambda' = \frac{\bar{h}'A}{\dot{m}'_f c_f} = \frac{(50.23)(31.4)}{(0.156)(1011.0)} = 10.0$$

Reduced period $$\Pi' = \frac{\bar{h}'AP'}{M_m c_m} = \frac{(50.23)(31.4)(3600)}{(4898.4)(920.0)} = 1.26$$

Cold period:

Reduced length $$\Lambda'' = \frac{\bar{h}''A}{\dot{m}''_f c_f} = \frac{(25.11)(31.4)}{(0.078)(1011.0)} = 10.0$$

Reduced period $$\Pi'' = \frac{\bar{h}''AP''}{M_m c_m} = \frac{(25.11)(31.4)(10800.0)}{(4898.0)(920.0)} = 1.88$$

The harmonic mean parameters are

$$\Lambda_H = \Lambda' = \Lambda'' = 10.0$$

$$\frac{1}{\Pi_H} = \frac{1}{2}\left(\frac{1}{\Pi'} + \frac{1}{\Pi''}\right)$$

$$= \frac{1}{2}\left(\frac{1}{1.26} + \frac{1}{1.88}\right) = 0.6628$$

$$\Pi_H = 1.51$$

The *reduced period/reduced length* ratios are

$$\frac{\Pi'}{\Lambda'} = \frac{1.26}{10.0} = 0.126$$

$$\frac{\Pi''}{\Lambda''} = \frac{1.88}{10.0} = 0.188$$

$$\frac{\Pi_H}{\Lambda_H} = \frac{1.51}{10.0} = 0.151$$

The degree of unbalance is

$$\beta = \frac{\min(\Pi'/\Lambda', \Pi''/\Lambda'')}{\max(\Pi'/\Lambda', \Pi''/\Lambda'')} = \frac{0.126}{0.188} = 0.67$$

The computed thermal effectiveness is

$$\Lambda' = \Lambda'' = 10.0$$

$$\Pi' = 1.26 \qquad \Pi'' = 1.88$$

$$\eta'_{\mathrm{REG}} = 0.947 \qquad \eta''_{\mathrm{REG}} = 0.635$$

The harmonic means are

$$\Lambda_H = 10.0 \qquad \Pi_H = 1.51$$

$$\eta^H_{\mathrm{REG}} = 0.831$$

The estimated values η'^*_{REG}, η''^*_{REG} of η'_{REG}, η''_{REG}, respectively, can be computed from η^H_{REG}:

$$\eta'^*_{\mathrm{REG}} = \frac{\Pi_H}{\Lambda_H}\frac{\Lambda'}{\Pi'}\eta^H_{\mathrm{REG}}$$

$$= \frac{(0.151)(0.831)}{0.126} = 0.996$$

$$\eta''^{*}_{\text{REG}} = \frac{\Pi_H \Lambda''}{\Lambda_H \Pi''} \eta^{H}_{\text{REG}}$$

$$= \frac{(0.151)(0.831)}{0.188} = 0.667$$

If thermal ratios accurate to ±0.04 are satisfactory for a particular requirement for this value of $\beta = 0.67$, then use of the symmetric regenerator with the harmonic means model is quite satisfactory, but it should be noted that the harmonic means model tends to overestimate performance.

Example 5.6 Example 5.5 is repeated with the length of the cold period $P'' = 9123.0$ s.

Cold period:

$$\text{Reduced period } \Pi'' = \frac{\bar{h}'' A P''}{M_m c_m}$$

$$= \frac{(25.11)(31.4)(9123.0)}{(4898.4)(920.0)} = 1.60$$

$$\frac{\Pi''}{\Lambda''} = \frac{1.60}{10.0} = 0.16$$

Degree of unbalance:

$$\beta = \frac{0.126}{0.16} = 0.78$$

Harmonic mean parameters:

$$\frac{1}{\Pi_H} = \frac{1}{2}\left(\frac{1}{1.26} + \frac{1}{1.60}\right)$$

$$\Pi_H = 1.41$$

$$\frac{\Pi_H}{\Lambda_H} = \frac{1.41}{10.0} = 0.141$$

Computed thermal effectiveness:

$$\Lambda' = \Lambda'' = 10.0$$

$$\Pi' = 1.26 \qquad \Pi'' = 1.60$$

$$\eta'_{\text{REG}} = 0.9126 \qquad \eta''_{\text{REG}} = 0.7187$$

Harmonic means:

$$\Lambda_H = 10.0 \qquad \Pi_H = 1.41$$

$$\eta^{H}_{\text{REG}} = 0.8311$$

Estimates of η'_{REG}, η''_{REG} from symmetric harmonic means model:

$$\eta'^{*}_{REG} = \frac{\Pi_H}{\Lambda_H} \frac{\Lambda'}{\Pi'} \eta^{H}_{REG}$$

$$= \frac{(0.141)(0.8311)}{0.126} = 0.930$$

$$\eta''^{*}_{REG} = \frac{\Pi_H}{\Lambda_H} \frac{\Lambda''}{\Pi''} \eta^{H}_{REG}$$

$$= \frac{(0.141)(0.8311)}{0.160} = 0.732$$

Here the thermal ratios η'^{*}_{REG}, η''^{*}_{REG} are within ± 0.02 of the correct values η'_{REG} and η''_{REG} for $\beta = 0.78$.

5.6 SOLUTION OF DIFFERENTIAL EQUATIONS

The cycle of regenerator operation consists of the hot period followed by the cold period. After a sufficiently large number of identical cycles, the temperature behavior of the regenerator becomes periodic, the period length being the duration of the cycle. At this stage, what has been referred to earlier in this chapter as steady-state behavior or cyclic equilibrium is reached, when the thermal performance is independent of the initial temperature conditions within the packing of the regenerator. However, if the running of a regenerator is disturbed by a step change in one or more operating parameters, for example, in hot inlet gas temperature or in gas flow rate on the hot and/or cold side, then the transient behavior of the regenerator can be observed until steady-state performance is reestablished or until another step change occurs.

The methods of solution of the regenerator problem, as formulated in Eqs. (5.7) and (5.8) with boundary conditions (for the reversals) specified in Eqs. (5.9) and (5.10), fall into two distinct classes. The open methods are those in which the gas and solid temperatures are evaluated in successive cycles until the mathematical model reaches steady-state performance. An arbitrary but if possible well-chosen solid temperature distribution is imposed as an initial condition for the problem.

The closed methods are those in which the reversal conditions (5.9) and (5.10) are employed immediately in formulating the mathematical model; the steady-state performance is computed directly, without any consideration being taken of any earlier transient cycles.

The closed methods of Hausen [5], Iliffe [2], and Nahavandi and Weinstein [3] embody the assumption that the inlet gas temperatures t'_{fi} and t''_{fi} do not vary with time in each period.

The great advantage of the closed methods lies in the direct calculation of cyclic equilibrium. In some rotary regenerators, periods of transience following alterations in operating conditions are relatively very short, and it is sufficient to predict only steady-state behavior. Certain other applications of regenerative heat exchangers involve rarely changing operating conditions; again, mathematical modeling of cyclic equilibrium performance is sufficient for the design of new units and the prediction of the performance of existing regenerators under possibly new thermal load conditions.

The open methods can be used to predict both transient and steady-state regenerator performance. The existing closed methods are mathematically complicated and are specific to the model embodied in Eqs. (5.7), (5.8), (5.9), and (5.10). In contrast, the open methods of the finite difference type can be readily adapted to cope with, for example, time-varying inlet gas temperatures, temperature-dependent thermal properties of gas and packing, and time-varying gas flow rates. However, it is sufficient at this stage to describe methods of solution of Eqs. (5.7) and (5.8).

It is most convenient to introduce the simplifying notation

$$F'\{\xi'\} = T'_m\{\xi', 0\}$$
$$F''\{\xi''\} = T''_m\{\xi'', 0\}$$

to represent the solid temperature distribution at the start of the hot and cold periods, respectively, using the [0, 1] temperature scale. Nusselt [6] showed that the solid temperature at position ξ' and time η' in the hot period, $T'_m\{\xi', \eta'\}$, is related to $F'\{\xi'\}$:

$$T'_m\{\xi', \eta'\} = 1 - \exp(-\eta')(1 - F'\{\xi'\}) + \int_0^{\xi'} \frac{iJ_1[2i\sqrt{(\xi' - \epsilon)\eta'}]}{\sqrt{(\xi' - \epsilon)\eta'}} \eta' \exp[-(\xi' - \epsilon) + \eta')](1 - F'\{\epsilon\})\, d\epsilon \tag{5.25}$$

For the cold period, the Nusselt equation is

$$T''_m\{\xi'', \eta''\} = \exp(-\eta'')F''\{\xi''\} \times \int_0^{\xi'} \frac{iJ_1[2i\sqrt{(\xi'' - \epsilon)\eta''}]}{\sqrt{(\xi'' - \epsilon)\eta''}} \eta'' \exp[-(\xi'' - \epsilon) + \eta'']F''\{\epsilon\}\, d\epsilon \tag{5.26}$$

This notation is simplified for $\eta' = \Pi'$ and $\eta'' = \Pi''$ as follows:

$$K\{\xi - \epsilon\} = -\frac{iJ_1[2i\sqrt{(\xi - \epsilon)\Pi}]}{\sqrt{(\xi - \epsilon)\Pi}} \Pi \exp[-(\xi - \epsilon) + \Pi]$$

This allows a simplified representation of the mathematical problem embodying the reversal conditions (5.9) and (5.10) using

$$1 - F''\left\{\Lambda''\left(1 - \frac{\xi'}{\Lambda'}\right)\right\} = \exp(-\Pi')(1 - F'\{\xi'\}) + \int_0^{\xi'} K'\{\xi' - \epsilon\}(1 - F'\{\epsilon\})\, d\epsilon \tag{5.27}$$

$$F'\left\{\Lambda'\left(1-\frac{\xi''}{\Lambda''}\right)\right\}=\exp(-\Pi'')F''\{\xi''\}$$

$$+\int_0^{\xi''} K''\{\xi''-\epsilon\}F''\{\epsilon\}\,d\epsilon \tag{5.28}$$

The solution of the original differential equations (5.7) and (5.8) for cyclic equilibrium (only) is thus reduced to the solution of a pair of integral equations (5.27) and (5.28). Iliffe [2] suggested that the integrals be represented approximately using Simpson's rule. The form of the solution employing this approach consists of numerical values of the temperatures $F'\{\xi'\}$ and $F''\{\xi''\}$ at the entrance and exit to the regenerator and at equally spaced positions down the length of the packing. It turns out that inherent difficulties, set out by Willmott and Thomas [7], are associated with this method.

Nahavandi and Weinstein [3] offered an alternative approach by representing the solid temperature distribution at the start of each period employing polynomials:

$$F'\{\xi'\}=\sum_{k=0}^{n} a_k(\xi')^k \tag{5.29}$$

$$F''\{\xi''\}=\sum_{k=0}^{n} b_k(\xi'')^k \tag{5.30}$$

These are substituted into the integral equations (5.27) and (5.28), whereupon the mathematical problem becomes one of determining the values of a_k and b_k ($k = 0, 1, 2, \ldots, n$). Again the integrals are evaluated employing methods of quadrature (Gaussian methods are recommended).

The open methods are quite different. Direct use is made of the differential equations (5.7) and (5.8) by representing them in finite difference form. Willmott [4] proposed to integrate Eqs. (5.7) and (5.8) using the trapezoidal rule. The gas and solid temperatures are evaluated at mesh points (r, s), that is, at a distance $r\,\Delta\xi$ from the gas entrance ($\Delta\xi$ = step length) and at a time $s\,\Delta\eta$ from the start of the period under consideration ($\Delta\eta$ = time step length):

$$t_f(r+1,s+1)=t_f(r,s+1)+\frac{\Delta\xi}{2}\left(\left.\frac{\partial t_f}{\partial\xi}\right|_{r+1}+\left.\frac{\partial t_f}{\partial\xi}\right|_{r}\right)$$

$$=t_f(r,s+1)+\frac{\Delta\xi}{2}\,[t_m(r+1,s+1)$$

$$-t_f(r+1,s+1)+t_m(r,s+1)$$

$$-t_f(r,s+1)] \tag{5.31}$$

$$t_m(r+1, s+1) = t_m(r+1, s) + \left(\frac{\Delta\eta}{2}\frac{\partial t_m}{\partial\eta}\bigg|_{s+1} + \frac{\partial t_m}{\partial\eta}\bigg|_{s}\right)$$

$$= t_m(r+1, s) + \frac{\Delta\eta}{2}\,[t_f(r+1, s+1)$$

$$- t_m(r+1, s+1) + t_f(r+1, s)$$

$$- t_m(r+1, s)] \tag{5.32}$$

Given any solid temperature distribution and the inlet gas temperature (which can of course vary with time) at that instant, Eq. (5.31) can be employed repeatedly and directly to compute the corresponding gas temperatures down the length of the regenerator for $r = 0, 1, 2, \ldots$.

$$t_f(r+1, s+1) = A_1 t_f(r, s+1) + A_2[t_m(r+1, s+1) + t_m(r, s+1)] \tag{5.33}$$

where $A_1 = \dfrac{2 - \Delta\xi}{2 + \Delta\xi}$

$$A_2 = \frac{\Delta\xi}{2 + \Delta\xi}$$

This solid temperature distribution is computed using a modified (by substitution for $t_f(r+1, s+1)$ from Eq. 5.33) form of Eq. (5.32):

$$t_m(r+1, s+1) = K_1 t_m(r+1, s) + K_2 t_f(r+1, s)$$

$$+ K_3 t_m(r, s+1) + K_4 t_f(r, s+1) \tag{5.34}$$

where $K_1 = \dfrac{B_1}{X}$ $\qquad K_2 = \dfrac{B_2}{X}$

$$K_3 = \frac{A_2 B_2}{X} \qquad K_4 = \frac{A_1 B_2}{X}$$

Expressions for B_1, B_2, and X are given below:

$$B_1 = \frac{2 - \Delta\eta}{2 + \Delta\eta}$$

$$B_2 = \frac{\Delta\eta}{2 + \Delta\eta}$$

$$X = 1 - A_2 B_2$$

In this way, Eqs. (5.7) and (5.8) can be integrated, first over a hot period followed by a cold period (here the reversal conditions are embodied much later in the mathematical process than for the closed methods) and, second, over successive cycles, either from an arbitrary starting position to cyclic equilibrium or over a period of transient performance under examination.

Needless to say, both the open and the closed methods generate the same solutions in terms of the computed thermal ratios and the solid temperature distributions at cyclic equilibrium for constant inlet gas temperatures. The model of the regenerator embodying the idealizing assumptions consists of Eqs. (5.7)-(5.10). Different methods of solving these equations have been discussed in this part of the chapter, but all relate to the same model. Discrepancies between theory and experiment or industrial practice will not lie in the methods (e.g., Iliffe or Willmott), but will result from the validity or otherwise of the idealizations built into the model.

In a later chapter we therefore turn to the representation of features not adequately represented by the linear model described here. In some applications of regenerators, for example, it is necessary to represent the temperature dependence of specific heat, especially at very low temperatures. Nevertheless, this linear model, which is comparatively easy to understand and to handle, provides a framework from which both early estimates of likely performance of regenerator systems can be made and from which more complex models may be developed.

REFERENCES

1. H. Hausen, *Wärmeübertragung im Gegenstrom, Gleichstrom und Kreuzstrom*, Springer-Verlag, Berlin, 1950.
2. C. E. Illiffe, "Thermal Analysis of Counterflow Regenerative Heat Exchanger," *J. Inst. Mech. Eng.*, vol. 159, 1948, pp. 363–372 (war emergency issue no. 44).
3. A. N. Nahavandi and A. S. Weinstein, "A Solution to the Periodic-Flow Regenerative Heat Exchanger Problem," *Appl. Sci. Res.*, vol. A10, 1961, pp. 335–348.
4. A. J. Willmott, "Digital Computer Simulation of a Thermal Regenerator," *Int. J. Heat Mass Transfer*, vol. 7, 1964, pp. 1291–1303.
5. H. Hausen, "Näherungsverfahren zur Berechnung des Wärmeaustauches in Regeneratoren," *Z. Angew. Math. Mech.*, vol. 2, 1931, pp. 105–114.
6. W. Nusselt, "Die Theorie des Winderhitzers," *Z. Ver. Deutsch. Ing.*, vol. 71, 1927, pp. 85–91.
7. A. J. Willmott and R. J. Thomas, "Analysis of the Long Contra-flow Regenerative Heat Exchanger," *J. Inst. Math. Appl.*, vol. 14, 1974, pp. 267–280.

CHAPTER

SIX

FINITE CONDUCTIVITY MODELS OF COUNTERFLOW REGENERATORS

6.1 INTRODUCTION

The regenerator theory presented so far has related to the simplest linear model. It has been assumed, for example, that the heat transfer coefficient $\bar{h}$ is a simple convective surface coefficient, and no reference has been made to the internal resistance to heat transfer of the storage material comprising the packing or "checkerwork" of the regenerator. This particular problem is now examined together with the general problem of representing unsteady heat conduction within the heat storing mass.

6.2 LUMPED HEAT TRANSFER COEFFICIENTS

The differential equations representing heat transfer in a regenerator were set out in Chap. 5. This model is frequently called the two-dimensional model.

$$\bar{h}A(t_m - t_f) = \dot{m}_f c_f L \frac{\partial t_f}{\partial x} + m_f c_f \frac{\partial t_f}{\partial \tau} \tag{5.1}$$

$$\bar{h}A(t_f - t_m) = M_m c_m \frac{\partial t_m}{\partial \tau} \tag{5.2}$$

The form of the heat transfer coefficient $\bar{h}$ employed in those equations was not discussed. However, it will be seen that in the case of rotary regenerators whose packing consists of wire mesh or thin sheets of material, there will be negligible temperature differences between the surface and the interior of the packing. Similarly, for a fixed bed regenerator, a metallic, even modestly thick-walled (around each channel) checkerwork can be considered to be of uniform temperature inside the solid at any level in the packing at any instant in either period of operation. In these cases, the sole effective resistance to heat transfer occurs at the surface of the regenerator channels, between gas and solid, and $\bar{h}$ will be just the convective heat transfer coefficient. The temperature t_m will refer to both the surface and interior of the channel walls at position x down the length of the regenerator at time τ from the start of the period considered. However, in a regenerator in which the channel walls are modestly thick and the solid material has a low thermal conductivity, both surface resistance to heat transfer and interior resistance will be equally important.

In electrical terms, it is possible to consider a pair of resistances in series. The current i (see Fig. 6.1) flowing through resistance R_1 is $i = (V_1 - V_2)/R_1$, and the same current flows through R_2 where $i = (V_2 - V_3)/R_2$. If R_3 is the combined resistance of R_1 and R_2, then $i = (V_1 - V_3)/R_3$.

Now,

$$V_1 - V_3 = (V_1 - V_2) + (V_2 - V_3)$$

that is, $iR_3 = iR_1 + iR_2$, or the well-known result

$$R_3 = R_1 + R_2$$

This is translated into heat transfer terms using heat flux as an analog to current and temperature as an analog to voltage. Let q = heat flux, t_s = surface solid temperature, and t_m = some internal solid temperature (yet to be defined).

The heat flux at the solid surface is $q = h(t_f - t_s)$; let G = the internal heat transfer coefficient, in which case the heat flux through the solid is $q = G(t_s - t_m)$. We denote $\bar{h}$ to be the overall or bulk heat transfer coefficient and specify by $\bar{h}$ the same heat flux $q = \bar{h}(t_f - t_m)$. In order to complete the analogy with electric resistances in series, we note that if $t_f \equiv V_1$, $t_s \equiv V_2$, and $t_m \equiv V_3$, then

$$\frac{1}{h} \equiv R_1 \qquad \frac{1}{G} \equiv R_2 \qquad \frac{1}{\bar{h}} \equiv R_3 \tag{6.1}$$

The resistance G^{-1} to heat transfer will be inversely proportional to the checkerwork thermal conductivity k_m and proportional to the semithickness w of the packing walls between channels. Hausen [1] developed a particular form for Eqs. (6.1), namely.

$$\frac{1}{\bar{h}} = \frac{1}{h} + \frac{w}{3k_m}\phi \tag{6.2}$$

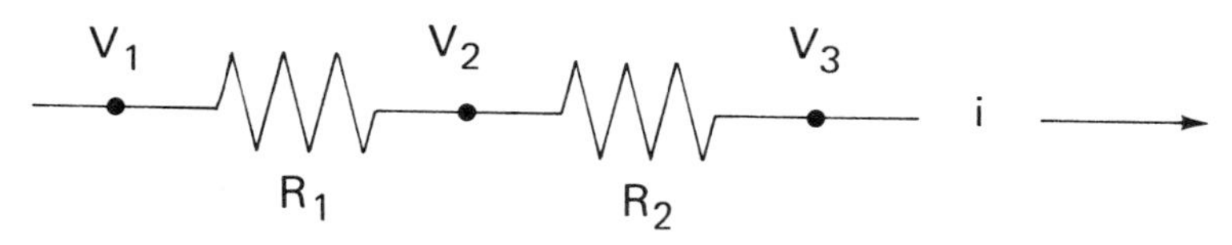

Figure 6.1 Electrical analogy of the summation of resistance to heat transfer.

If y is the direction into the checkerwork perpendicular to gas flow, then the solid temperature $t_m(x, \tau)$ may be defined to be a spatial average of the solid temperature $t_m(x, y, \tau)$ using Eq. (6.3),

$$t_m(x, \tau) = \frac{1}{w} \int_0^w t_m(x, y, \tau)\, dy \tag{6.3}$$

Referring again to the electrical analogy, it will be seen that

$$\bar{h}(t_m(x, \tau) - t_f(x, \tau)) = h(t_s(x, \tau) - t_f(x, \tau)) \tag{6.4}$$

The inclusion of the factor ϕ was in fact a later refinement of this summation of "resistances." It is included for these reasons. Once a period of operation is established, it can usually (although not always!) be assumed that the heat flux q at the surface remains constant and that there is a parabolic distribution of temperature inside the solid material in the direction y, perpendicular to gas flow. For these steady heating/cooling conditions, Hausen elegantly developed Eq. (6.2) without the factor ϕ as a simple summation of resistances. At the reversals, however, the parabolic profile must be inverted from its steady heating to its steady cooling shape or vice versa (see Fig. 6.2).

The summation of the "resistances" h^{-1} and $3k_m/w$ in Eq. (6.2) implies that the solid temperature increases/decreases monotonically from t_s to t_m when the "voltage difference," $t_s - t_m$, is used to calculate the current heat flux. But at the reversals, the parabolic profile is badly distorted; the factor ϕ attempts to average out over the whole cycle the effects of these distortions by modifying the bulk heat transfer coefficient $\bar{h}$, which is then applied over the whole period under consideration.

Hausen [1] proposed that

$$\phi = 1 - \frac{w^2}{15\alpha_m}\left(\frac{1}{P'} + \frac{1}{P''}\right)$$

$$\text{if} \qquad \frac{w^2}{\alpha_m}\left(\frac{1}{P'} + \frac{1}{P''}\right) \leqslant 5 \tag{6.5}$$

$$\text{and} \qquad \phi = \frac{2.142}{\sqrt{0.3 + 4w^2(1/P' + 1/P'')/2\alpha_m}}$$

$$\text{if} \qquad \frac{w^2}{\alpha_m}\left(\frac{1}{P'} + \frac{1}{P''}\right) > 5 \tag{6.6}$$

The elegance of this approach lies in the fact that the differential equations

$$\text{Fluid:} \qquad \frac{\partial t_f}{\partial \xi} = t_m - t_f \tag{5.7}$$

$$\text{Storage material:} \qquad \frac{\partial t_m}{\partial \eta} = t_f - t_m \tag{5.8}$$

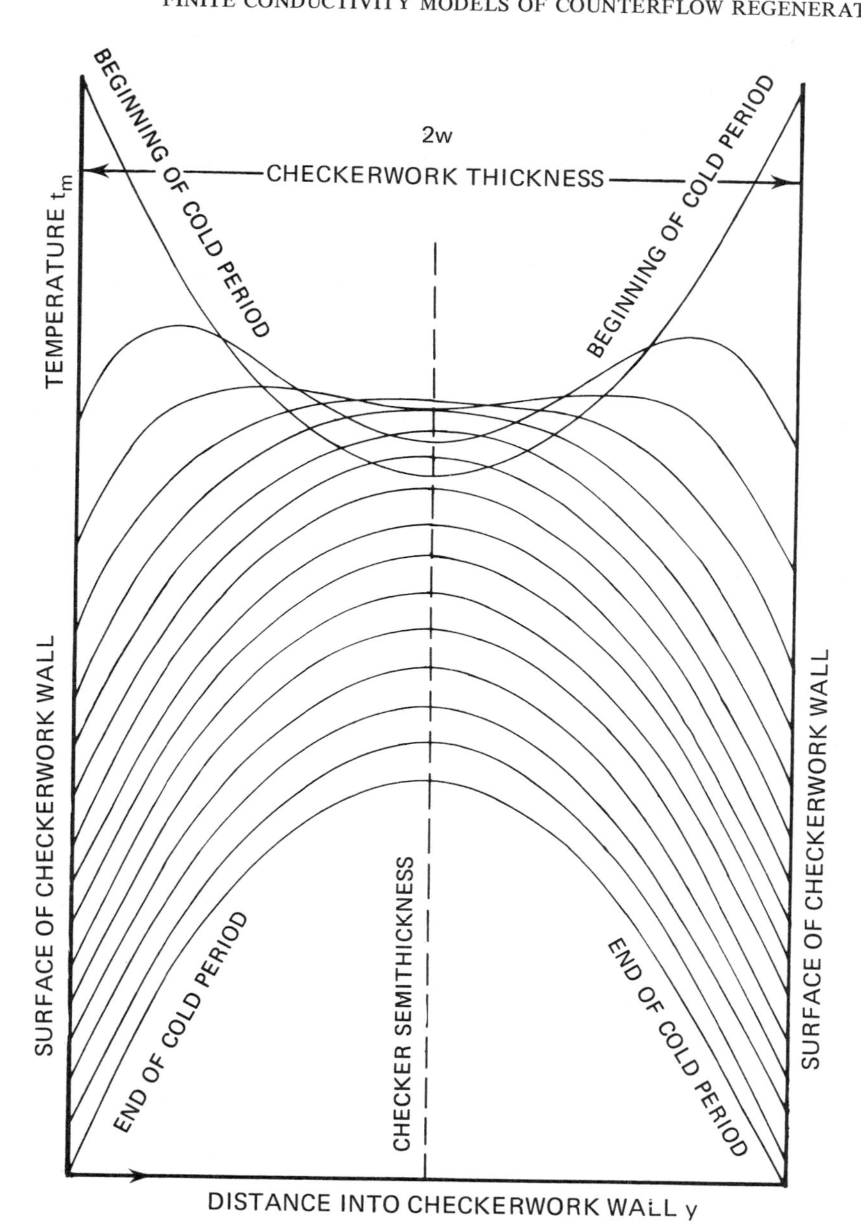

Figure 6.2 Chronological variation of solid temperature in a regenerator.

set out in Chap. 5, together with the derived dimensionless parameters, the reduced length Λ and the reduced period Π, remain exactly the same in form whether the bulk heat transfer coefficient $\bar{h}$ or the surface heat transfer coefficient h (internal resistance to heat transfer ignored) is used. This considerably widens the applicability of Eqs. (5.7) and (5.8) in predicting probable regenerator behavior. The dimensions of the two-dimensional model are the x and τ used in Eqs. (5.1) and (5.2); the corresponding parameters in Eqs. (5.7) and (5.8) are ξ and η.

In most circumstances, the use of the bulk heat transfer coefficient is quite acceptable. The possible circumstances under which the Hausen form of ϕ might not be used will be discussed in a later section.

Example 6.1 A regenerator constructed of ceramic material has the following dimensions:

Heating surface area $A = 132.4\ \text{m}^2$
Heating storing mass $M_m = 6399.6$ kg
Semithickness of checkers $w = 0.026$ m

The packing thermal properties are

Density $\rho_m = 1850.0\ \text{kg/m}^3$
Conductivity $k_m = 1.38$ W/m °C
Specific heat $c_m = 1050.0$ J/kg °C
Gas specific heat $c_f = 1100.0$ J/kg °C
Gas mass flow rate $\dot{m}_f = 1.25$ kg/s
Surface heat transfer coefficient $h = 20.2\ \text{W/m}^2\ {}^\circ\text{C}$

We consider two cases:

1. Length of hot and cold periods = 1200.0 s.
2. Length of hot and cold periods = 3600.0 s.

The diffusivity of the solid is α_m:

$$\alpha_m = \frac{k_m}{c_m \rho_m}$$

$$= \frac{1.38}{(1.05 \times 10^3)(1.85 \times 10^3)}$$

$$= 7.104 \times 10^{-7}\ \text{m}^2/\text{s}$$

The Hausen ϕ factor can now be computed.

Case 1, $P' = P'' = 1200.0$ s:

$$\frac{w^2}{\alpha_m}\left(\frac{1}{P'} + \frac{1}{P''}\right) = Z \text{ (say)}$$

$$= \frac{(2.6)(5.2)(10^{-4})(2)}{(7.104)(10^{-7})(1.2)(10^3)}$$

$$= \frac{(2.6)(5.2)(2)}{(7.104)(1.2)} = 1.586 < 5$$

It follows that ϕ is computed using

$$\phi = 1 - \frac{Z}{15} = 0.894$$

Case 2, $P' = P'' = 3600.0$ s:

$$Z = \frac{(2.6)(5.2)(2)}{(7.104)(3.6)} = 0.529 < 5$$

$$\phi = 1 - \frac{Z}{15} = 0.965$$

Note that ϕ is smaller for case (1) and therefore more influential on the lumped heat transfer coefficient $\bar{h}$ for the shorter cycle times. This coefficient is calculated using

$$\frac{1}{\bar{h}} = \frac{1}{h} + \frac{w}{3k_m}\phi \tag{6.2}$$

Case 1:

$$\frac{1}{\bar{h}} = \frac{1}{20.2} + \frac{(0.026)(0.894)}{(3.0)(1.38)}$$

$$= 0.0495 + 0.00561$$

$$= 0.0551$$

$$\bar{h} = 18.15$$

Case 2:

$$\frac{1}{\bar{h}} = \frac{1}{20.2} + \frac{(0.026)(0.965)}{(3.0)(1.38)}$$

$$= 0.0495 + 0.00606$$

$$= 0.0555$$

$$\bar{h} = 17.997$$

The dimensionless parameters Λ and Π (symmetric case) can be evaluated:

$$\Lambda = \frac{\bar{h}A}{\dot{m}_f c_f} \qquad \Pi = \frac{\bar{h}AP}{M_m c_m}$$

Case 1:

$$\Lambda = \frac{(18.15)(132.4)}{(1.25)(1100.0)}$$

$$= 1.748$$

$$\Pi = \frac{(18.15)(132.4)(1200.0)}{(6399.6)(1050.0)}$$

$$= 0.429$$

Case 2:

$$\Lambda = \frac{(17.997)(132.4)}{(1.25)(1100.0)}$$

$$= 1.733$$

$$\Pi = \frac{(17.997)(132.4)(3600.0)}{(6399.6)(1050.0)}$$

$$= 1.277$$

The corresponding thermal ratios have been computed (using Nahavandi and Weinstein's method, discussed in Chap. 5) and are as follows:

	Case 1	Case 2
Reduced length Λ	1.748	1.733
Reduced period Π	0.429	1.277
Thermal ratio η_{REG}	0.465	0.449

We shall now calculate* the reduced length, reduced period, and thermal ratio for this problem using only the surface heat transfer coefficient h instead of the lumped coefficient $\bar{h}$.

	Case 1	Case 2
Reduced length using surface heat transfer coefficient	1.945	1.945
Reduced period using surface heat transfer coefficient	0.468	1.433
Thermal ratio	0.491	0.475
Thermal ratio calculated using lumped heat transfer coefficient $\bar{h}$	0.465	0.449

*Regenerator calculations have been undertaken for this text, using a computer program embodying the Nahavandi and Weinstein method.

It is useful to compare these differences in thermal ratio for a high-temperature regenerator where $t'_{fi} = 1500°\text{C}$ and $t''_{fi} = 100°\text{C}$, the inlet temperatures in the hot and cold periods, respectively. Denote the corresponding time mean exit gas temperatures to be t'_{fo} and t''_{fo}. Then

$$t'_{fo} = t'_{fi} - \eta_{\text{REG}}(t'_{fi} - t''_{fi})$$

$$t''_{fo} = t''_{fi} + \eta_{\text{REG}}(t'_{fi} - t''_{fi})$$

Case 1 computed with lumped heat transfer coefficient $\bar{h}$:

$$t'_{fo} = 1500.0 - (0.465)(1400.0)$$

$$= 849.0°\text{C}$$

$$t''_{fo} = 100.0 + (0.465)(1400.0)$$

$$= 751.0°\text{C}$$

Case 1 computed using surface instead of lumped heat transfer coefficient:

$$t'_{fo} = 1500.0 - (0.491)(1400.0)$$

$$= 812.6°\text{C}$$

$$t''_{fo} = 100.0 + (0.491)(1400.0)$$

$$= 787.4°\text{C}$$

This means that for this case an error of about 36°C in the exit temperatures would be introduced unless the resistance to heat transfer within the heat storing material is embodied within the heat transfer coefficient $\bar{h}$. The reader is invited to repeat this comparison for case 2.

6.3 FURTHER CONSIDERATIONS OF THE EFFECT OF CHECKERWORK CONDUCTIVITY

We have discussed the employment of lumped heat transfer coefficients in Eqs. (5.1) and (5.2). We now come to modifications in the assumptions built into the regenerator model that involve alterations to the mathematical formulation of the problem. Both involve the representation of heat conduction within the solid packing. Perhaps in an artificial, but what turns out to be an acceptable and convenient way, longitudinal conduction in the solid (i.e., in the direction of gas flow) is studied separately from latitudinal conduction (i.e., in a direction perpendicular to gas flow).

In metallic regenerators, the thermal conductivity of the packing is so good that the contribution of internal resistance to heat transfer built into the bulk heat transfer coefficient h,

$$\frac{1}{\bar{h}} = \frac{1}{h} + \frac{w}{3k_m}\phi \qquad (6.2)$$

may be negligible and $\bar{h} \to h$. On the other hand, longitudinal conduction may be so important as to necessitate its inclusion in a mathematical model of such a regenerator.

In thick-walled ceramic regenerators, longitudinal conduction will be negligible but the internal resistance to heat transfer within the checkerwork may be significant, as has been seen in Example 6.1. In this case, the problem is to identify under what circumstances the use of the bulk heat transfer coefficient is acceptable and when a more elaborate model must be employed.

6.4 LONGITUDINAL CONDUCTION IN THE REGENERATOR PACKING

A possible measure of the potential effect of longitudinal conduction down the length of the packing of a regenerator is the ratio γ,

$$\gamma = \frac{\text{Longitudinal heat conduction}}{\text{Heat input/extracted by gas}}$$

The temperature gradient down the length of the regenerator can be represented very approximately by $(t'_{fi} - t'_{fo})/L$ for the hot period and $(t''_{fo} - t''_{fi})/L$ for the cold period, assuming that the solid temperature gradient is roughly parallel to the gradient in the gas. Note that t'_{fo} and t''_{fo} are the hot and cold period average exit gas temperatures; in the fixed bed regenerator these are chronological averages over the periods concerned, whereas for the rotary regenerator they are the temperatures of the mixed gases coming from the several sectors of the rotating checkerwork.

The volume of the checkerwork is equal to the product of the heating surface area A and the semithickness w of the walls surrounding the checkerwork channels. The effective cross-sectional area of the solid packing is therefore Aw/L. It follows that the rate of heat conduction in the checkerwork is approximately $Awk_m(t'_{fi} - t'_{fo})/(L^2)$ in the hot period and $Awk_m(t''_{fo} - t''_{fi})/(L^2)$ in the cold period.

The average rate of heat input to the regenerator is $\dot{m}'_f c'_f(t'_{fi} - t'_{fo})$ for the hot period, whereas $\dot{m}''_f c''_f(t''_{fo} - t''_{fi})$ is the average rate at which heat is extracted from the regenerator in the cold period.

Two factors γ' and γ'' can now be derived:

$$\gamma' = \frac{\text{Longitudinal heat conduction}}{\text{Rate of heat input}}$$

$$= \frac{Awk_m(t'_{fi} - t'_{fo})}{2L^2} \, \frac{1}{\dot{m}'_f c'_f(t'_{fi} - t'_{fo})}$$

$$= \frac{Awk_m}{\dot{m}'_f c'_f L^2} \tag{6.7}$$

Similarly,

$$\gamma'' = \frac{Awk_m}{\dot{m}''_f c''_f L^2} \tag{6.8}$$

For the case where $\gamma = \gamma' = \gamma''$, Tipler [2] mentioned a value of $\gamma = 10^{-2}$ in the case of regenerators used in gas turbines and considered that longitudinal conduction could be neglected. Willmott [3] obtained a value of the order of 10^{-5} for γ for a medium-size Cowper stove and therefore ignored longitudinal conduction effects in this case.

This same γ was called the *conduction parameter* in the classic paper by Bahnke and Howard [4]. They suggested that if $\Delta E/E$ is defined for the balanced case ($\eta_{\mathrm{REG}} = \eta'_{\mathrm{REG}} = \eta''_{\mathrm{REG}}$) by

$$\frac{\Delta E}{E} = \frac{\eta_{\mathrm{REG}}(\text{no conduction}) - \eta_{\mathrm{REG}}(\text{conduction})}{\eta_{\mathrm{REG}}(\text{no conduction})}$$

then the decrease in thermal effectiveness caused by longitudinal conduction can be represented by the very simple equation

$$\frac{\Delta E}{E} = \gamma \tag{6.9}$$

for reduced length $\Lambda > 20$ and $\gamma < 0.1$.

Bahnke and Howard present a whole set of results computed from regenerator simulations for both balanced and unbalanced regenerators, from which it can be concluded that:

1. $\Delta E/E$ increases with the conduction factor γ and with reduced length Λ.
2. $\Delta E/E$ decreases as the ratio of reduced length to reduced period, namely, as Λ/Π increases.
3. For an unbalanced regenerator ($\dot{m}'_f c'_f P' \neq \dot{m}''_f c''_f P''$), γ has a negligible effect on $\Delta E/E$ (in this case $\Delta E/E$ is taken to refer to the cold-side thermal ratios).

However, the Bahnke and Howard paper lacks a formal description of the mathematical model describing regenerator behavior with longitudinal conduction. It will be shown that certainly the first of Bahnke and Howard's results can be anticipated from such a formal idealization.

To embody longitudinal conduction effects, Eq. (5.2) is modified:

$$\frac{\partial t_m}{\partial \tau} = \frac{\bar{h}A}{M_m c_m}(t_f - t_m) + \alpha_m \frac{\partial^2 t_m}{\partial x^2} \tag{6.10}$$

Equation (5.1) is modified to neglect the effect of the thermal capacity of the gas resident in the regenerator at any instant:

$$\frac{\partial t_f}{\partial x} = \frac{\bar{h}A}{\dot{m}_f c_f L}(t_m - t_f) \tag{6.11}$$

The simplifying parameters ξ and η are introduced:

$$\xi = \frac{\bar{h}A}{\dot{m}_f c_f L} x \qquad \eta = \frac{\bar{h}A}{M_m c_m} \tau \tag{6.12}$$

from which it can be concluded that

$$\frac{\partial^2 t_m}{\partial x^2} = \left(\frac{\bar{h}A}{\dot{m}_f c_f L}\right)^2 \frac{\partial^2 t_m}{\partial \xi^2}$$

and
$$\frac{\partial t_m}{\partial \eta} = \frac{M_m c_m}{\bar{h}A} \frac{\partial t_m}{\partial \tau}$$

$$= \frac{M_m c_m}{\bar{h}A}\left[\frac{\bar{h}A}{M_m c_m}(t_f - t_m) + \alpha_m \left(\frac{\bar{h}A}{\dot{m}_f c_f L}\right)^2 \frac{\partial^2 t_m}{\partial \xi^2}\right]$$

$$= t_f - t_m + \psi \frac{\partial^2 t_m}{\partial \xi^2} \tag{6.13}$$

where
$$\psi = \frac{M_m c_m \alpha_m \bar{h}A}{(\dot{m}_f c_f L)^2} \tag{6.14}$$

It turns out that ψ can be rearranged by noting that the mass M_m of the checkerwork is equal to the checkerwork volume times the density ρ_m, that is, $M_m = Aw\rho_m$. The diffusivity α_m is equal to $k_m/(\rho_m c_m)$. It follows that

$$\psi = \frac{\bar{h}A}{\dot{m}_f c_f} \frac{k_m}{\rho_m c_m} \frac{Aw\rho_m c_m}{L^2 \dot{m}_f c_f}$$

$$= \Lambda \frac{k_m A w}{L^2 \dot{m}_f c_f} = \gamma \Lambda \tag{6.15}$$

The differential equations (6.10) and (6.11) are reduced in form:

$$\frac{\partial t_m}{\partial \eta} = t_f - t_m + \gamma \Lambda \frac{\partial^2 t_m}{\partial \xi^2} \tag{6.16}$$

$$\frac{\partial t_f}{\partial \xi} = t_m - t_f \tag{6.17}$$

Clearly, the influence of the conductivity term in Eq. (6.16) will increase with the factor γ and the reduced length Λ; this was observed in the results presented by Bahnke and Howard.

The further significance of this representation lies in the fact that the construction and operation of the regenerator can be summarized in just six factors (and the two inlet gas temperatures):

	Hot period	Cold period
Reduced length	Λ'	Λ''
Reduced period	Π'	Π''
Conduction factor	γ'	γ''

Example 6.2 Examine the likely longitudinal conduction in the Feolite regenerator described in Chap. 5. The relevant data are as follows:

Heating surface area $A = 5.8\ \text{m}^2$
Semithickness of packing $w = 0.04$ m
Length of regenerator $L = 5.8$ m
Thermal conductivity of Feolite $k_m = 2.1$ W/m °C
Gas mass flow rate $\dot{m}_f = 0.156$ kg/s
Gas specific heat $c_f = 1011.0$ J/kg °C
(symmetric case $\gamma = \gamma' = \gamma''$)

$$\gamma = \frac{Awk_m}{\dot{m}_f c_f L^2}$$

$$= \frac{(5.8)(0.04)(2.1)}{(0.156)(1011.0)(5.8)^2}$$

$$= 9.2 \times 10^{-5}$$

Clearly this is a case where longitudinal conduction is not important. Even if the regenerator were made of mild steel (conductivity k_m approximately 52.0 W/m °C) or copper (conductivity 385.0 W/m °C), the value of γ for this regenerator would change insufficiently.

Mild steel:

$$\gamma = \frac{(9.2)(52.0)}{2.1} 10^{-5}$$

$$= 2.3 \times 10^{-3}$$

Copper:

$$\gamma = \frac{(9.2)(385.0)}{2.1} 10^{-5}$$

$$= 1.69 \times 10^{-2}$$

Since the reduced length Λ for this configuration is only 1.847, this will have a negligible multiplicative effect on γ in the factor $\psi = \gamma\Lambda$ in Eq. (6.16).

Example 6.3 Bahnke and Howard suggest that longitudinal conductivity becomes important if $\gamma > 0.1$. What sort of regenerator configuration, made of mild steel, under the same flow rate operating conditions specified in Example 6.2, would exhibit significant longitudinal conduction effects?

$$\gamma = \frac{Awk_m}{\dot{m}_f c_f L^2}$$

If the heating surface area A is increased by heightening the regenerator (increasing L), this will result in a decrease in γ. The alternative is to increase the surface area A by widening the regenerator, which would effect an increase in γ. The last alternative is to increase the thickness of the walls of the packing. If this alone is considered, then for $\gamma = 2.3 \times 10^{-1}$, when conductivity starts to be important, the steel walls would need to be 8.0 m thick! If the width of the regenerator was increased 10-fold, then the steel would need to be only 0.8 m thick! In other words, for this type of fixed bed system, longitudinal conduction is negligible, unless extremely thick-walled and very broad regenerators are considered.

6.5 NOTE ON NUMERICAL SOLUTION OF EQUATIONS INCLUDING LONGITUDINAL CONDUCTION

A simple approach is to represent Eq. (6.17) in finite difference form using the methodology described in Chap. 5, Eq. (5.31); this facilitates the calculation of the gas temperatures down the length of the regenerator at any instant in time. Equation (6.16) is integrated explicitly:

$$t_m(r, s+1) = t_m(r, s) + \Delta\eta \left.\frac{\partial t_m}{\partial \eta}\right|_{r,\,s} \tag{6.18}$$

where $\Delta\eta$ is the time step in the finite difference scheme. This can be expanded:

$$t_m(r, s+1) = t_m(r, s) + \Delta\eta \left[t_f(r, s) - t_m(r, s) + \frac{\gamma\Lambda}{(\Lambda\xi)^2} \left(t_m(r+1, s) - 2t_m(r, s) + t_m(r-1, s) \right) \right] \tag{6.19}$$

If it is assumed that there are no positive or negative heat losses at the ends of the regenerator, then at the entrance ($r = o$), we have

$$t_m(o, s+1) = t_m(o, s) + \Delta\eta \left[t_f(o, s) - t_m(o, s) + \frac{2\gamma\Lambda}{(\Delta\xi)^2} \left(t_m(1, s) - t_m(o, s) \right) \right] \tag{6.20}$$

and at the exit ($r = M$),

$$t_m(M, s+1) = t_m(M, s) + \Lambda\eta \left[t_f(M, s) - t_m(M, s) + \frac{2\gamma\Lambda}{(\Delta\xi)^2} \left(t_m(M-1, s) - t_m(M, s) \right) \right] \tag{6.21}$$

At any instant $s + 1$ during the hot/cold period of operation simulated, $t_m(r, s+1)$ is computed using Eqs. (6.19), (6.20), and (6.21); the gas temperatures are then computed using the difference representation of Eq. (6.17).

Note that this method will not be stable if $\gamma\Lambda\ \Delta\eta/(\Delta\xi)^2 > \frac{1}{2}$. Suppose that $\Lambda = 10$, $\gamma = 0.1$, and $\Delta\xi = 1.0$; then $0.1 \times 10 \times \Delta\eta/1.0 > \frac{1}{2}$. For $\Pi = 1$, this means that two steps must be taken to avoid instability, whereas 10 to 20 time steps is usual for this sort of simulation. However, for a small reduced length, say, $\Lambda = 1, \gamma = 0.1$, $\Delta\xi = 0.1$, $0.1 \times 1 \times \Delta\eta/(0.1)^2 > \frac{1}{2}$ implies that at least 20 time steps for $\Pi = 1$ must be taken if stability of the numerical scheme is to be preserved.

For a practical program, it is likely that an implicit scheme of the Crank-Nicolson type would be used, thereby avoiding these problems of stability. This is not discussed here.

6.6 LATITUDINAL CONDUCTION IN THE REGENERATOR CHECKERWORK

Most mathematical models employed in industrial practice for the prediction of thermal regenerator performance assume either that the internal resistance to heat transfer within the checkerboard is negligible ($\bar{h} = h$), or that this resistance can be represented satisfactorily using a lumped heat transfer coefficient, as described in Sec. 6.2:

$$\frac{1}{\bar{h}} = \frac{1}{h} + \frac{w}{3k_m}\,\phi \tag{6.2}$$

In this section is described a model, known as the three-dimensional model, in which the heat transfer within the checkerwork in a direction y, perpendicular to the direction x of gas flow, is represented explicitly by the diffusion equation

$$\frac{\partial t_m}{\partial \tau} = \alpha_m \frac{\partial^2 t_m}{\partial y^2} \tag{6.22}$$

The checkerwork is considered to be composed of plain walls separated from one another by the channels through which the hot and cold gases flow (see Fig. 6.3).

At the surface of the solid, $x = 0$, the heat flux across the surface from the gas is represented by Eq. (6.23),

$$\left.\frac{\partial t_m}{\partial y}\right|_{y=0} = \frac{h}{k_m}(t_s - t_f) \tag{6.23}$$

where t_s is the surface solid temperature. From the symmetry of the heat transfer characteristics of the plain wall system, it may be deduced that the heat flux through the middle of the wall in the direction y is zero:

$$\left.\frac{\partial t_m}{\partial y}\right|_{y=w} = 0 \tag{6.24}$$

The heat gained/lost by the gas passing through the regenerator and the gas currently

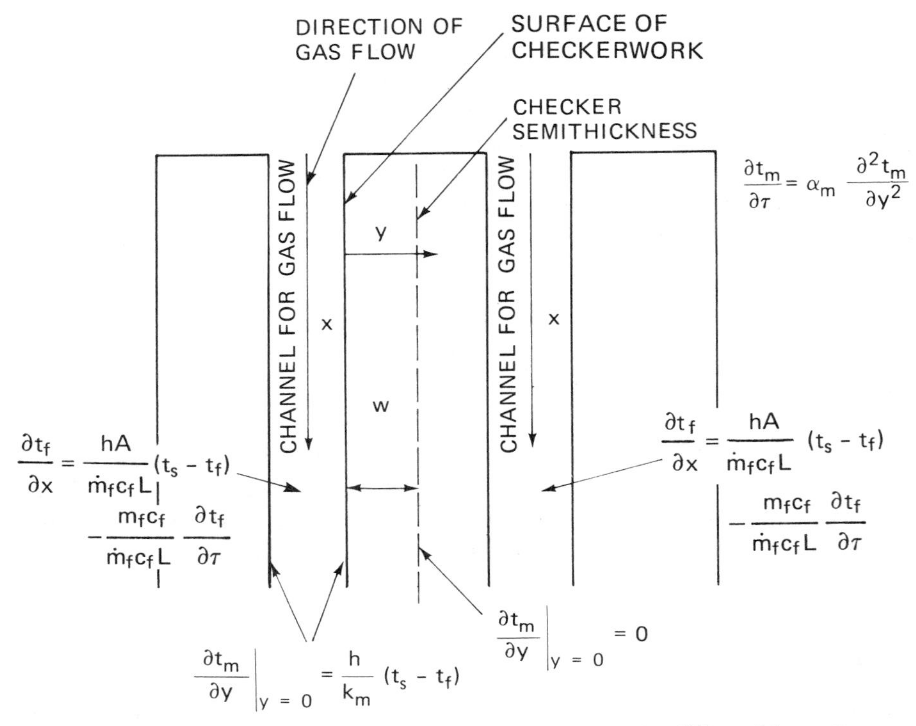

Figure 6.3 Illustration of checkerwork cross section and descriptive differential equations.

resident in the checkerwork channels is calculated using the differential equation (6.25):

$$hA(t_s - t_f) = \dot{m}_f c_f L \frac{\partial t_f}{\partial x} + m_f c_f \frac{\partial t_f}{\partial \tau} \tag{6.25}$$

This system of differential equations can be simplified using the substitutions

$$\xi = \frac{hA}{\dot{m}_f c_f L} x \qquad Y = \frac{y}{w} \qquad Z = \frac{\alpha_m}{w^2}\left(\tau - \frac{m_f}{\dot{m}_f L} x\right) \tag{6.26}$$

whereupon they take the form

$$\frac{\partial t_m}{\partial Z} = \frac{\partial^2 t_m}{\partial Y^2} \tag{6.27}$$

$$\frac{\partial t_f}{\partial \xi} = t_s - t_f \tag{6.28}$$

The dimensions of the three-dimensional model are x, y, and τ, which correspond to the three dimensionless parameters ξ, Y, and Z. The boundary conditions represented previously by Eqs. (6.23) and (6.24) then become

$$\left.\frac{\partial t_m}{\partial Y}\right|_{Y=0} = \mathrm{Bi}(t_s - t_f) \tag{6.29}$$

$$\left.\frac{\partial t_m}{\partial Y}\right|_{Y=1} = 0 \tag{6.30}$$

For each hot and cold period of the cycle, it is possible to define the descriptive parameters reduced time Ω and reduced length Λ:

$$\Omega = \frac{\alpha_m}{w^2}\left(P - \frac{m_f}{\dot{m}_f}\right) \tag{6.31}$$

$$\Lambda = \frac{hA}{\dot{m}_f c_f} \tag{6.32}$$

The overall effect of the surface heat flux relative to the internal resistance to heat transfer within the checkerwork is measured by the Biot modulus Bi:

$$\mathrm{Bi} = \frac{hw}{k_m} \tag{6.33}$$

The regenerator and its mode of operation can thus be summarized for this model using just six factors:

	Hot period	Cold period
Reduced time	Ω'	Ω''
Reduced length	Λ'	Λ''
Biot modulus	Bi'	Bi''

6.7 RELATIONSHIP BETWEEN THE TWO-DIMENSIONAL AND THREE-DIMENSIONAL MODELS

In the linear model set out in Chap. 5, two dimensions are used, namely, $\bar{\xi}$ and $\bar{\eta}$ (we use the bar to denote that $\bar{\xi}$ and $\bar{\eta}$ embody the bulk heat transfer coefficient $\bar{h}$), and this gives rise to the parameters reduced length $\bar{\Lambda}$ and reduced period $\bar{\Pi}$, where

$$\bar{\Lambda} = \frac{\bar{h}A}{\dot{m}_f c_f} \tag{6.34}$$

$$\bar{\Pi} = \frac{\bar{h}A}{M_m c_m}\left(P - \frac{m_f}{\dot{m}_f}\right) \tag{6.35}$$

It has been noted that this is called the two-dimensional model as distinct from the three-dimensional model which involves the dimensions ξ, Y, and Z (see Eqs. 6.26) and the parameters reduced time Ω, reduced length Λ, and Biot modulus Bi.

In both models, the time variation of the inlet gas temperatures for both hot and cold periods must be specified; most, if not all previous authors have restricted their considerations to constant inlet gas temperatures.

The question arises how adequate the two-dimensional model is to represent the thermal behavior of a regenerator; the answer revolves around the use of the lumped heat transfer coefficient $\bar{h}$.

The Hausen form of $\bar{h}$ is developed from the assumption that the mean solid temperature t_m varies linearly with time in both hot and cold periods:

$$t_m(x, \tau) = \frac{1}{w}\int_0^w t_m(x, y, \tau)\, dy \tag{6.36}$$

In the middle of the regenerator, this assumption is generally true. If the entrance gas temperature is constant in both periods, the closer to the regenerator entrance, the more nonlinear are the chronological variations of solid temperature. The nonvarying entrance gas temperatures can be treated as propagating this nonlinear behavior down the regenerator.

Hausen [1] suggested that a factor K/K_o (whose original use does not concern us here) might be used to measure the extent of the effect of these nonlinearities. Note that $0 < K/K_o < 1$, and the smaller the value of K/K_o, the greater the effect of the nonlinearities. When $\dot{m}_f' c_f' P' = \dot{m}_f'' c_f'' P''$ (the balanced case when $\eta_{\text{REG}} = \eta_{\text{REG}}' = \eta_{\text{REG}}''$), then

$$\frac{K}{K_o} = \frac{\eta_{\text{REG}}}{1 - \eta_{\text{REG}}}\left(\frac{1}{\bar{\Lambda}'} + \frac{1}{\bar{\Lambda}''}\right) \tag{6.37}$$

From Fig. 6.4, it will be observed that K/K_o increases with harmonic mean reduced length $\bar{\Lambda}_H$,

$$\frac{1}{\bar{\Lambda}_H} = \frac{1}{2}\left(\frac{1}{\bar{\Lambda}'} + \frac{1}{\bar{\Lambda}''}\right) \tag{6.38}$$

and decreases with $\bar{\Pi}_H$,

$$\frac{1}{\bar{\Pi}_H} = \frac{1}{2}\left(\frac{1}{\bar{\Pi}'} + \frac{1}{\bar{\Pi}''}\right) \tag{6.39}$$

In other words, the larger the regenerator the less the effect of the nonlinearities, whereas the longer the reduced period the greater the time available in each cycle for the nonlinearities to penetrate the regenerator packing.

Presented in Tables 6.1, 6.2, and 6.3 are comparative figures for thermal ratio computed using the two-dimensional model (with bulk heat transfer coefficients) and the three-dimensional model described earlier in this chapter. Consideration is restricted to the symmetric case ($\Lambda' = \Lambda''$, $\Omega' = \Omega''$, $\text{Bi}' = \text{Bi}''$). It will be observed the greater the value of K/K_o, the closer will be the values of η_{REG} obtained from the two- and three-dimensional models.

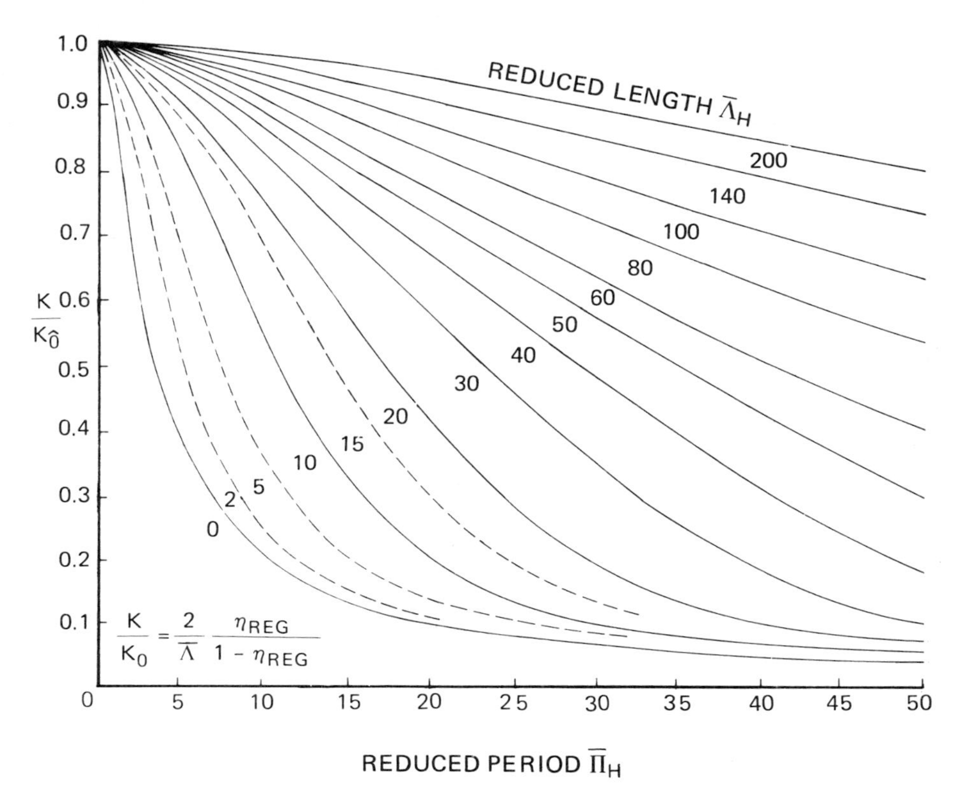

Figure 6.4 Hausen K/K_o curves.

It will be recalled from Sec. 6.2 that the ϕ factor in the bulk heat transfer coefficient attempts to average out the effects of the inversions of the temperature profiles at the beginning of each period over the whole cycle of regenerator operation. In the same way, the thermal ratios η'_{REG} and η''_{REG} average out over each period the cumulative temperature performance of the regenerator. As one might expect, therefore, even for low values of ϕ (< 0.7), the correspondence between the two- and three-dimensional thermal ratios is good for $K/K_o > 0.8$. Low values of ϕ correspond to relatively large durations of the reversal effect on the internal solid temperature profile in each period of operation.

However, the Hausen factor ϕ makes no attempt to deal with differences in the time variations of gas and solid temperature predicted by the two- and three-dimensional models.

We deal first, however, with the relationship between the two- and three-dimensional descriptive parameters.

Table 6.1 Values of thermal ratio η_{REG} (for $\bar{\Lambda}' = \bar{\Lambda}''$ and $\bar{\Pi}' = \bar{\Pi}''$)

(*a*) Computed using Willmott's two-dimensional method, and
(*b*) computed by the three-dimensional method, $\phi = 0.9$, $\Omega = 1.333$

Reduced length, $\bar{\Lambda}$		Reduced period, $\bar{\Pi}$	
	1	2	3
1. (*a*)	0.3221	0.2930	0.2559
(*b*)	0.3193	0.2849	0.2437
2. (*a*)	0.4912	0.4664	0.4305
(*b*)	0.4890	0.4590	0.4169
3. (*a*)	0.5937	0.5757	0.5477
(*b*)	0.5924	0.5702	0.5360
4. (*a*)	0.6622	0.6490	0.6282
(*b*)	0.6613	0.6453	0.6194
5. (*a*)	0.7109	0.7012	0.6856
(*b*)	0.7105	0.6988	0.6794
6. (*a*)	0.7474	0.7400	0.7280
(*b*)	0.7473	0.7385	0.7239
7. (*a*)	0.7758	0.7699	0.7605
(*b*)	0.7758	0.7691	0.7580
8. (*a*)	0.7984	0.7936	0.7861
(*b*)	0.7986	0.7933	0.7848
9. (*a*)	0.8169	0.8129	0.8068
(*b*)	0.8172	0.8130	0.8064
10. (*a*)	0.8322	0.8289	0.8238
(*b*)	0.8326	0.8293	0.8241

Hausen proposed that

$$\phi = 1 - \frac{w^2}{15\alpha_m}\left(\frac{1}{P'} + \frac{1}{P''}\right)$$

if $$\frac{w^2}{\alpha_m}\left(\frac{1}{P'} + \frac{1}{P''}\right) \leqslant 5 \tag{6.5}$$

otherwise

$$\phi = \frac{2.142}{\sqrt{0.3 + w(1/P' + 1/P'')}} \tag{6.6}$$

or, employing the dimensionless parameters set out previously,

$$\phi = 1 - \frac{1}{15}\left(\frac{1}{\Omega'} + \frac{1}{\Omega''}\right)$$

if $$\frac{1}{\Omega'} + \frac{1}{\Omega''} \leqslant 5 \tag{6.40}$$

otherwise

$$\phi = \frac{2.142}{\sqrt{0.3 + 2(1/\Omega' + 1/\Omega'')}} \tag{6.41}$$

An immediate relationship between ϕ and Ω' and Ω'' is provided.

Reduced length is considered by examining the form of the bulk heat transfer coefficient:

$$\frac{1}{\bar{\Lambda}} = \frac{\dot{m}_f c_f}{\bar{h}A} = \frac{\dot{m}_f c_f}{A}\left(\frac{1}{h} + \frac{w\phi}{3k_m}\right)$$

$$= \frac{1}{\Lambda}\left(1 + \frac{\mathrm{Bi}}{3}\right) \tag{6.42}$$

Reduced period is considered by examining the form of $\bar{h}$ in the same way:

$$\frac{1}{\bar{\Pi}} = \frac{1}{\Omega}\left(\frac{1}{\mathrm{Bi}} + \frac{\phi}{3}\right) \tag{6.43}$$

Table 6.2 Values of thermal ratio η_{REG} (for $\bar{\Lambda}' = \bar{\Lambda}''$ and $\bar{\Pi}' = \bar{\Pi}''$)

(*a*) Computed using Willmott's two-dimensional method, and
(*b*) computed by the three-dimensional method, $\phi = 0.8$, $\Omega = 0.666$

Reduced length, $\bar{\Lambda}$		Reduced period, $\bar{\Pi}$	
		1	2
1.	(*a*)	0.3221	0.2930
	(*b*)	0.3127	0.2667
2.	(*a*)	0.4912	0.4664
	(*b*)	0.4836	0.4404
3.	(*a*)	0.5937	0.5757
	(*b*)	0.5887	0.5553
4.	(*a*)	0.6622	0.6490
	(*b*)	0.6589	0.6346
5.	(*a*)	0.7109	0.7012
	(*b*)	0.7089	0.6914
6.	(*a*)	0.7474	0.7400
	(*b*)	0.7463	0.7337
7.	(*a*)	0.7758	0.7699
	(*b*)	0.7752	0.7655
8.	(*a*)	0.7984	0.7936
	(*b*)	0.7982	0.7909
9.	(*a*)	0.8169	0.8169
	(*b*)	0.8170	0.8170
10.	(*a*)	0.8322	0.8322
	(*b*)	0.8326	0.8326

Table 6.3 Values of thermal ratio η_{REG} (for $\bar{\Lambda}' = \bar{\Lambda}''$ and $\bar{\Pi}' = \bar{\Pi}''$)

(*a*) Computed using Willmott's two-dimensional method, and
(*b*) computed by the three-dimensional method, $\phi = 0.7$, $\Omega = 0.444$

Reduced length, $\bar{\Lambda}$		Reduced period, $\bar{\Pi}$
1.	(*a*)	0.3221
	(*b*)	0.3027
2.	(*a*)	0.4912
	(*b*)	0.4749
3.	(*a*)	0.5937
	(*b*)	0.5822
4.	(*a*)	0.6622
	(*b*)	0.6543
5.	(*a*)	0.7109
	(*b*)	0.7056
6.	(*a*)	0.7474
	(*b*)	0.7438
7.	(*a*)	0.7758
	(*b*)	0.7734
8.	(*a*)	0.7984
	(*b*)	0.7969
9.	(*a*)	0.8169
	(*b*)	0.8161
10.	(*a*)	0.8322
	(*b*)	0.8320

For a given two-dimensional regenerator configuration (symmetric case), the corresponding three-dimensional configuration is obtained in the following way.

From the value of ϕ, the reduced time $\Omega = \Omega' = \Omega''$ is calculated:

$$\Omega = \frac{2}{15(1-\phi)} \qquad \text{if } \Omega \geqslant 0.4 \tag{6.44}$$

$$= \frac{4}{0.3-(2.142/\phi)^2} \qquad \text{if } \Omega < 0.4 \tag{6.45}$$

The Biot modulus may now be computed:

$$\text{Bi} = \left(\frac{\Omega}{\Pi} - \frac{\phi}{3}\right)^{-1} \tag{6.46}$$

from which it will be observed that certain two-dimensional configurations are in fact meaningless in that the corresponding value of Bi for the three-dimensional model is negative. Finally, the three-dimensional reduced length may be calculated:

$$\Lambda = \bar{\Lambda}\left(1 + \text{Bi}\,\frac{\phi}{3}\right) \tag{6.47}$$

6.8 POSSIBLE IMPROVEMENTS IN THE LUMPED HEAT TRANSFER COEFFICIENT

In order to understand the limitations of the Hausen form of the bulk heat transfer coefficient in the two-dimensional regenerator model, and in particular the form of ϕ employed there, it is useful to recall how this coefficient was developed. At each and every level in the regenerator, Hausen [5] assumed that the time variation of solid temperature was linear:

$$\frac{\partial^2 t_m}{\partial y^2} = \frac{1}{\alpha_m}\frac{\partial t_m}{\partial \tau} = \text{a constant } \kappa \tag{6.48}$$

Integrating in the direction y, perpendicular to gas flow, and noting that

$$\left.\frac{\partial t_m}{\partial y}\right|_{y=w} = 0 \tag{6.49}$$

yields the relation

$$\frac{\partial t_m}{\partial y} = \kappa(y - w) \tag{6.50}$$

Further integration of Eq. (6.50) provides

$$t_m(y) = t_s + \tfrac{1}{2}\kappa y(y - 2w) \tag{6.51}$$

where t_s is the surface solid temperature. This last equation illustrates the parabolic form of the internal solid temperature profile. If we define $\bar{t}_m$ to be the mean solid temperature at any instant, then

$$\bar{t}_m = \frac{1}{w}\int_0^w t_m(y)\,dy = \frac{1}{w}\int_0^w t_s + \frac{1}{2}\,[\kappa y(y - 2w)]\,dy \tag{6.52}$$

or

$$\bar{t}_m - t_s = \frac{1}{w}\int_0^w \frac{1}{2}\,\kappa y(y - 2w)\,dy \tag{6.53}$$

Equation (6.53) then becomes

$$t_s - \bar{t}_m = \frac{\kappa w^2}{3} = \frac{w^2}{3\alpha_m}\frac{\partial \bar{t}_m}{\partial \tau} \tag{6.54}$$

Equation (5.2) can be modified using the surface heat transfer coefficient:

$$\frac{\partial \bar{t}_m}{\partial \tau} = \frac{hA}{M_m c_m}(t_f - t_s) \tag{6.55}$$

Equation (6.54) may be developed:

$$t_s - \bar{t}_m = \frac{w^2}{3\alpha_m}\frac{hA}{M_m c_m}(t_f - t_s) \tag{6.56}$$

Noting that $M_m = Aw\rho_m$ and $\alpha_m = k_m/\rho_m c_m$, Eq. (6.56) may be modified:

$$t_s - \bar{t}_m = (t_f - \bar{t}_m) - (t_f - t_s) = \frac{hA}{3k_m}(t_f - t_s) \tag{6.57}$$

If the lumped heat transfer coefficient is defined very simply by

$$\bar{h}(t_f - \bar{t}_m) = h(t_f - t_s) \tag{6.58}$$

then Eq. (6.57) becomes

$$\frac{1}{\bar{h}} = \frac{1}{h} + \frac{w}{3k_m} \tag{6.59}$$

The underlying assumption of this development is found in Eq. (6.48), namely, that the time variation of solid temperature is purely linear. In fact, reversals take place during which the parabolic profile denoted by Eq. (6.51) is inverted as a result of the changes in sign of κ (Eq. 6.48) (see Fig. 6.2).

Hausen [1] therefore modified Eq. (6.57) and introduced the factor ϕ. He considered that the effect of the parabolic profile inversions could be averaged out over the cycle; Eq. (6.57) is represented in time-average form:

$$\frac{1}{P' + P''}\int_0^{P'+P''} (t_s - t_m)\, d\tau = \frac{1}{P' + P''}\int_0^{P'+P''} \frac{h2w}{k_m}(t_f - t_s)\phi\, d\tau \tag{6.60}$$

The form of ϕ that Hausen derived has been specified previously. The Hausen approach basically assumes that the time-average heat fluxes calculated using h and $\bar{h}$ are equal:

$$\frac{1}{P' + P''}\int_0^{P'+P''} h(t_f - t_s)\, d\tau = \frac{1}{P' + P''}\int_0^{P'+P''} \bar{h}(t_f - t_m)\, d\tau \tag{6.61}$$

From Eq. (6.60) we have

$$\phi = \frac{1}{P' + P''}\int_0^{P'+P''} \frac{k_m}{h2w}\frac{t_s - t_m}{t_f - t_s}\, d\tau \tag{6.62}$$

Hausen's analytical results for ϕ have been confirmed by Butterfield, Schofield, and Young [6], who solved Eq. (6.48) numerically by applying alternately positive and negative (but equal in magnitude) heat fluxes $h(t_f - t_s)$ over successive hot and cold periods until cyclic equilibrium was established.

This numerical method points to a possibly more sophisticated approach. Noting that the Biot modulus $\mathrm{Bi} = hw/k_m$, an "instantaneous" value of $\phi(\tau)$ can be defined:

$$\phi(\tau) = \frac{6}{\mathrm{Bi}}\left(\frac{t_s(\tau) - \bar{t}_m(\tau)}{t_f(\tau) - t_s(\tau)}\right) \tag{6.63}$$

The variations of $t_s(\tau)$ and $\bar{t}_m(\tau)$ are computed by solving numerically:

$$\frac{\partial t_m}{\partial \tau} = \alpha_m \frac{\partial^2 t_m}{\partial y^2} \tag{6.22}$$

$$\left.\frac{\partial t_m}{\partial y}\right|_{y=w} = 0 \tag{6.24}$$

$$\left.\frac{\partial t_m}{\partial y}\right|_{y=0} = \frac{h}{k_m}(t_f - t_s) = \pm q \tag{6.23}$$

in successive hot and cold periods, cycled until equilibrium is established. The value of

$$\frac{1}{P' + P''} \int_0^{P'+P''} \phi(\tau)\, d\tau$$

corresponds to the Hausen value of ϕ.

Comparatively little work has been published in this possible use of a time-dependent ϕ. What is proposed here is that the two-dimensional model should embody a time-dependent bulk heat transfer coefficient $\bar{h}(\tau)$, where

$$\frac{1}{\bar{h}(\tau)} = \frac{1}{h} + \frac{w}{3k_m}\phi(\tau) \tag{6.64}$$

Why should such a modification be undertaken? After all, the thermal effectiveness η_{REG} predicted by both the two- and three-dimensional models of a regenerator do correspond fairly closely over a wide range of parameters.

It turns out that the correspondence between the time variations of thermal performance are sometimes significantly different between those predicted by the two-dimensional and three-dimensional models. For example, the exit gas temperature in the cold period varies in a fairly linear way using the two-dimensional model, whereas there is a steep decline, for the three-dimensional model, in exit gas temperature (as the parabolic profile inverts) before linear variation is established (see Fig. 6.5).

A measure of the discrepancy between the two models is given by

$$\psi = \frac{A_2 - B_2}{A_3 - B_3}$$

where

A_2 = initial exit gas temperature (2-D model)

A_2 = initial exit gas temperature (3-D model)

B_2 = final exit gas temperature (2-D model)

B_3 = final exit gas temperature (3-D model)

For the symmetric case ($\bar{\Lambda}' = \bar{\Lambda}''$, $\bar{\Pi}' = \bar{\Pi}''$), Fig. 6.6 displays ψ as a function of the ϕ factor. It is where ψ is unacceptably low that the two-dimensional model using time-varying $\phi(\tau)$ should be particularly useful in predicting time variations in thermal behavior.

The likely requirements of industry for regenerator predictive models are the development of nonlinear models specific to the application under consideration. It is most undesirable to use the three-dimensional model in such circumstances; the

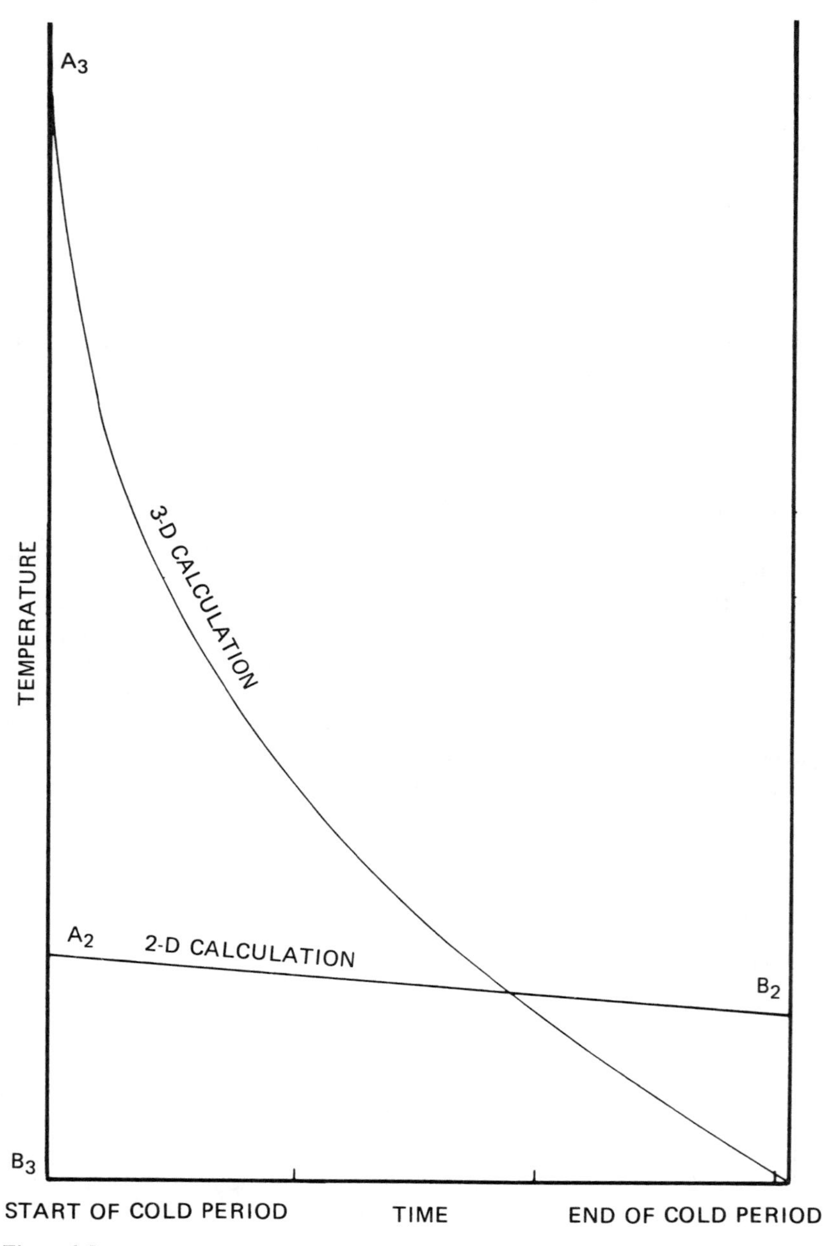
A3
3-D CALCULATION
TEMPERATURE
A2
2-D CALCULATION
B2
B3
START OF COLD PERIOD
TIME
END OF COLD PERIOD

Figure 6.5

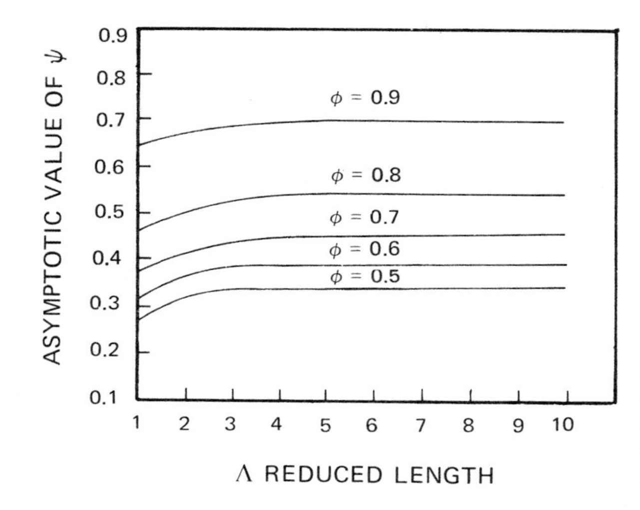

Figure 6.6 Relationship between ψ, the ratio of (max. exit temp./min. exit temp.) 2-D to the (max. exit temp./min. exit temp.) 3-D and reduced length $\bar{\Pi} = 1$.

computing will be complex to develop and program, as well as consume an unnecessary amount of computer time. In these circumstance, the refinements of the two-dimensional model we propose here may be well worth while.

REFERENCES

1. H. Hausen, "Vervollständigte Berechnung des Wärmeaustausches in Regeneratoren," *Z. Ver. Deutsch. Ing.*, Beiheft Verftk No. 2, 1942, pp. 31-43.
2. W. Tipler, "A Simple Theory of the Heat Regenerator," Shell Technical Report ICT/14, 1947.
3. A. J. Willmott, "The Regenerative Heat Exchanger Computer Representation," *Int. J. Heat Mass Transfer*, vol. 12, 1969, pp. 997-1014.
4. G. D. Bahnke and C. P. Howard, "The Effect of Longitudinal Heat Conduction on Periodic-Flow Heat Exchanger Performance," *J. Eng. Power*, April 1964, pp. 105-120.
5. H. Hausen, "Berechnung der Steintemperature in Winderhitzern," *Arch. Eiseniittwes.*, vol. 10, 1938/1939, pp. 474-480.
6. P. Butterfield, J. S. Schofield, and P. A. Young, "Hot Blast Stoves," *J. Iron Steel Inst.*, vol. 201, 1963, pp. 497-508.

CHAPTER

SEVEN

NONLINEAR MODELS OF COUNTERFLOW REGENERATORS

7.1 INTRODUCTION

Many likely requirements of industry for regenerator predictive models include the development of nonlinear models specific to the application under consideration. In certain very-low-temperature applications, it may be necessary to accommodate within the model the temperature variation of gas specific heat. Simulations of the regenerators used to preheat the air for ironmaking and for zinc and lead smelting blast furnaces (called Cowper stoves) must allow for the time variation of gas mass flow rate and the accompanying variations in heat transfer coefficient. Models of high-temperature regenerators fired by the combustion products of a fuel gas or a liquid fuel–these include glass furnace regenerators and Cowper stoves–must attempt some representation of radiative as well as convective heat transfer between gas and solid in the hot period of the cycle of regenerator operation. Such nonlinear models have as a basis the two-dimensional model discussed in Chap. 5, together with the representation of resistance to heat transfer internal to the regenerator packing, discussed in detail in Chap. 6. It is most unlikely that the three-dimensional model would be necessary, still less practicable in these circumstances. The influence of longitudinal conductivity is negligible and is not considered.

The form of the lumped heat transfer coefficient $\bar{h}$ described in Chap. 6 enables the continued use of Eqs. (5.1) and (5.2) even for models in which the heat transfer coefficients and the thermal properties of both gas and solid are permitted to be dependent on gas/solid temperature and therefore vary spatially and with time in the hot and cold period. The chronological variation of gas mass flow rate can be similarly represented.

The linear model set out in Chap. 5 is an approximation to a very wide spectrum of problems; the regenerator is specified by four dimensionless parameters, reduced length Λ' and Λ'' and reduced period Π' and Π'', together with hot and cold period inlet gas temperatures, possibly varying with time in a prescribed fashion. For constant inlet gas temperature conditions in each period, many authors have presented graphical solutions of the differential equations, in the form of the thermal effectiveness η'_{REG} and η''_{REG} as functions of reduced length and reduced period.

There are no corresponding general solutions for nonlinear models, simply because there are an infinite number of possible nonlinear models, each tailored to the particular environment in which the regenerator is required to operate, each incorporating those nonlinear features necessary in the model if a good enough correspondence between the theoretical model and possible practical measurements on a physical regenerator system is to be achieved.

In this chapter, therefore, the means will be discussed whereby the finite difference representation of Eqs. (5.1) and (5.2) can be manipulated in order that different nonlinear features may be embodied in the model.

7.2 MATHEMATICAL REPRESENTATION

It will be recalled that for the linear model, it proved useful to introduce the dimensionless parameters

$$\xi = \frac{\bar{h}A}{\dot{m}_f c_f L} x \tag{7.1}$$

$$= \frac{\bar{h}A}{M_m c_m} \left(\tau - \frac{m_f}{\dot{m}_f L} x\right) \tag{7.2}$$

since Eqs. (5.1) and (5.2) become simplified:

$$\frac{\partial t_f}{\partial \xi} = t_m - t_f \tag{7.3}$$

$$\frac{\partial t_m}{\partial \eta} = t_f - t_m \tag{7.4}$$

Employing the trapezoidal rule, Eq. (7.3) became

$$t_f(r+1, s+1) = t_f(r, s+1) + \frac{\Delta\xi}{2} [t_m(r+1, s+1) - t_f(r+1, s+1) + t_m(r, s+1) - t_f(r, s+1)] \tag{7.5}$$

Expanding this slightly, we have

$$t_f(r+1, s+1) = t_f(r, s+1) + \tfrac{1}{2}\{\Delta\xi_{r+1,s+1}[t_m(r+1, s+1) - t_f(r+1, s+1)] + \Delta\xi_{r,s+1}[t_m(r, s+1) - t_f(r, s+1)]\} \tag{7.6}$$

We suffix $\Delta\xi$ to become $\Delta\xi_{r,s}$ to embody nonlinear features, that $\bar{h}$, $\dot{m}_f$, and c_f may be functions of gas/solid temperature and therefore dependent on space and time. More specifically,

$$\Delta\xi_{r,s} = \frac{\bar{h}[r,s]A}{\dot{m}_f[s]c_f[r,s]L}\,\Delta x \tag{7.7}$$

where Δx is the step length in the finite difference scheme.

Equation (7.4) is a little more complicated, and it is necessary to expand $\Delta\eta_{r,s}$ as a total derivative from Eq. (7.2):

$$\Delta\eta_{r,s} = \left.\frac{\partial\eta}{\partial\tau}\right|_{r,s}\Delta\tau + \left.\frac{\partial\eta}{\partial x}\right|_{r,s}\Delta x \tag{7.8}$$

from which it follows that

$$\Delta\eta_{r,s} = \frac{\bar{h}[r,s]S}{M_m c_m[r,s]}\left(\Delta\tau - \frac{m_f}{\dot{m}_f[s]L}\,\Delta x\right) \tag{7.9}$$

where $\Delta\tau$ is the step length in time.

Equation (7.4) can now be expanded in finite difference form:

$$\begin{aligned} t_m(r+1, s+1) = t_m(r+1, s) + \tfrac{1}{2}\{\Delta\eta_{r+1,s+1}[t_f(r+1, s+1) \\ - t_m(r+1, s+1)] + \Delta\eta_{r+1,s}[t_f(r, s+1) - t_m(r, s+1)]\} \end{aligned} \tag{7.10}$$

Equation (7.6) can now be condensed in an analogous manner to that used in Chap. 5:

$$\begin{aligned} t_f(r+1, s+1) = A1_{r,s+1}t_f(r, s+1) + A2_{r+1,s+1}t_m(r+1, s+1) \\ + A3_{r,s+1}t_m(r, s+1) \end{aligned} \tag{7.11}$$

where $$A1_{r,s} = \frac{2 - \Delta\xi_{r,s}}{2 + \Delta\xi_{r+1,s}} \tag{7.12}$$

$$A2_{r,s} = \frac{\Delta\xi_{r,s}}{2 + \Delta\xi_{r,s}} \tag{7.13}$$

$$A3_{r,s} = \frac{\Delta\xi_{r,s}}{2 + \Delta\xi_{r+1,s}} \tag{7.14}$$

Similarly Eq. (7.10) can be condensed:

$$\begin{aligned} t_m(r+1, s+1) = B1_{r+1,s}t_m(r+1, s) + B2_{r+1,s+1}t_f(r+1, s+1) \\ + B3_{r+1,s}t_f(r+1, s) \end{aligned} \tag{7.15}$$

where $$B1_{r,s} = \frac{2 - \Delta\eta_{r,s}}{2 + \Delta\eta_{r,s+1}} \tag{7.16}$$

$$B2_{r,s} = \frac{\Delta\eta_{r,s}}{2 + \Delta\eta_{r,s}} \tag{7.17}$$

$$B3_{r,s} = \frac{\Delta\eta_{r,s}}{2 + \Delta\eta_{r,s+1}} \tag{7.18}$$

Substituting for $t_f(r+1, s+1)$ in Eq. (7.15), we obtain

$$t_m(r+1, s+1) = K1_{r+1,s} t_m(r+1, s) + K2_{r+1,s} t_f(r+1, s) \\ + K3_{r,s+1} t_m(r, s+1) + K4_{r,s+1} t_f(r, s+1) \tag{7.19}$$

where $$K1_{r,s} = \frac{B1_{r,s}}{1 - A2_{r,s+1} B2_{r,s+1}} \tag{7.20}$$

$$K2_{r,s} = \frac{B3_{r,s+1}}{1 - A2_{r,s+1} B2_{r,s+1}} \tag{7.21}$$

$$K3_{r,s} = \frac{A3_{r,s} B2_{r+1,s}}{1 - A2_{r+1,s} B2_{r+1,s}} \tag{7.22}$$

$$K4_{r,s} = \frac{A1_{r+1,s} B2_{r+1,s}}{1 - A2_{r+1,s} B2_{r+1,s}} \tag{7.23}$$

At first sight, this representation with its plethora of subscripts r, s may seem overawing! However, notice that, provided that all the $A1$, $A2$, $A3$, $B1$, $B2$, $B3$, $K1$, $K2$, $K3$, and $K4$ can be computed for all the required nodes r, s over the space-time finite difference mesh, Eqs. (7.11) (for integrating down the length of the regenerator at any time s), (7.15) (for integrating with time at the regenerator entrance), and (7.19) (for integrating with time over the remainder of the mesh) retain their simple linear form.

This approach enables a computer program to tackle regenerator simulations embodying nonlinear features to be developed fairly rapidly. Indeed, it is possible to provide a generalized computer program implementing the algorithm set out in Eqs. (7.11), (7.15), and (7.19), applied suitably to the hot and cold periods of the cycle. The particular nonlinear features under consideration are then "localized" into the coefficients $A1$, $A2$, $A3$, $B1$, $B2$, $B3$, $K1$, $K2$, $K3$, and $K4$, being incorporated into the general program by the provision of special subroutines that evaluate these coefficients.

Such an approach is worth considering if it is desired to explore by computer simulation of regenerator performance which nonlinear features have a significant effect and should therefore always be represented as nonlinear features in the model. For a specific program, however, such an approach would prove less efficient than a tailor-made computer program; for example, a program embodying time-varying gas flow rate and thus time-varying convective heat transfer coefficient does not require spatial variations (change in subscript r) in $\Delta\xi_{r,s}$ or $\Delta\eta_{r,s}$. Nevertheless, this simple approach enables most if not all nonlinear features to be embodied in a regenerator program.

At the end of Chap. 6, we discussed the possible requirement of a time-varying ϕ factor within the lumped heat transfer coefficient:

$$\frac{1}{\bar{h}(\tau)} = \frac{1}{h(\tau)} + \frac{w}{3k_m}\,\phi(\tau) \tag{7.24}$$

Such a chronologically varying ϕ, as well as a varying surface heat transfer coefficient, can be built into the nonlinear modeling approach that has been described.

7.3 QUASI-LINEARIZATION OF NONLINEAR MODELS

The linear forms of Eqs. (7.11), (7.15), and (7.19) depend on the prior knowledge of $\Delta\xi_{r,s}$ and $\Delta\eta_{r,s}$ over the space-time mesh. Of course, in general this information is not available; $\Delta\xi_{r,s}$ and $\Delta\eta_{r,s}$ may well be functions of the gas temperature $t_f(r, s)$ and/or the solid temperature $t_m(r, s)$, and it is the purpose of solving the differential equations (5.1) and (5.2) using this scheme to compute these temperatures.

The integration of the differential equations over successive cycles of operation can be regarded for the kth cycle as the operation of a function $F^{(k)}$ on a vector of solid temperatures $\mathbf{v}^{(k)}$; the elements of the vector $\mathbf{v}^{(k)}$ consist of the inlet and exit solid temperatures and a number (defined by the number of spatial nodes in the finite difference mesh) of equally spaced intermediate solid temperatures at the start of the kth cycle. This integration can then be represented as

$$\mathbf{v}^{(k+1)} = F^{(k)}(\mathbf{v}^{(k)}) \tag{7.25}$$

The exact form of $F^{(k)}$ will be determined by the values of $\Delta\xi_{r,s}$ and $\Delta\eta_{r,s}$ applicable to the kth cycle, values of which are not available at the start of the kth cycle.

At cyclic equilibrium, the vector $\mathbf{v}$ at the start of a cycle is unaltered by the operation of the function F:

$$\mathbf{v} = F(\mathbf{v}) \tag{7.26}$$

In the method proposed nearly 10 years ago for variable gas flow regenerator problems [1], it is suggested that integration is progressed over a set of artificial cycles: we denote the "artificiality" by using $\mathbf{u}^{(k)}$ to denote the vector of solid temperatures at the start of the kth cycle instead of $\mathbf{v}^{(k)}$. Although the values of $\Delta\xi_{r,s}$ and $\Delta\eta_{r,s}$ are unknown for the current kth cycle, they can be computed for the previous $(k-1)$th cycle. These artificial cycles can be represented by Eq. (7.27):

$$\mathbf{u}^{(k+1)} = F^{(k-1)}(\mathbf{u}^{(k)}) \tag{7.27}$$

It turns out in practice that the natural cycling of the regenerator toward equilibrium carries along the convergence of $F^{(k)}$ to F and $\mathbf{u}^{(k)}$ to $\mathbf{v}$, that is,

$$\lim_{k\to\infty} F^{(k)} = F$$

$$\lim_{k\to\infty} \mathbf{u}^{(k)} = \mathbf{v} \tag{7.28}$$

Initial estimates of $F^{(0)}$ and $\mathbf{u}^{(1)}$ can be obtained using the linear model for which estimated average constant values of $\Delta\xi$ and $\Delta\eta$ are used.

It must be stressed that Eq. (7.27) represents the simulation of a purely artificial kth cycle; in no way can a sequence of such cycles be used as a simulation of any transient regenerator performance. Once cyclic equilibrium has been achieved and $F^{(k)} = F^{(k-1)} = F$, however, representation of regenerator performance is achieved. The feature of this approach is that the linearity of Eqs. (7.11), (7.15), and (7.19) is preserved; the procedure is sometimes called quasi-linearization.

7.4 METHODS FOR SIMULATING NONSTEADY-STATE PERFORMANCE OF REGENERATORS

It is sometimes required to simulate regenerator performance during the time following one or more step changes in operating conditions and to follow the performance of the regenerator before cyclic equilibrium is established again. Such nonsteady-state behavior is sometimes referred to as transient behavior.

In this section, a possible method for computing the simulation of regenerator transient performance is briefly discussed.

At any internal point $(r+1, s+1)$ on the space-time mesh, the solid temperature is computed using Eq. (7.19). Subsequently the gas temperature is evaluated by using Eq. (7.11).

Both these equations involve prior knowledge of $t_m\{r+1, s+1\}$ and $t_f\{r+1, s+1\}$ to facilitate the necessary evaluation of $\Delta\xi_{r+1,s+1}$ and $\Delta\eta_{r+1,s+1}$. Estimates $t_m^*\{r+1, s+1\}$ and $t_f^*\{r+1, s+1\}$ can be obtained by linear or quadratic extrapolation from previously computed values of t_m and t_f at "earlier" nodes of the finite difference mesh. Using these estimates, approximate values of $A1, A2, A3, K1, K2, K3$, and $K4$ can be computed and revised values of $t_m\{r+1, s+1\}$ and $t_f\{r+1, s+1\}$ obtained using Eqs. (7.19) and (7.11). This process is continued iteratively (usually no more than two or three iterations are required) until the computed values of $t_m\{r+1, s+1\}$ and $t_f\{r+1, s+1\}$ converge. The values of t_m and t_f are then computed at the next node, and so on over the whole mesh covering the cycle of operation under consideration.

The integration then continues. At the expense of local iterations at each node on the space-time net, the integration

$$\mathbf{v}^{(k+1)} = F^{(k)}(\mathbf{v}^{(k)}) \tag{7.29}$$

over successive cycles is achieved and the simulation of transient performance of the regenerator is made possible. The artificiality of the cycles of the quasi-linear regenerator model is eliminated, at the expense of the computational economy (for cyclic equilibrium problems) provided by the quasi-linearization.

7.5 VARIABLE GAS FLOW PROBLEMS IN REGENERATORS

A system of fixed bed regenerators delivers a preheated gas whose temperature varies in sawtooth fashion (see Fig. 7.1). During each cold period, the exit gas temperature

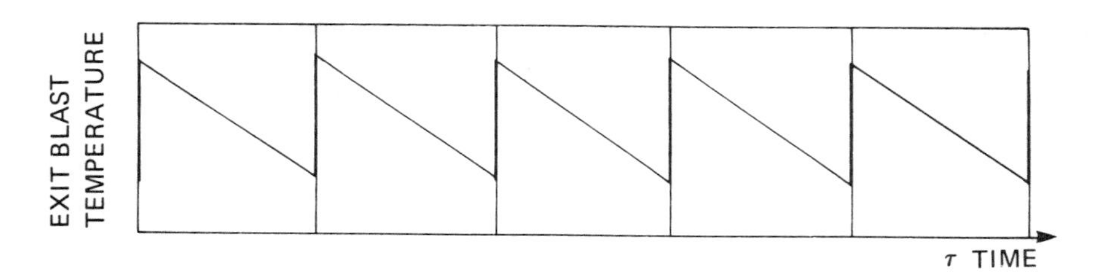

Figure 7.1 Variation of blast temperature from Cowper stove without bypass main.

steadily declines with time until a reversal occurs when a fresh regenerator is brought into operation and the temperature of the gas presented by the system of regenerators rises sharply before starting to decline again.

Cowper stoves are fixed bed regenerators used to preheat the blast for the iron-making and zinc smelting processes. It is necessary to eliminate the sawtooth effect (Fig. 7.1) to provide stable smelting furnace operation. This is achieved normally by running the stoves with a "bypass main" (see Fig. 7.2) or in "staggered parallel" (see Fig. 7.3).

In steady-state operation, the Cowper stove system delivers preheated blast at a constant flow rate of $\hat{\dot{m}}_f$ at a constant blast temperature t_B''. The flow of air is divided at a valve between the stove (in its cold period of operation) and the bypass main, or between two stoves, for staggered parallel operation, both stoves in their cold periods of operation but out of phase by half a period.

The valve position is adjusted continuously by a control system in such a way that the blast temperature t_B'' is held constant. This imposes a steady thermal load on the Cowper stove system equal to $\hat{\dot{m}}_f'' c_f''(t_B'' - t_{fi}'')$. The time variation of gas mass flow rate for bypass main operation is illustrated in Fig. 7.4. If the system could be perfectly controlled, the local flow rates at any instant are determined by the heat balance equations (7.30) or (7.31) and (7.32).

Bypass main:

$$\hat{\dot{m}}_f'' c_f''(t_B'' - t_{fi}'') = \dot{m}_f''(\tau) c_f''(t_{fo}''(\tau) - t_{fi}'') \tag{7.30}$$

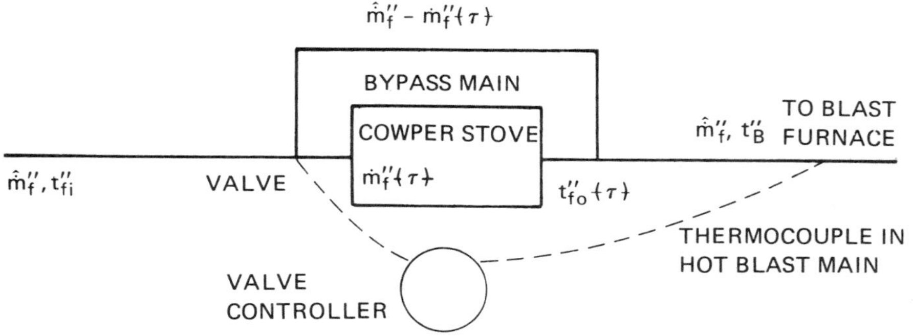

Figure 7.2 Blast temperature control system, bypass main.

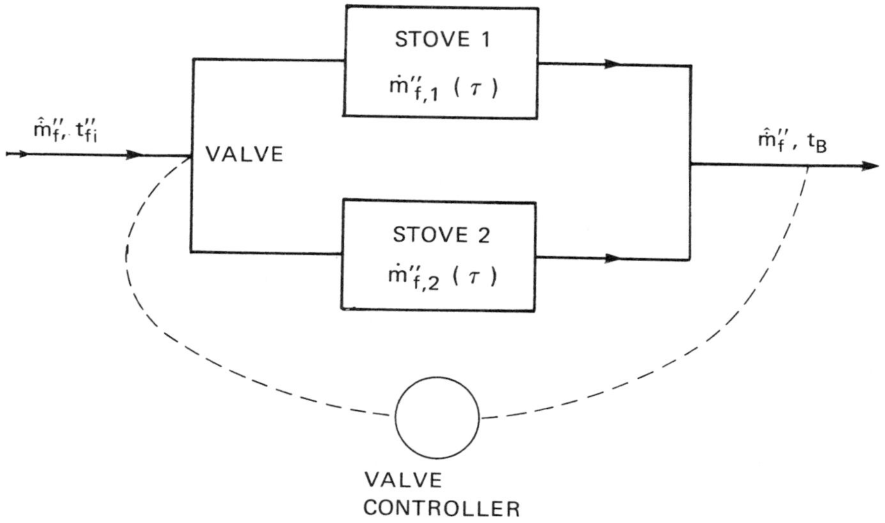

Figure 7.3 Blast temperature control system, staggered parallel.

Staggered parallel:

$$\hat{\dot{m}}_f'' c_f''(t_B'' - t_{fi}'') = \dot{m}_{f,1}''(\tau) c_f''(t_{fo,1}''(\tau) - t_{fi}'') + \dot{m}_{f,2}''(\tau) c_f''(t_{fo,2}''(\tau) - t_{fi}'') \quad (7.31)$$

$$\hat{\dot{m}}_f'' = \dot{m}_{f,1}'' + \dot{m}_{f,2}'' \quad (7.32)$$

Note that $\dot{m}_{f,1}''$ and $\dot{m}_{f,2}''$ are the flow rates of blast passing through the pair of regenerators, numbered 1 and 2, which are operating together in staggered parallel.

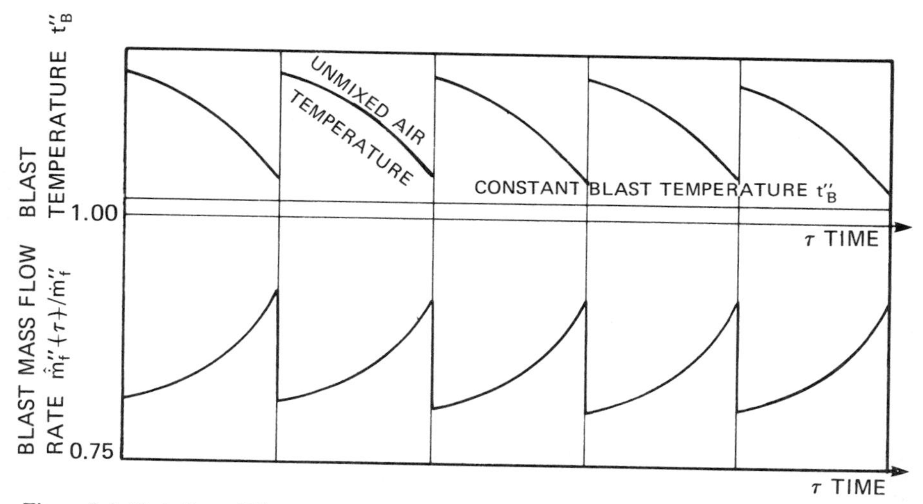

Figure 7.4 Variation of blast temperature and flow rate bypass main system.

In the hot period, it might be idealized in the model that the hot gas passes through the stove at a constant flow rate $\hat{\dot{m}}'$ with inlet temperature t'_{fi}. For transient performance simulations, the hot period gas flow and inlet temperature may well vary from one period to the next over a succession of cycles of regenerator operation.

The computer simulation of such Cowper stoves embodies several distinct yet closely intertwined procedures.

First, if the variation of the flow rate $\dot{m}''(\tau)$ for a particular stove is specified, the variation of the convective heat transfer coefficient $h''(\tau) = C\dot{m}''(\tau)^a$, typically $0.8 \leqslant a \leqslant 1.0$) and thus the lumped heat transfer coefficient $\bar{h}''(\tau)$ can be calculated, from which can be computed the corresponding values of $\Delta\xi_{r,s}$ (Eq. 7.7) and $\Delta\eta_{r,s}$ (Eq. 7.9). The cold period can then be simulated by solving Eqs. (5.1) and (5.2) employing the scheme of Eqs. (7.11), (7.15), and (7.19).

Second, the mechanism of controlling the variation of flow rate must be embodied in the simulation that enables the mixed blast temperature t''_B to be held constant. Now rather different approaches are required to deal with cyclic equilibrium as contrasted with transient performance problems.

For stationary performance, the quasi-linearization technique may be employed; the variation of flow rate $\dot{m}''(\tau)$ for the kth cycle is computed from the temperature performance of the regenerator in the $(k - 1)$th cycle. The calculation of regenerator performance involves not only evaluating how the gas and solid temperatures vary over an equilibrium cycle, but also how the gas flow $\dot{m}''_f(\tau)$ changes with time over the cold period of the same cycle. It is necessary to impose an additional boundary condition for the problem, and for bypass main operation it is common to specify that the ratio of the air flow rate passing through the regenerator at the end of the cold period and the total air flow rate delivered to the furnace, $\dot{m}''_f(P'')/\hat{\dot{m}}''_f$ is equal to some constant K (typically $0.8 \leqslant K < 1.0$). In this case, the mixed blast temperature t''_B is unknown and is calculated by the simulation. Alternatively, it is possible to specify in advance the required value of t''_B and the value of $\dot{m}''_f(P'')/\hat{\dot{m}}''_f$ at the end of the cold period and to calculate, using the simulation, which combinations of gas flow $\dot{m}'_f$ in the hot period and cycle duration will allow such cold period thermal demands to be met.

Where $\dot{m}''_f(P'')/\hat{\dot{m}}''_f = K$ is specified in advance, then given the computed exit air temperature at the end of the cold period, $t''_{fo}(P'')$ for the $(k-1)$th cycle, the possible blast temperature t''_B can be computed using a slightly rearranged form of Eq. (7.30):

$$t''_B = t''_{fi} + K(t''_{fo}(P'') - t''_{fi}) \tag{7.33}$$

Using this t''_B, held constant for the whole cold period, and the variation of exit air temperature $t''_{fx}(\tau)$ for the $(k - 1)$th cycle, an "estimated" variation $\dot{m}''_f(\tau)$ for the remainder of the cold period of the kth cycle is obtained using Eq. (7.30).

Instead of fixing K in advance, we have indicated that it is equally possible to set t''_B as the required blast temperature. If t''_B is set too high, however, the computed value of $\dot{m}''_f(P'')/\hat{\dot{m}}''_f$ will be greater than one! If t''_B is set too low, the value of $\dot{m}''_f(P'')/\hat{\dot{m}}''_f$ will be less than a possible target value. Either case will call for an adjustment of hot gas flow rate $\dot{m}'_f$ and/or the cycle time in the simulation. The simulation of staggered parallel operation follows along the same lines.

Transient performance is much more difficult to represent; each and every cycle must be simulated during the transient phase. This requires that the variation of air flow $\dot{m}_f''(\tau)$ must be computed as well as the gas and solid temperature fluctuations over all the cold periods simulated. Usually the blast temperature t_B'' is set in advance; at each iteration involving Eqs. (7.19) and (7.11) described earlier, successive estimates of the instantaneous value of $\dot{m}_f''(\tau)$, namely, $\dot{m}_f''(s+1)$, are employed until the computed values of $t_m(r+1, s+1)$ and $t_f(r+1, s+1)$ together with $\dot{m}_f''(s+1)$ converge to their final values.

However, there is a further problem when nonstationary performance, for example, of staggered parallel operation is considered. When a change in thermal load occurs, each of the stoves will be at a different phase of its operation and it is thus necessary to simulate all three or, more usually, four stoves separately. For cyclic equilibrium problems, provided that all three or four stoves are identical, only one stove need be simulated.

Razelos and Benjamin [2] present an interesting example calculation of a Cowper stove installation, together with graphical results typical of those possible using this simulation method. Figures 7.5 and 7.6 contrast the air flow variations through each stove in the cold period between bypass main (sometimes called serial) and staggered parallel operation. They present exit temperatures on the dimensionless temperature scales T_{fo}' and T_{fo}'' defined by Eqs. (5.11) and (5.12). The important dimensions of the stove installation are as follows:

Heating surface area $A = 32{,}940\ \text{m}^2$
Gas inlet temperature $t_{fi}' = 1440°\text{C}$
Air inlet temperature $t_{fi}'' = 93°\text{C}$

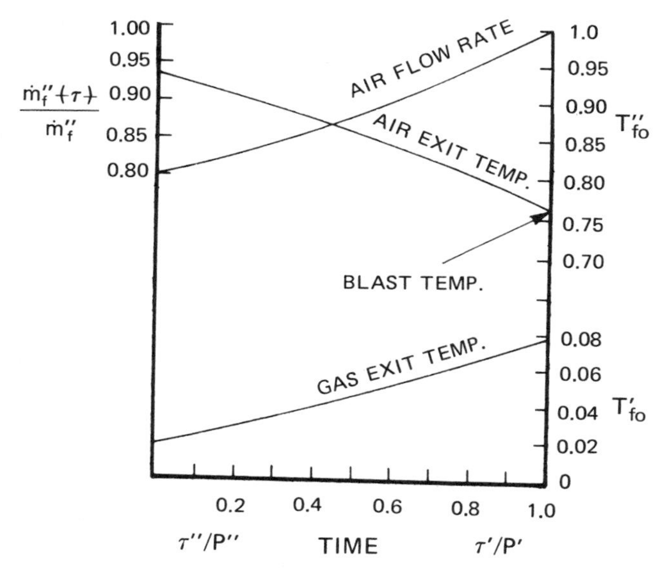

Figure 7.5 Variation of dimensionless exit blast temperature, air flow rate, and hot period exit gas temperature for bypass main operation.

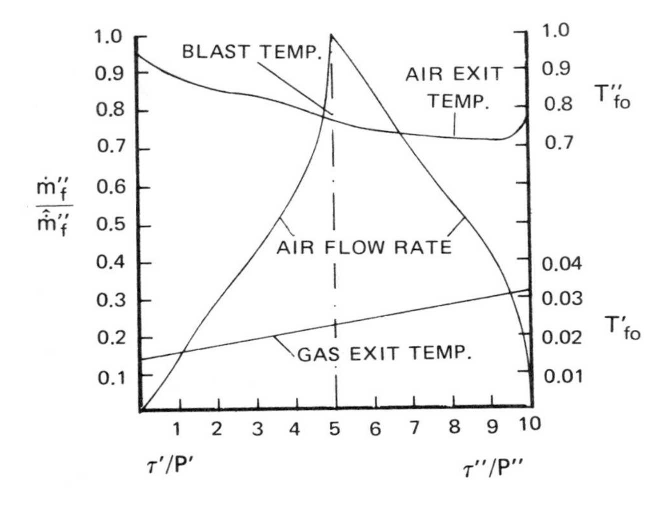

Figure 7.6 Variation of dimensionless exit blast temperature, air flow rate, and hot period exit gas temperature for staggered parallel operation.

Required blast flow rate $\hat{\dot{m}}_f'' = 59\ \text{m}^3/\text{s}$
Desired blast temperature $t_B'' = 1093°\text{C}$
Time on blast $P'' = 5400$ s
Maximum gas exit temperature $t_{fo}' \leqslant 343°\text{C}$

Gas flow rate:
3-stove bypass main arrangement $\dot{m}_f' = 15.23\ \text{m}^3/\text{s}$
Time on gas $P' = 14{,}400$ s

Gas flow rate:
4-stove staggered parallel arrangement $\dot{m}_f' = 31.46\ \text{m}^3/\text{s}$
Time on gas $P'' = 3600$ s

7.6 RADIATIVE HEAT TRANSFER BETWEEN GAS AND SOLID SURFACE IN REGENERATORS

In certain high-temperature regenerators, including Cowper stoves and glass furnace regenerators, the heat exchanger is fired by a freshly combusted gas containing significant proportions of carbon dioxide and water vapor. The channels of the checkerwork of such regenerators are sufficiently wide to avoid blockage by deposited dirt of other carryover solid material in the hot period of operation. However, the radiative heat flux between carbon dioxide/water vapor and the surface of the checkerwork is proportional to what is called the "beam length," which in this case is the width of the checkerwork channels. These factors therefore combine to make radiative heat transfer as important as, and in some cases significantly more important than, convective heat transfer between gas and solid surface in the hot period of operation.

Depending on the relative importances of convective and radiative heat transfer between gas and solid in the particular circumstances of regenerator construction and operation under consideration, different approximate representations of the radiative effects can be employed.

However, although radiation between gas and solid is proportional to $t_f^4 - t_s^4$, where t_s is surface solid temperature and both temperatures are absolute, the underlying philosophy adapted here is to maintain the "linearity" of the regenerator model described in Chap. 5 and earlier in this chapter.

The heat flux due to gas-solid radiation is q_R, where

$$q_R = \sigma \left(\frac{\epsilon_s + 1}{2}\right)(\epsilon t_f^4 - \alpha t_s^4) \tag{7.34}$$

where ϵ_s is the emissivity of the solid surface, ϵ is the emissivity of the gas and α is the absorbivity of the gas. Now ϵ and α are proportional to the percentage of carbon dioxide in the gas (water vapor is treated in a similar manner) and to the beam length; both are functions of the gas temperature.

The value of ϵ can be obtained from the Hottel charts (see Fig. 7.7), given the product of the beam length (channel width) L_o and the proportion of carbon dioxide in the gas, and given the gas temperature. In circumstances where $t_f > t_s$, certainly when $t_f/t_s > 1.25$, the value of α can be extracted approximately from the same charts, using surface solid temperature instead of gas temperature. In the hot period, t_f is greater than t_s.

A more accurate formula is given by Hottel for α:

$$\alpha = \left(\frac{t_f}{t_s}\right)^{0.65} \epsilon\left(\frac{t_s L_o t_s}{t_s}\right)$$

where the emissivity ϵ is obtained from Fig. 7.7 using the surface temperature and the product of L_o and t_s/t_f. This and the possible inclusion of the effects of water vapor are discussed in [3].

The nonlinear form of Eq. (7.34) is approximated by an "equivalent" radiative heat transfer coefficient h_R; the corresponding heat flux at the solid surface is represented by

$$q_R = h_R(t_f - t_s) \tag{7.35}$$

from which it follows that at any instant in the hot period at a particular position in the regenerator,

$$h_R = \sigma \left(\frac{\epsilon_s + 1}{2}\right)\left(\frac{\epsilon t_f^4 - \alpha t_s^4}{t_f - t_s}\right) \tag{7.36}$$

and the local "bulk" heat transfer coefficient $\bar{h}$ is

$$\frac{1}{\bar{h}} = \frac{1}{h + h_R} + \frac{w\phi}{3\lambda} \tag{7.37}$$

By taking a global (i.e., in space and time over the hot period) average of h_R or an estimate of this average, and applying this to the whole of hot period operation, the simple linear model can be retained as described in Chap. 5. In this case, the hot period

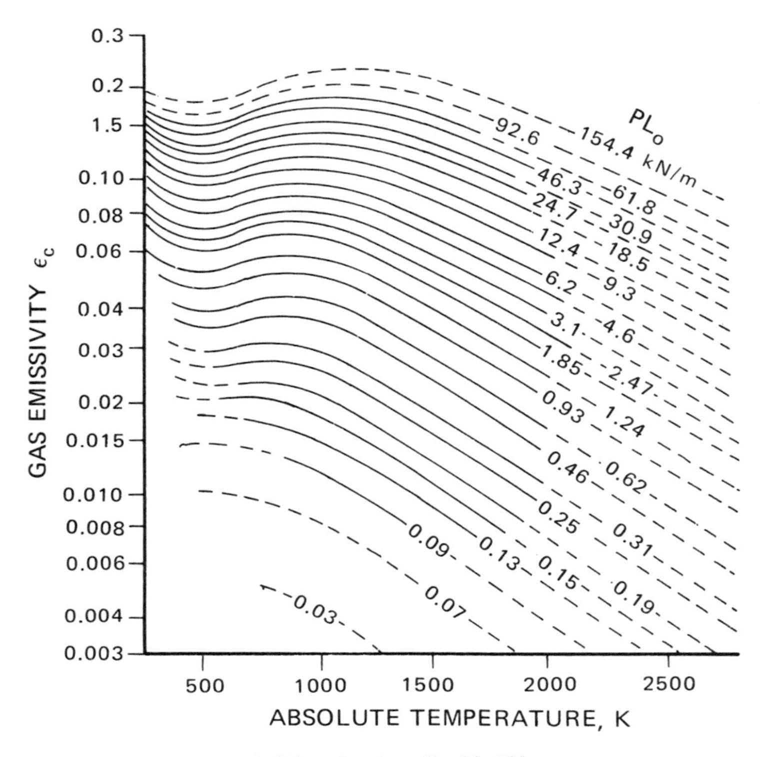

Figure 7.7 Effective emissivity of carbon dioxide [3].

dimensionless parameters Λ' and Π' embody the average value of h_R through Eq. (7.37).

However, a more precise representation can be achieved using the quasi-linear model. Here h_R is permitted to vary spatially and chronologically as a function of gas and solid temperature (7.36), noting also that ϵ and α are themselves functions of gas and solid temperature, respectively. The local bulk heat transfer coefficients $\bar{h}_{r,s}$ and surface coefficients $(h + h_R)_{r,s}$ computed in preparation for integration of the differential equations (5.11) and (5.21) over the kth cycle, together with the gas and mean solid temperatures computed for the kth cycle, are combined to extract the local "surface" solid temperature $t_s\{r, s\}$ at all the nodes on the finite difference net. Equation (6.4) is employed to yield

$$t_s\{r, s\} = t_f\{r, s\} + \frac{h_{r,s}}{(h + h_R)_{r,s}} [t_m\{r, s\} - t_f\{r, s\}] \tag{7.38}$$

These values of $t_s\{r, s\}$, together with the values of $t_f\{r, s\}$, are then employed to compute a revised local value of the "radiative heat transfer coefficient" h_R using Eq. (7.36) from which the local value of the bulk heat transfer coefficient $\bar{h}$ is calculated from Eq. (7.37). Revised values of $\Delta\xi_{r,s}$ and $\Delta\eta_{r,s}$ are now computed over the nodes of the net, and the quasi-linear integration for the $(k + 1)$th cycle may commence.

The accuracy of this approach depends on the reliability of Eq. (7.38) and the form of the Hausen phi factor ϕ employed in Eq. (7.37). In Chap. 6 it was noted that a precise calculation of surface solid temperature as needed for Eq. (7.36) may require the heat transfer within the checkerwork to be represented explicitly within the three-dimensional model of the regenerator. Alternatively, it may be sufficient to employ the time-varying ϕ factor within the two-dimensional model and thereby yield a more than adequate representation of gas-solid surface radiative effects in a regenerator. Little or no work has been reported in the literature on this problem, despite the importance of radiative heat transfer in certain high-temperature regenerators.

REFERENCES

1. A. J. Willmott, "Simulation of a Thermal Regenerator Under Conditions of Variable Mass Flow," *Int. J. Heat Mass Transfer*, vol. 11, 1968, pp. 1105–1116.
2. P. Razelos and M. K. Benjamin, "Computer Model of Thermal Regenerators with Variable Mass Flow Rates," *Int. J. Heat Mass Transfer*, vol. 21, 1978, pp. 735–743.
3. H. C. Hottel and R. B. Egbert, "Radient Heat Transmission from Water Vapor," *Trans. AIChE*, vol. 38, pp. 531–568.

CHAPTER

EIGHT

IMPROVED COMPUTATIONAL METHODOLOGY FOR REGENERATORS

8.1 INTRODUCTION

In this chapter, the discussion of the computational methodology mentioned initially in Chap. 5 is extended. Concern focuses on the differential equations describing regenerator behavior, namely,

$$\bar{h}A(t_m - t_f) = \dot{m}_f c_f L \frac{\partial t_f}{\partial x} + m_f c_f \frac{\partial t_f}{\partial \tau} \tag{5.1}$$

and $$\bar{h}A(t_f - t_m) = M_m c_m \frac{\partial t_m}{\partial \tau} \tag{5.2}$$

with the reversal conditions (5.3) and (5.4). Reference will be made largely to the form of these equations that uses the dimensionless parameters ξ and η [see the defining equations (5.5) and (5.6)] and the form employing the [0, 1] temperature scale. These equations become

$$\frac{\partial T_f}{\partial \xi} = T_m - T_f \tag{5.13}$$

$$\frac{\partial T_m}{\partial \eta} = T_f - T_m \tag{5.14}$$

where T_f and T_m are defined by Eqs. (5.11) and (5.12). Examined first are the open methods, in which the gas and solid temperatures are evaluated in successive cycles until the regenerator model attains cyclic equilibrium. The open methods are largely schemes of the finite difference type, which have the advantage that they can be adapted to deal with nonlinear problems, as discussed in Chap. 7. The closed methods are those in which the steady-state performance is computed directly by embodying the reversal conditions (5.9) and (5.10) in a formulation of the simulation of a thermal regenerator as a boundary value problem. Although the closed methods have the great advantage in calculating directly the steady-state performance of the regenerator, certain inherent difficulties can arise and these will be mentioned. Nevertheless, in developing the closed methods, advantage is taken of the integral equations (5.25) and (5.26) proposed originally by Nusselt [1]; it is possible to use these integral equations in an open method of solution and thus avoid the instabilities of a closed scheme. Be it sufficient to say in this introduction that these instabilities manifest themselves in the ill-conditioning of the simultaneous linear equations generated in the closed schemes.

For many linear problems, however, the closed schemes are the most computationally economical. Although these closed methods have been discussed in the literature, few details of how such methods can be implemented on a digital computer have been published. This matter will be discussed here. Even so, it is also necessary to describe how an open scheme can be accelerated toward cyclic equilibrium, since under the very circumstances in which the open methods, without acceleration, require many cycles to reach equilibrium, the closed methods tend to break down, obviating any possible advantage in these conditions such closed methods might be expected to exhibit.

8.2 OPEN METHODS: NUMERICAL SOLUTION OF THE REGENERATOR EQUATIONS

In a numerical solution of Eqs. (5.13) and (5.14), the gas and solid temperatures are computed spatially, at the entrance and exit to the regenerator and at a number of equally spaced intermediate positions, and chronologically at equal intervals of time. The mesh of gas and solid temperatures is shown in Fig. 8.1; the distance step length is $\Delta\xi$, whereas the step length in time is $\Delta\eta$.

In Chap. 5, mention is made of the Willmott [2] integration scheme whereby the nondimensional equation (5.33) is reformulated:

$$T_f(r+1, s+1) = A_1 T_f(r, s+1) + A_2[T_m(r, s+1) + T_m(r+1, s+1)] \tag{8.1}$$

with
$$A_1 = \frac{2 - \Delta\xi}{2 + \Delta\xi} \tag{8.2}$$

$$A_2 = \frac{\Delta\xi}{2 + \Delta\xi} \tag{8.3}$$

At any time $s+1$, if the solid temperatures are known, the gas temperatures down the

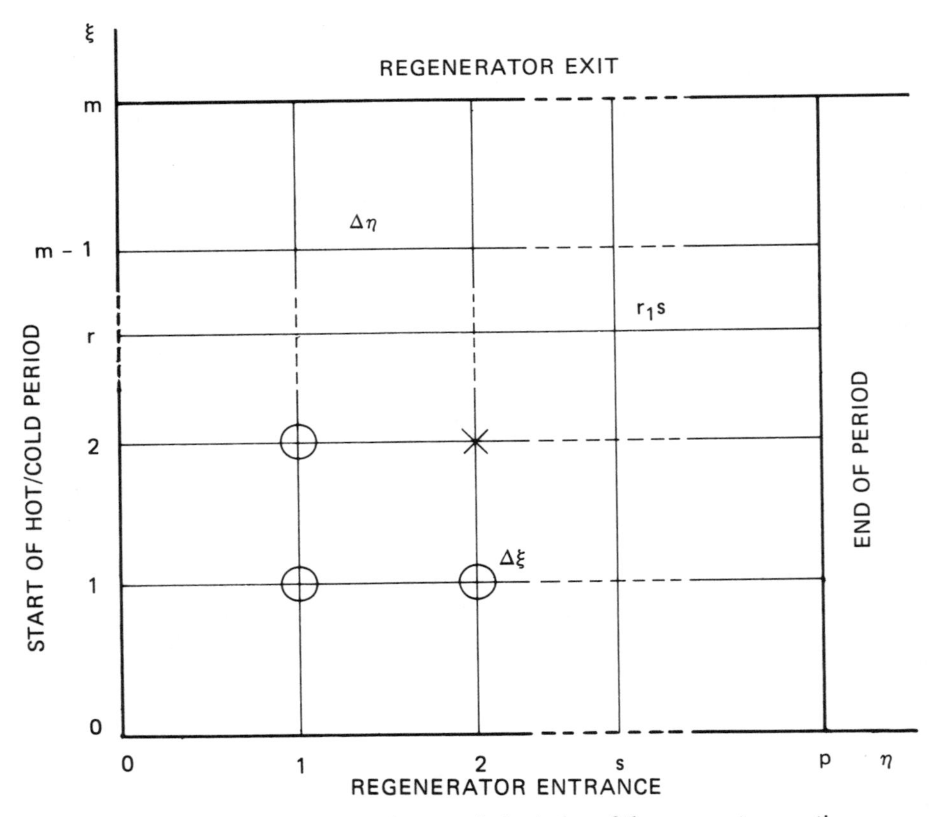

Figure 8.1 Finite difference mesh for the numerical solution of the regenerator equations.

length of the regenerator can be computed from the prespecified inlet gas temperature by applying Eq. (8.1) for $r = 0, 1, 2, \ldots, m-1$. In particular, at the start of a period where the solid temperatures are known as being those at the end of the preceding period, the initial gas temperature distribution can be calculated in this way.

Certain problems arise in implementing this scheme in FORTRAN, because this language does not allow zero subscripts for any array, in particular the arrays of gas and solid temperatures defined by, for example,

```
DIMENSION TF(30), TS(30)
```

This means that the regenerator entrance must be denoted by JR = 1 and the exit by JR = M1; it is convenient to denote M + 1 by M1, thereby retaining the nomenclature M of the numerical method for m. The gas temperatures TF can now be computed from the solid temperatures at the beginning of the period by the following simple program:

```
    DO 140 JR = 2,M1
    TF(JR) = A1*TF(JR - 1) + A2*(TM(JR) + TM(JR - 1))
140 CONTINUE
```

Note it is possible to retain the notation for $A1$ and $A2$ [see Eqs. (8.2) and (8.3)] within such a computer program.

At any particular instant $(r+1, s+1)$, for example, (2, 2) in Fig. 8.1, the gas and solid temperatures are known at $(r+1, s)$, (r, s), and $(r, s+1)$, denoted by circles on the mesh at (2, 1), (1, 1), and (1, 2). In the Willmott scheme, the solid temperature T_m is computed using (8.4):

$$T_m\{r+1, s+1\} = K_1 T_m\{r+1, s\} + K_2 T_f\{r+1, s\} + K_3 T_m\{r, s+1\} + K_4 T_f\{r, s+1\} \tag{8.4}$$

It is to be noted that

$$K_1 = \frac{B_1}{X} \tag{8.5}$$

$$K_2 = \frac{B_2}{X} \tag{8.6}$$

$$K_3 = \frac{A_2 B_2}{X} \tag{8.7}$$

$$K_4 = \frac{A_1 B_2}{X} \tag{8.8}$$

and that the expressions of B_1 or B_2 and X are as follows:

$$B_1 = \frac{2 - \Delta\eta}{2 + \Delta\eta} \tag{8.9}$$

$$B_2 = \frac{\Delta\eta}{2 + \Delta\eta} \tag{8.10}$$

$$X = 1 - A_2 B_2 \tag{8.11}$$

Having evaluated $T_m\{r+1, s+1\}$ using Eq. (8.4), it is now possible to find $T_f\{r+1, s+1\}$ by using Eq. (8.1).

The solid temperature at the entrance can be evaluated using Eq. (8.12) at each instant in time where

$$T_m\{0, s+1\} = B_1 T_m\{0, s\} + B_2[T_f\{0, s\} + T_f\{0, s+1\}] \tag{8.12}$$

This can be simplified by noting that the entrance gas temperature $T_f\{0, s\}$ for $s = 0, 1, 2, \ldots, p$ is equal to unity with the hot period and zero in the cold period.

Once the initial gas and solid temperatures at the start of a period have been established using Eq. (8.1) in the manner described, integration at the successive intervals of time consists of the following stages:

1. Evaluation of the entrance solid temperature using Eq. (8.13*h*) or (8.13*c*):

Hot period: $$T_m\{0, s+1\} = 2B_2 + B_1 T_m\{0, s\} \tag{8.13h}$$

Cold period: $$T_m\{0, s+1\} = B_1 T_m\{0, s\} \tag{8.13c}$$

2. Alternate evaluation of the solid and the gas temperatures down the length of the regenerator by applying Eq. (8.4) and then Eq. (8.1).

Perhaps the most important feature of this scheme lies in the fact that it is only necessary to store the gas and solid temperatures at any instant in time. There is no need to store, in a computer program, two-dimensional arrays of numbers corresponding to the gas and solid temperatures at all positions on the space-time mesh illustrated in Fig. 8.1. This advantage arises because Eq. (8.4) requires only temperatures at $(r + 1, s)$ and $(r, s + 1)$; the temperatures at (r, s) are not needed. As a consequence, when $T_m(r + 1, s + 1)$ and $T_f(r + 1, s + 1)$ are evaluated, they can be stored in the locations previously occupied by $T_m(r + 1, s)$ and $T_f(r + 1, s)$, respectively. In Fig. 8.2 it will be seen that prior to the evaluation of $T_m(r + 1, s + 1)$, the first r elements of each array and TF and TM (corresponding to T_f and T_m) hold temperature for time $s + 1$; the remaining temperatures $r + 1, r + 2, \ldots, ml$ $(ml = m + 1)$ refer to the previous time position s. Next the $(r + 1)$th element of TM is overwritten by the solid temperature at time $s + 1$; similarly, the $(r + 1)$th element of TF subsequently holds the gas temperature for time $s + 1$. The final state of the vectors of temperature TF and TM are then ready for the evaluation of the gas and solid temperatures at $r + 2, s + 1$.

By denoting as the number of steps in time required for the integration by NP, this process is readily programed in FORTRAN:

```
      (Hot period)
      DO 175 JS = 1, NP
      TM(1) = 2.0*B2 + B1*TM(1)
      DO 175 JR = 1, M
      TM(JR + 1) = K1*TM(JR + 1) + K2*TF(JR + 1)
                 + K3*TM(JP) + K4*TF(JR)
      TF(JR + 1) = A1*TF(JR) + A2*(TM(JR + 1) + TM(JR))
  175 CONTINUE
```

(Note that it is necessary to specify K1, K2, K3, and K4 to be real.)

Such a computer program may be required to record the time variation of the gas and perhaps solid temperatures at certain positions in the regenerator, usually at the exit and entrance, and possibly in the middle of the checkerwork. This information can be held in separate arrays within the computer program. Similarly, it is often required to display the spatial distribution of gas and solid temperature at the start of each period of the cycle of operation, and this can be achieved in the same way.

This method of regenerator simulation is so simple and so economical that it can be implemented even for a programmable hand calculator with modest storage facilities. It is more usual to program the method in a high-level language such as FORTRAN; the calculations can then be performed on most computers.

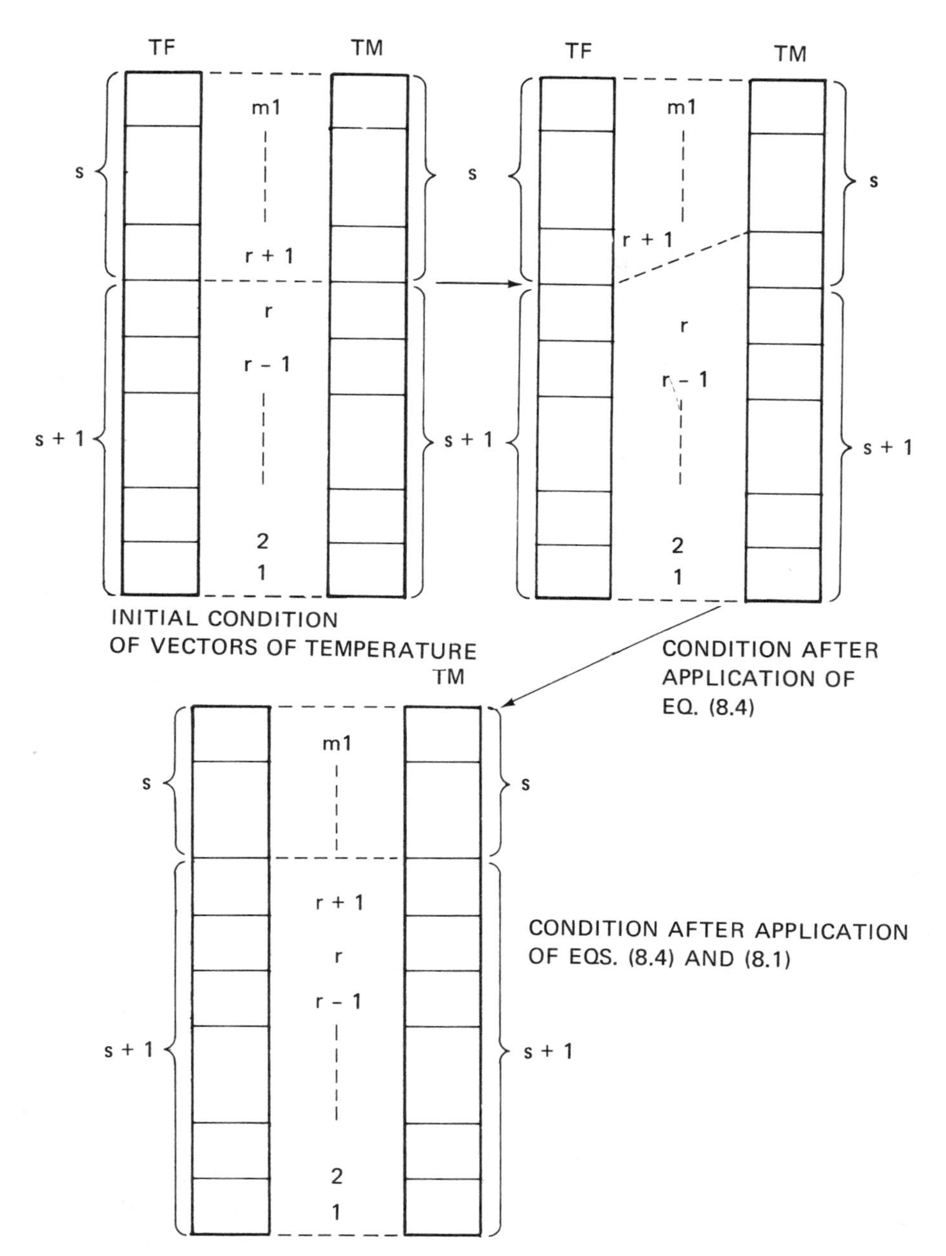

Figure 8.2 Temperature data structures for computer program implementing the Willmott method.

Other methods of regenerator calculation have been proposed, in particular by Hausen [3] and Allen [4], working independently. They suggested a scheme of calculation whereby the gas and solid temperatures could be evaluated independently. The basis of their scheme is Eq. (8.14):

$$T_f(r+1, s+1) = A_1^* T_f(r, s+1) + A_2^* T_f(r+1, s) + A_3^* T_f(r, s) \tag{8.14}$$

The constants A_1^*, A_2^*, and A_3^* are defined in the following way:

$$A_1^* = \frac{2 + \Delta\eta - \Delta\xi}{2 + \Delta\eta + \Delta\xi} \tag{8.15}$$

$$A_2^* = \frac{2 - \Delta\eta + \Delta\xi}{2 + \Delta\eta + \Delta\xi} \tag{8.16}$$

$$A_3^* = \frac{2 - \Delta\eta - \Delta\xi}{2 + \Delta\eta + \Delta\xi} \tag{8.17}$$

In this method, it is no longer possible to overwrite the contents of the location (in a computer program) holding initially $T_f(r + 1, s)$, by the value of $T_f(r + 1, s + 1)$ calculated using Eq. (8.14). The value of $T_f(r + 1, s)$ will be required, when multiplied by A_3^*, in the calculation of $T_f(r + 2, s + 1)$ at the next stage of the calculation. It is possible to discard the value of $T_f(r + 1, s)$ by the time $T_f(r + 3, s + 1)$ is required to be calculated, and because this is so, an algorithm can be devised using only one spatial array of gas temperatures. Such an algorithm is not as compact as that for the Willmott scheme [2], and requires a separate similar scheme to (8.14) whereby the solid temperatures may be computed.

8.3 CONTROL OF TRUNCATION ERROR

The truncation error associated with the Hausen-Allen scheme is of order $(\Delta\xi)^2\Delta\eta + \Delta\xi(\Delta\eta)^2$. This means that if the step length in distance is adjusted alone, the scheme is of first-order accuracy; that is, if we halve the distance step length, the truncation error is only halved, and similarly for adjustments in time step length alone. The method is rendered to a third-order accuracy only if $\Delta\xi$ and $\Delta\eta$ are changed together, that is, if $\Delta\xi$ is set equal to $k\,\Delta\eta$, in which case the truncation error becomes of order $(k^2 + k)(\Delta\eta)^3$. Halving the step length, in both distance and time, now results in the truncation error being reduced to an eighth of its previous value.

The truncation error of the Willmott scheme is of order $(\Delta\xi)^3$ for Eq. (8.1) but only of order $\Delta\eta(\Delta\xi)^3 + (\Delta\eta)^3$ for Eq. (8.4). This method is rendered to third-order accuracy overall when $\Delta\xi$ and $\Delta\eta$ are changed together, so that the truncation error is of order $(\Delta\eta)^3(1 + k^3\,\Delta\eta)$ for Eq. (8.4). In this respect the Willmott and Hausen-Allen schemes have no advantage over each other.

Any computer program that implements either scheme and that seeks to control the truncation error must adjust both $\Delta\xi$ and $\Delta\eta$ simultaneously. A flow diagram illustrating such a process is given in Fig. 8.3. A possible scheme is

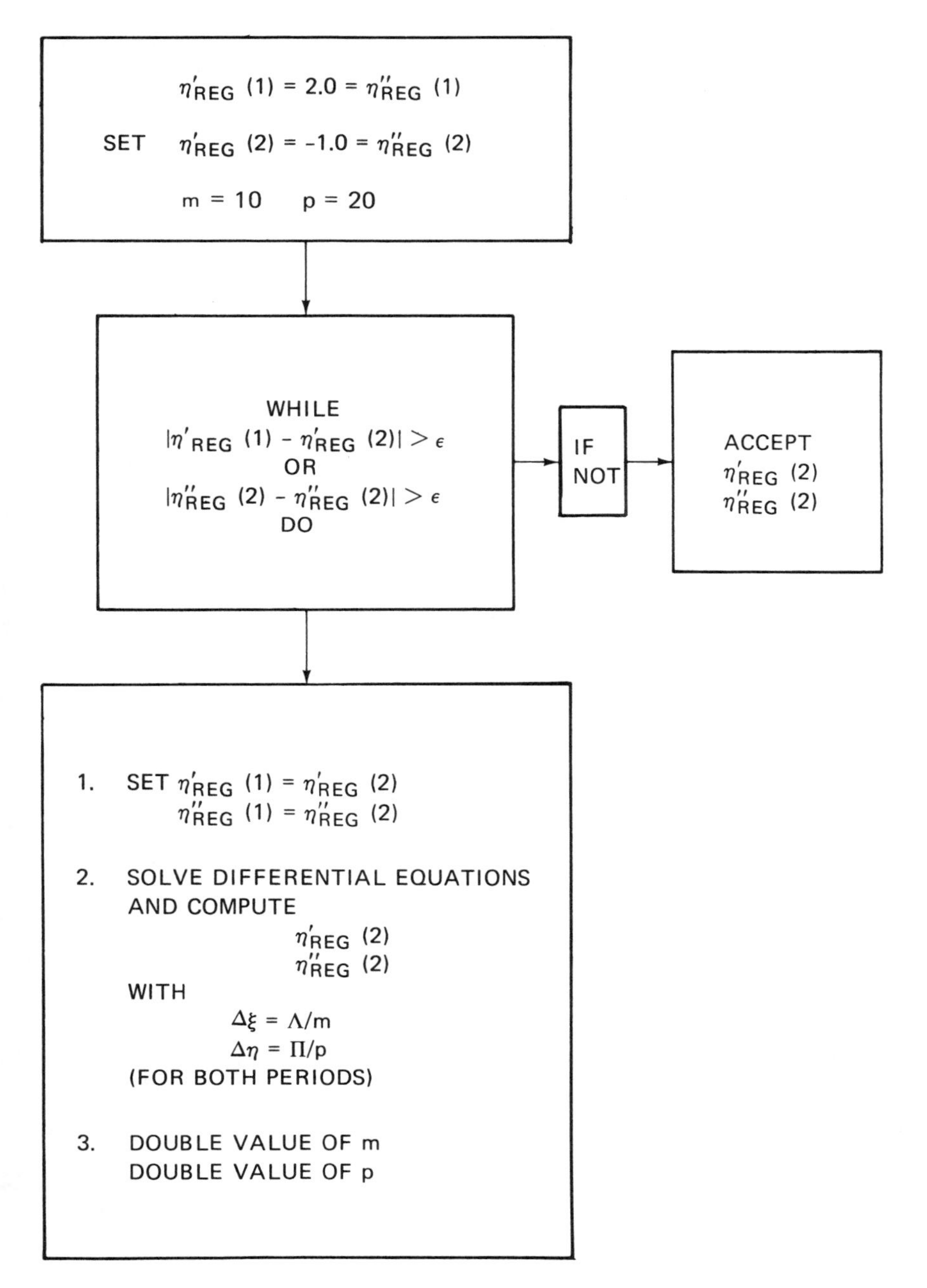

Figure 8.3 Regenerator calculations with truncation error control.

to specify in advance how accurate, for example, a thermal ratio is required of a regenerator calculation. The differential equations are first solved with chosen step lengths, $\Delta\xi$ and $\Delta\eta$ ($\Delta\eta$ can be different for hot and cold periods); the calculation is repeated with $\frac{1}{2}\,\Delta\xi$ and $\frac{1}{2}\,\Delta\eta$ as step lengths. The thermal ratios computed by the two calculations are compared. If we are willing to accept a thermal ratio accurate to $\pm\epsilon = \pm 0.0001$ (for example), then, if the two hot thermal ratios and the two cold thermal ratios are within ± 0.0001 of one another, the most accurate thermal ratio computed with $\frac{1}{2}\,\Delta\xi$ and $\frac{1}{2}\,\Delta\eta$ is accepted. Otherwise, the step lengths are halved yet again and the process shown in Fig. 8.3 continues until a sufficiently accurate solution is obtained.

In Chap. 10 a discussion of how any regenerator simulation can be deemed to have reached cyclic equilibrium is presented. It is enough to say here that the exit temperatures of the gas at the end of the hot periods (or cold periods) in successive cycles must be at least as close to one another as $\pm\epsilon$ for cyclic equilibrium to be judged to have been attained. In this way the accuracy requirements met by the control of truncation error are also satisfied by the cyclic equilibrium conditions.

8.4 OPEN METHODS: INTEGRAL METHODS FOR SIMULATION OF THE REGENERATOR

Nusselt [1] and subsequently Nahavandi and Weinstein [5] represented the solution of Eqs. (5.13) and (5.14) in integral form:

$$T''_m(\xi'', \eta'') = e^{-\eta''} F''(\xi'') - \int_0^{\xi''} \frac{iJ_1 2i\sqrt{(\xi'' - \epsilon)\eta''}}{\sqrt{(\xi'' - \epsilon)\eta''}}\, \eta'' e^{-\eta'' + \xi'' - \epsilon} F''(\epsilon)\, d\epsilon \tag{8.18}$$

for the cold period and

$$T'_m(\xi', \eta') = 1 - e^{-\eta'}(1 - F'(\xi')) + \int_0^{\xi'} \frac{iJ_1 2i\sqrt{(\xi' - \epsilon)\eta'}}{\sqrt{(\xi' - \epsilon)\eta'}} \times \eta' e^{-\eta' + \xi' - \epsilon}(1 - F'(\epsilon))\, d\epsilon \tag{8.19}$$

for the hot period. We present the equations in this order because (8.18) is the slightly simpler in form. Equations (8.18) and (8.19) embody a simplifying notation: it is specified that the solid matrix temperature distribution at the start of the hot period is $F'(\xi')$ and at the start of the cold period is $F''(\xi'')$. Thus

$$F'(\xi') = T'_m(\xi', 0) \tag{8.20}$$

$$F''(\xi'') = T''_m(\xi'', 0) \tag{8.21}$$

$T_m(\xi, \eta)$ is the solid temperature at position ξ down the regenerator and at time η. The primes and double primes in Eqs. (8.18), (8.19), (8.20), and (8.21) distinguish between the hot and cold period, respectively.

The solid temperature distribution at the end of the cold period at time $\eta'' = \Pi''$

is

$$T''_m(\xi'', \Pi'') = e^{-\Pi''} F''(\xi'') + \int_0^{\xi''} K''(\xi'' - \epsilon) F''(\epsilon)\, d\epsilon \tag{8.22}$$

Similarly, the solid temperature at the finish of the hot period at $\eta' = \Pi'$ is

$$T'_m(\xi', \Pi') = 1 - e^{-\Pi'}(1 - F'(\xi)) + \int_0^{\xi'} K'(\xi' - \epsilon)(1 - F'(\epsilon))\, d\epsilon \tag{8.23}$$

where a further simplifying notation has been introduced through the identity (8.24):

$$K(\xi - \epsilon) = \frac{-iJ_1 2i\sqrt{(\xi - \epsilon)\Pi}}{\sqrt{(\xi - \epsilon)\Pi}} e^{-(\Pi + \xi - \epsilon)} \tag{8.24}$$

Equation (8.22) represents an integration of the differential equations through the cold period, and Eq. (8.23) through the hot period. Provided that adequate note is taken of the reversal conditions, the alternate application of (8.22) and (8.23) represents an open method of simulating a thermal regenerator.

The reversal condition incorporates the ideas that distance is always measured from the gas entrance in both periods, even though the gases flow through the regenerator in opposite directions in counterflow mode, and that the temperature at any position in the heat storing matrix at the end of one period is equal to that at the same position at the commencement of the next period. The same position measured to be distance ξ' from the gas entrance in the hot period is measured to be ξ'' from the gas entrance in the cold period. In counterflow mode of regenerator operation, ξ' is related to ξ'' by Eq. (8.25):

$$\frac{\xi'}{\Lambda'} = \frac{1 - \xi''}{\Lambda''} \tag{8.25}$$

The reversal boundary conditions take the form

$$T'_m(\xi', \Pi') = T''_m\left(\Lambda''\left\{\frac{1 - \xi'}{\Lambda'}\right\}, 0\right) \tag{8.26}$$

$$T''_m(\xi'', \Pi'') = T'_m\left(\Lambda'\left\{\frac{1 - \xi''}{\Lambda''}\right\}, 0\right) \tag{8.27}$$

so that the initial solid temperature distributions $F''(\xi'')$ and $F'(\xi')$ in Eqs. (8.22) and (8.23) can be found using Eqs. (8.28) and (8.29):

$$F''\left(\Lambda''\left\{1 - \frac{\xi'}{\Lambda'}\right\}\right) = T'_m(\xi', \Pi') \tag{8.28}$$

and

$$F'\left(\Lambda'\left\{1 - \frac{\xi''}{\Lambda''}\right\}\right) = T''_m(\xi'', \Pi'') \tag{8.29}$$

In order to integrate Eqs. (8.22) and (8.23), it is suggested that the Iliffe [6] scheme be used, although it was devised originally for a closed method. By the application of Iliffe's scheme to Eqs. (8.22) and (8.23), given the functions of $F'(\xi')$ and $F''(\xi'')$, the form of $T'_m(\xi', \Pi')$ and $T''_m(\xi'', \Pi'')$ can be found.

It is convenient to define F'_k and F''_k to be $F'(k\ \Delta\xi')$ and $F''(k\ \Delta\xi'')$, respectively; these are the temperatures of the solid at the start of the hot and cold periods at the entrance and exit to the regenerator and at intermediate levels spaced $\Delta\xi'$ or $\Delta\xi''$ apart. Figure 8.1 illustrates this arrangement. In this definition of F'_k and F''_k, $k = 0$ denotes the gas entrance, and $k = N$ (where $N\ \Delta\xi' = \Lambda'$ and $N\ \Delta\xi'' = \Lambda''$) denotes the gas exit.

At the regenerator entrance, $\xi' = 0$ or $\xi'' = 0$ and

$$T''_m(0, \Pi'') = e^{-\Pi''} F''_o \tag{8.30}$$

and

$$T'_m(0, \Pi') = 1 - e^{-\Pi'}(1 - F'_o) \tag{8.31}$$

Elsewhere, at a position $\xi' = k\ \Delta\xi'$ or $\xi'' = k\ \Delta\xi''$, the integral equations take the form

$$T''_m(k\ \Delta\xi'', \Pi'') = e^{-\Pi''} F''_k + \int_0^{k\ \Delta\xi''} K''(k\ \Delta\xi'' - \epsilon) F''(\epsilon)\, d\epsilon \tag{8.32}$$

$$T'_m(k\ \Delta\xi', \Pi') = e^{-\Pi'}(1 - F'_k) + \int_0^{k\ \Delta\xi'} K'(k\ \Delta\xi' - \epsilon)(1 - F'(\epsilon))\, d\epsilon \tag{8.33}$$

Iliffe introduced the further simplifying notation:

$$K(i\ \Delta\xi - j\ \Delta\xi) = K_{i-j}$$

In the case where k is even, Iliffe proposed to approximate the integrals by Simpson's rule, in which case Eq. (8.32) becomes

$$\begin{aligned} T''_m(k\ \Delta\xi'', \Pi'') = e^{-\Pi''} F''_k + \frac{\Delta\xi''}{3} (K''_k F''_o + 4K''_{k-1} F''_1 \\ + 2K''_{k-2} F''_2 + \cdots + 4K''_1 F''_{k-1} + K''_o F''_k) \end{aligned} \tag{8.34}$$

In the case where $k = 3$, the "three-eighths rule" is applied to yield

$$T''_m(3\ \Delta\xi'', \Pi'') = e^{-\Pi''} F''_3 + \frac{3\ \Delta\xi''}{8} (K''_3 F''_o + 3K''_2 F''_1 + 3K''_1 F''_2 + K''_o F''_3) \tag{8.35}$$

In other cases where k is odd, the integral is split up:

$$\begin{aligned} \int_0^{k\ \Delta\xi''} F''(\epsilon) K''(k\ \Delta\xi'' - \epsilon)\, d\epsilon = \int_0^{3\ \Delta\xi''} F''(\epsilon) K''(k\ \Delta\xi'' - \epsilon)\, d\epsilon \\ + \int_{3\ \Delta\xi''}^{k\ \Delta\xi''} F''(\epsilon) K''(k\ \Delta\xi'' - \epsilon)\, d\epsilon \end{aligned} \tag{8.36}$$

The first integral in Eq. (8.36) represents the three-eighths rule, whereas Simpson's rule is applied to the second integral. Iliffe described a special procedure to deal with the case where $k = 1$.

Exactly analogous replacements can be made to the hot period equation (8.23). In this way, by integrating numerically Eqs. (8.23) for the hot period and then (8.22) for the cold period, successive cycles of regenerator performance can be simulated. The method is more complicated than Willmott's method described in Sec. 8.3; it is, however, much faster because in order to evaluate $T'_m(k\ \Delta\xi', \Pi')$ or $T''_m(k\ \Delta\xi'', \Pi'')$ for $k = 0, 1, 2, \ldots, N$ ($N\ \Delta\xi = \Lambda$ for hot and cold periods), it is not necessary to evaluate the gas and solid temperatures at intermediate positions in time. The values of the temperatures at the end of a period are calculated directly from the temperatures at the beginning of the period under consideration, and these are given by the reversal conditions (8.28) and (8.29).

It is also easier to be able to control truncation error, since this is only a function of $(\Delta\xi)$. A method analogous to that used for truncation error control described in Sec. 8.2.1 can be used here.

When the overall performance is required, including the detailed space and time variations of gas and solid temperature, a possible approach is to cycle the regenerator model to equilibrium using the integral equations discussed here; when equilibrium is reached, it is then possible to switch to the Willmott scheme (Sec. 8.3.1) to compute all the temperature values required.

8.4.1 Evaluation of the *K* Function

Both integrals (8.22) and (8.23) require a function $K(x)$ to be evaluated, where

$$K(x) = e^{-\Pi - x}\sqrt{\frac{\Pi}{x}}\, iJ_1(2i\sqrt{\Pi x}) \tag{8.37}$$

Iliffe [6] suggests that the first-order Bessel function J_1 be evaluated using a series expansion. Alternatively, it is possible to express the Bessel function with complex argument as an integral:

$$iJ_1(2i\Pi x) = \frac{1}{\pi}\int_0^{\pi} e^{-2x\Pi\cos t}\cos t\, dt \tag{8.38}$$

and
$$K(x) = \frac{-1}{\pi}\int_0^{\pi}\sqrt{\frac{\Pi}{x}}\, e^{-2\sqrt{\Pi x}\cos t - x - \Pi}\cos t\, dt \tag{8.39}$$

It can be demonstrated that

$$\lim_{x \to 0} K(x) = \Pi e^{-\Pi} \tag{8.40}$$

It is convenient to evaluate the integral (8.39) numerically, and we suggest using the 10-point Gaussian quadrature method. An analysis of the function $K(x)$ reveals that for $0 \leqslant x \leqslant 0.0001$, the limiting value (8.40) can be assumed with a relative error of less than 0.02% for $0.25 \leqslant \Pi \leqslant 5$.

8.4.2 Calculation of Regenerator Effectiveness

When the temperature distributions $F'(\xi')$ and $F''(\xi'')$ are alone known for cyclic equilibrium, it is necessary to evaluate the thermal ratio (regenerator effectiveness) using the form shown by Iliffe to be

$$\eta_{\text{REG}} = \frac{\Lambda''}{\Pi''}\left(\frac{1}{\Lambda''}\int_0^{\Lambda''} F''(\xi'')\,d\xi'' - \frac{1}{\Lambda'}\int_0^{\Lambda'} F'(\xi')\,d\xi'\right) \tag{8.41}$$

The integrals can be evaluated numerically. If $N\,\Delta\xi' = \Lambda'$ and $N\,\Delta\xi'' = \Lambda''$, and N is even, then

$$\frac{1}{\Lambda''}\int_0^{\Lambda''} F''(\xi'')\,d\xi'' = \frac{1}{N\,\Delta\xi''}\,\frac{\Delta\xi''}{3}\,(F_0'' + 4F_1'' + 2F_2'' + \cdots + 4F_{N-1}'' + F_N'')$$

$$= \frac{1}{3N}(F_0'' + 4F_1'' + 2F_2'' + \cdots + 4F_{N-1}'' + F_N) \tag{8.42}$$

with the other integral treated in the same way.

8.5 CLOSED METHODS

8.5.1 Further Consideration of Reversal Conditions for the Symmetric Case

The thermal regenerator problem is simplified by consideration of the symmetric case where $\Lambda = \Lambda' = \Lambda''$ and $\Pi = \Pi' = \Pi''$. Here the temperature performance of the solid in the hot period is exactly symmetric to that in the cold period at cyclic equilibrium. In these particular circumstances, the reversal condition can be written

$$T_m'(\xi', 0) = 1 - T_m''(\xi'', 0) \tag{8.43}$$

using the [0, 1] temperature scale. This means that the problem can be reduced to the "single-period" boundary value problem. The reversal condition can then be written

$$T_m'(\xi', 0) + T_m'(\Lambda - \xi', \Pi) = 1 \tag{8.44}$$

8.5.2 Development of the Closed Methods

Upon application of the reversal conditions (8.26) and (8.27), the solid temperature distributions $F'(\xi')$ and $F''(\xi'')$ at the start of the hot period and the start of the cold period, respectively, are related by the integral equations

$$F'\left(\Lambda'\left(1 - \frac{\xi''}{\Lambda''}\right)\right) = e^{-\Pi''}F''(\xi'') + \int_0^{\xi''} K''(\xi'' - \epsilon)F''(\epsilon)\,d\epsilon \tag{8.45}$$

and
$$1-F''\left\{\Lambda''\left(1-\frac{\xi'}{\Lambda'}\right)\right\}=e^{-\Pi'}[1-F'\{\xi'\}]$$
$$+\int_0^{\xi'} K'\{\xi'-\epsilon\}[1-F'\{\epsilon\}]\,d\epsilon \tag{8.46}$$

For the symmetric case, application of the reversal condition (8.44) for the cold period yields the integral equation

$$F''\{\Lambda-\xi\}+e^{-\Pi}F''\{\xi\}+\int_0^{\xi} K\{\xi-\epsilon\}F''(\epsilon)\,d\epsilon=1 \tag{8.47}$$

In this way, the integral equations (8.45) and (8.46) (or 8.47) incorporate the reversal conditions directly; by solving these equations, the cyclic equilibrium performance of the regenerator can be computed immediately.

8.5.3 The Method of Nahavandi and Weinstein: Symmetric Case

In the Nahavandi and Weinstein approach [5], the temperature distribution $F\{\xi\}$ [omitting now the double primes in Eq. (8.47)] is represented by an approximating polynomial

$$F\{\xi\}=\sum_{j=0}^{n} a_j\xi^j$$

Equation (8.47) now takes the form

$$\sum_{j=0}^{n} a_j(\Lambda-\xi)^j+e^{-\Pi}\sum_{j=0}^{n} a_j\xi^j+\sum_{j=0}^{n} a_j\int_0^{\xi}\epsilon^j K\{\xi-\epsilon\}\,d\epsilon=1 \tag{8.48}$$

By applying Eq. (8.48) at $n+1$ distinct values ξ_i $(i=0,1,2,\ldots,n)$ of ξ on the range $0\leqslant\xi\leqslant\Lambda$, a set of $n+1$ simultaneous linear equations is obtained in the $n+1$ unknown coefficients $a_0, a_1, a_2, \ldots, a_n$. These equations are written in the form

$$R\mathbf{a}=\mathbf{1} \tag{8.49}$$

where $\mathbf{a}$ is the column vector $\{a_0, a_1, a_2, \ldots, a_n\}$ and $\mathbf{1}$ is the column vector $\{1, 1, 1, \ldots, 1\}$. R is a $(n+1)\times(n+1)$ matrix, the (i,j)th element of which is

$$(\Lambda-\xi_i)^j+e^{-\Pi}\xi_i^j+\int_0^{\xi_i}\epsilon^j K\{\xi_1-\epsilon\}\,d\epsilon$$

for $0\leqslant i,j\leqslant n$. The evaluation of the K function has been discussed in Sec. 8.4.1.

It will be noted in Chap. 9 that with a very slight modification, a computer program implementing the method of Nahavandi and Weinstein [5] for counterflow regenerators can be adapted to deal with parallel-flow regenerators.

Regenerator effectiveness (thermal ratio) can be readily computed using Eq. (8.41), which for the symmetric regenerator takes the form

$$\eta_{\text{REG}} = \frac{1}{\Pi} \int_0^{\Lambda} (F''(\xi) - F'(\xi))\, d\xi \tag{8.50}$$

Applying the reversal condition (8.47), the expression (8.50) for the thermal ratio takes the form

$$\eta_{\text{REG}} = \frac{1}{\Pi} \int_0^{\Lambda} (F''(\xi) + F''(\Lambda - \xi) - 1)\, d\xi \tag{8.51}$$

Upon substituting the polynomial expansion, the integral

$$\eta_{\text{REG}} = \frac{1}{\Pi} \int_0^{\Lambda} \left\{ \sum_{j=0}^{n} [a_j(\xi^j + (\Lambda - \xi)^j] - 1 \right\} d\xi \tag{8.52}$$

can be evaluated explicitly using

$$\eta_{\text{REG}} = \frac{1}{\Pi} \sum_{j=0}^{n} \left(\frac{a_j \Lambda^{j+1}}{j+1} - \Lambda \right) \tag{8.53}$$

given the values of the coefficients $\{a_0, a_1, a_2, \ldots, a_n\}$ obtained by solving the linear equations (8.49).

8.5.4 The Method of Nahavandi and Weinstein: Unsymmetric Case

For the unsymmetric case it is necessary to develop both the temperature distributions $F'(\xi)$ and $F''(\xi)$ in the form

$$F'(\xi') = \sum_{j=0}^{n} a_j'(\xi')^j \tag{8.54}$$

$$F''(\xi'') = \sum_{j=0}^{n} a_j''(\xi'')^j \tag{8.55}$$

Equation (8.45) now becomes, for the cold period,

$$\sum_{j=0}^{n} a_j' \left[\Lambda' \left(1 - \frac{\xi''}{\Lambda''} \right) \right]^j = e^{-\Pi''} \sum_{j=0}^{n} a_j''(\xi'')^j + \int_0^{\xi''} K(\xi'' - \epsilon) \sum_{j=0}^{n} a_j''(\xi'')^j \, d\epsilon \tag{8.56}$$

and Eq. (8.46) can be expanded in a similar way. Equation (8.56) is applied at $n+1$ distinct positions ξ_i'' $(i = 0, 1, 2, \ldots, n)$ on the range $0 \leqslant \xi'' \leqslant \Lambda''$; the expanded form Eq. (8.46) is applied at another $n+1$ positions ξ_i' $(i = 0, 1, 2, \ldots, n)$ on $0 \leqslant \xi' \leqslant \Lambda'$.

Together these yield a set of $2n + 2$ linear equations which can be solved for the coefficients

$$\{a'_0, a'_1, a'_2, \ldots, a'_n, a''_0, a''_1, a''_2, \ldots, a''_n\}$$

Having computed these coefficients, the thermal ratio can be evaluated using a developed form of Eq. (8.41), employing the same approach whereby Eq. (8.53) is obtained from Eq. (8.51) for the symmetric case.

8.5.5 Choice of Data Points ξ_i

Consideration is restricted to the symmetric case without loss of generality; the methods discussed are equally applicable to unsymmetric cases.

Nahavandi and Weinstein [5] chose to solve the set of simultaneous linear equations

$$R\mathbf{a} = \mathbf{1} \tag{8.49}$$

for values $\xi_i = i\Lambda/n$, for $i = 0, 1, 2, \ldots, n$ that are equally spaced apart.

In the middle of a regenerator, the temperature of the solid varies almost linearily with ξ and η in both periods. Nonlinear temperature behavior is propagated from the entrances to the regenerator by the constant inlet gas temperatures in both periods. This suggests that the data points ξ_i should be more closely clustered in some way around the entrances than in the middle of the regenerator. From a mathematical point of view, it is known that if it is required to interpolate a polynomial $Q(x)$ of degree n by another polynomial $P(x)$ of degree $n - 1$, the maximum value of the error $|Q(x) - P(x)|$ is minimized if the interpolation points are chosen to be the Chebyshev points.

For the solution of Eq. (8.49) it is recommended that such Chebyshev points ξ_i be used where

$$\xi_i = \frac{\Lambda}{2}\left(1 - \cos\frac{i\pi}{n}\right) \qquad i = 0, 1, 2, \ldots$$

These Chebyshev points are clustered around the regenerator entrances in a manner illustrated for $n = 8$ in Fig. 8.4. Note, however, that for $n = 1$ and $n = 2$ the Chebyshev points and the equally spaced points coincide.

The thermal ratio of a regenerator of specified reduced length Λ and reduced period Π is first computed assuming $F''(\xi)$ to be a linear or perhaps a quadratic function of ξ. The computation is repeated using successively higher powers in the polynomial expansion of ξ representing $F''(\xi)$. This iterative process is said to have "converged" when two consecutive evaluations of η_{REG} is less than some prescribed level of ϵ, say, $\epsilon = 10^{-4}$ or 10^{-5}.

It turns out that use of the Chebyshev data points reduces by one, or at most two, the degree of the power expansion $\Sigma\, a_j \xi^j$ required for the convergence of the thermal ratio, compared with the equal spacing of the data points. Although in one sense this is only a marginal gain, nonetheless, the method of evaluation of the

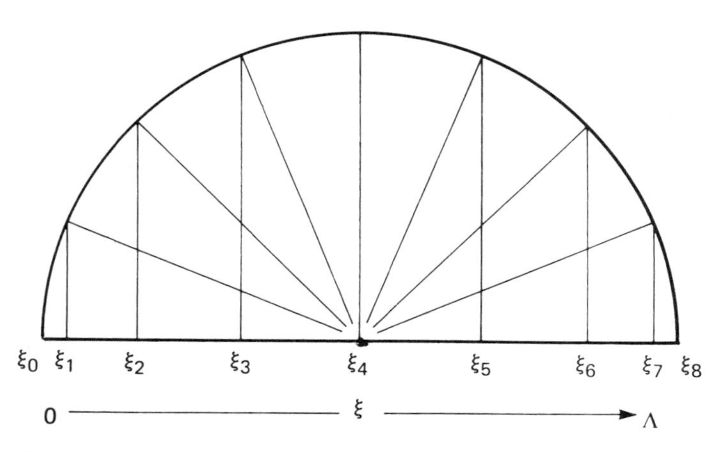

Figure 8.4 Chebyshev distribution of data points.

Chebyshev points is not complicated, and full advantage should be taken of this modification.

In another sense, the gain is considerable. It will be recalled that the (i, j)th element of the matrix R is

$$(\Lambda - \xi_i)^j + e^{-\Pi}\xi_i^j + \int_0^{\xi_i} \epsilon^j K(\xi_i - \epsilon)\, d\epsilon$$

for $0 \leqslant i, j \leqslant n$. Representation of the temperature distribution $F''(\xi)$ by a polynomial of degree 8, for example, in a regenerator of reduced length $\Lambda = 10$, involves elements of the matrix of order at least 10^8 in size. This can cause difficulties, which cannot be discussed here, in solving the linear equation (8.49) as a result of some of the elements of the matrix R being very large and others being very small. By reducing the degree of the polynomial needed, through the use of the Chebyshev data points, the range of the size of the elements of the matrix R can be limited and the effect of such difficulties restricted. Even so, the Nahavandi and Weinstein method, like the closed method of Iliffe [6] to be discussed shortly, should not be employed for large reduced lengths, say, $\Lambda > 30$, the upper limit being determined by the size of the numbers that can be accommodated by the computer used for the calculations.

8.5.6 The Closed Method of Iliffe

Without loss of generality, we restrict our considerations to the symmetric case, and in particular to the integral equation (8.47)

$$F''(\Lambda - \xi) + e^{-\Pi}F''(\xi) + \int_0^{\xi} K(\xi - \epsilon)F''(\epsilon)\, d\epsilon = 1 \tag{8.47}$$

The integral is replaced by the approximations (with the same notation) discussed in

Sec. 8.3. Using Simpson's rule, Eq. (8.47) becomes, for k even,

$$F_{N-k} + e^{-\Pi}F_k + \frac{\Delta\xi}{3}(K_kF_0 + 4K_{k-1}F_1 + 2K_{k-2}F_2 + \cdots + 4K_1F_{k-1} + K_0F_k) = 1 \qquad (8.57)$$

In the case where $k = 3$, the three-eighths rule is applied to yield

$$F_{N-3} + e^{-\Pi}F_3 + \tfrac{3}{8}(K_3F_0 + 3K_2F_1 + 3K_1F_2 + K_0F_3) = 1 \qquad (8.58)$$

Again, where k is odd, the integral is split up:

$$\int_0^{k\,\Delta\xi} F''(\epsilon)K(k\,\Delta\xi - \epsilon)\,d\epsilon = \int_0^{3\,\Delta\xi} F''(\epsilon)K(k\,\Delta\xi - \epsilon)\,d\epsilon + \int_{3\,\Delta\xi}^{k\,\Delta\xi} F''(\epsilon)K(k\,\Delta\xi - \epsilon)\,d\epsilon \qquad (8.59)$$

The first integral is represented using the three-eighths rule, whereas Simpson's rule is applied to the second. The special procedure to deal with the case where $k = 1$, described by Illife [6], is employed.

In this manner, the integral equation (8.47) is replaced by a set of simultaneous linear equations of the form

$$S\mathbf{F} = \mathbf{1} \qquad (8.60)$$

where S is a $(N + 1) \times (N + 1)$ square matrix, $\mathbf{F}$ is the column vector $\{F_0, F_1, F_2, \ldots, F_N\}$ of temperatures and $\mathbf{1}$ is again the vector $\{1, 1, \ldots, 1\}$. This set of simultaneous linear equations is solved for the temperatures $\{F_0, F_1, F_2, \ldots, F_N\}$, $N\,\Delta\xi = \Lambda$.

Application of quadrature methods of the Newton-Cotes type to the integral equation in the Iliffe approach automatically implies that the temperatures $F''(\xi)$ must be computed at positions $\xi = k\,\Delta\xi$, for $k = 0, 1, 2, \ldots, N$ equally spaced apart. In contrast, the form of Eq. (8.48) in Nahavandi and Weinstein's method implies no particular prepositioning of the values of ξ_i $(i = 0, 1, 2, \ldots, n)$ on the range $0 < \xi_i < \Lambda$ except that all $n + 1$ positions should be distinct. Alternative displays of the data points have been discussed in Sec. 8.4.5.

In both the Iliffe and the Nahavandi and Weinstein methods, the integral

$$\int_0^{\xi} F''(\epsilon)K(\xi - \epsilon)\,d\epsilon$$

is replaced by the summation of a quadrature formula. Although methods of quadrature might be applied in the Iliffe method that are alternatives to those set out in the Iliffe paper, such methods would, necessarily, be of the Newton-Cotes type where the data points ξ_i are equally spaced. However, quadrature methods of either the Newton-Cotes or the Gauss type can be adopted when the Nahavandi and Weinstein method is used. In this sense, therefore, there is an extra degree of flexibility associated with the Nahavandi and Weinstein method.

The major disadvantage of the Iliffe method, however, is that the set of equations (8.60) becomes ill-conditioned (the matrix S becomes almost singular) as the ratio of reduced length to reduced period Λ/Π increases, the more so the larger the reduced length. The matrix S is perturbed by the truncation errors associated with numerical quadrature representation, for example, Simpson's rule, of integrals. These perturbations give rise, when Eqs. (8.60) are ill-conditioned, to errors in the computed values of the elements of the vector of temperature $\mathbf{F}$; these can be reduced in the following way. As the number of levels increases, the step length between levels decreases, as does the magnitude of the truncation errors and the perturbations of the matrix S. As the number of levels increases, however, so the number of simultaneous linear equations also increases; in practice, it turns out that it may be necessary to perform computations on a matrix as large as order 40×40 to minimize the effect of this ill-conditioning. These problems are discussed in detail by Willmott and Thomas [7]. In general, therefore, it is suggested that the Nahavandi and Weinstein method be used if a closed method is to be preferred. We discuss here the Iliffe method, not only because of its historical importance in the solution of the regenerator problem, but because the Iliffe scheme in its open form, discussed in Sec. 8.4, provides a rapid means of cycling a regenerator model to equilibrium. Further, the Iliffe open scheme does not suffer from any form of ill-conditioning, as is experienced with the closed form.

8.6 NUMERICAL ACCELERATION OF REGENERATOR SIMULATIONS

Willmott and Kulakowski [8] demonstrated how to reduce the number of cycles simulated when calculating the cyclic equilibrium behavior of a regenerator by an open method. Such an open method can be regarded as an iterative process that converges to the periodic behavior of the regenerator at equilibrium. The convergence is accelerated by interrupting the iterations from time to time and, employing a modified form of the Aitken Δ^2 process, projecting the thermal state of the regenerator toward what it will be at equilibrium without simulating all the intermediate transient cycles. It will be shown that the more severe the inertia of the system, the more effective is the procedure employed to achieve this acceleration.

Although calculations on linear problems are discussed here, Willmott and Kulakowski extended the acceleration methodology to more realistic models, as discussed in Chap. 7, in which nonlinear features are included such as the temperature dependence of specific heat, the effect of radiative heat transfer between solid and hot gas containing carbon dioxide and water vapor, or the chronological variation of gas flow rate. This points to a further advantage of the method: the additional calculation involved by introducing further nonlinear features into the model of a regenerator with large inertia can be undertaken if few iterations are required when the acceleration method is applied. This is especially useful when the mathematical model is incorporated within an optimization scheme. Such calculations require many regenerator simulations from which the "best" design or set of operating conditions can be selected relative to a chosen cost function.

The cycle-by-cycle simulation of regenerator behavior can be considered as an iterative process. The open solution of Eqs. (5.13) and (5.14) can be regarded as an operation G on the packing temperature distribution at the start of the hot period. A simplifying notation is introduced: the initial packing temperature (at the start of the hot period) is denoted by $\mathbf{F}^{(0)} = F^{(0)}(x)$; that after the nth cycle is given by

$$\mathbf{F}^{(n)} = F^{(n)}\{x\} = G\{F^{(n-1)}\{x\}\} \tag{8.61}$$

where $$F^{(n)}\{x\} = T'_n\{x, 0\} = T''_{n-1}\{L - x, P''\} \tag{8.62}$$

At cycle equilibrium, the packing temperature distribution is $F^{(\infty)} = F^{(\infty)}\{x\}$, where

$$F^{(\infty)}\{x\} = G\{F^{(\infty)}\{x\}\} \tag{8.63}$$

Equation (8.61) represents an iterative process generating from $\mathbf{F}^{(0)}$ a sequence of distributions

$$\{\mathbf{F}^{(k)}\} \qquad k = 1, 2, \ldots$$

which converge to a limit

$$\lim_{k\to\infty} \mathbf{F}^{(k)} = \mathbf{F}^{(\infty)} \tag{8.64}$$

Numerical methods exist for extrapolating from a number of terms in such a sequence $\{F^{(k)}\}$ to an estimate of the limit. By halting the iterative process (8.61) from time to time and by using such estimates of the limit as fresh initial conditions to the process, the number of iterations necessary to reach convergence can be significantly reduced. There is, however, the danger that the estimated limit can be grossly wrong and the modified sequence may then diverge instead of converging. Thus, with any such extrapolation, steps must be taken to ensure that the sequence still converges.

Based on the assumption that such a sequence converges exponentially, Aitken [9] developed the classical extrapolation method. This can be applied to the spatial temperature distributions $\mathbf{F}^{(n)}$, $\mathbf{F}^{(n-1)}$, and $\mathbf{F}^{(n-2)}$ from which can be estimated $\mathbf{F}^{(\infty)}$; this estimate is $\mathbf{F}^{(n)*}$ and is calculated by

$$\mathbf{F}^{(n)*} = \mathbf{F}^{(n)} - \frac{(\mathbf{F}^{(n)} - \mathbf{F}^{(n-1)})^2}{\mathbf{F}^{(n)} - 2\mathbf{F}^{(n-1)} + \mathbf{F}^{(n-2)}} \tag{8.65}$$

This distribution $F^{(n)*}$, which is hopefully closer to limit $F^{(\infty)}$, can then be taken as a fresh initial condition in the solution of the descriptive differential equations (5.13) and (5.14). The algorithm is Steffensen's development [10] of the Aitken acceleration procedure.

In the numerical model, the packing temperature distribution is a set of discrete values

$$F^{(n)} = \{f_i^{(n)}\} \qquad i = 0, 1, 2, \ldots, N$$

corresponding to the temperatures at the entrance and exit to the regenerator and at $N-1$ equally spaced intermediate positions along the length of the regenerator, on the finite difference mesh. See Fig. 8.1. The Aitken formula (8.65) is written in more detailed form

$$f_i^{(n)*} = f_i^{(n)} - \frac{(f_i^{(n)} - f_i^{(n-1)})^2}{f_i^{(n)} - 2f_i^{(n-1)} + f_i^{(n-2)}} \qquad i = 0, 1, 2, \ldots, N \tag{8.66}$$

It is conditional that an iterative process of the form

$$\mathbf{F}^{(n)}\{x\} = G\{\mathbf{F}^{(n-1)}\{x\}\} \tag{8.61}$$

will converge only if the absolute value of the first derivative of G is less than one. Since G is not an explicit function but represents implicitly the solution of the differential equations (5.13) and (5.14) over one complete cycle, local values $K_i^{(n)}$ of this first derivative of G can only be estimated:

$$K_i^{(n)} = \frac{f_i^{(n)} - f_i^{(n-1)}}{f_i^{(n-1)} - f_i^{(n-2)}} \qquad i = 0, 1, 2, \ldots, N \tag{8.67}$$

Should $|K_i^{(n)}| \geqslant 1$, the danger exists that the iterative process may become divergent.

The detailed form (8.66) of the Aitken formula can be specified in terms of $K_i^{(n)}$, namely,

$$f_i^{(n)*} = f_i^{(n)} + \frac{K_i^{(n)}}{1 - K_i^{(n)}} (f_i^{(n)} - f_i^{(n-1)}) \qquad i = 0, 1, 2, \ldots, N \tag{8.68}$$

Because the physical regenerator system will always move toward cyclic equilibrium, the convergence of the simulative iteration (8.61) is assured. Inspection of Eq. (8.68), however, indicates that values of $K_i^{(n)}$ close to one may generate a badly deformed temperature distribution $\{f_i^{(n)*}\}$, $i = 0, 1, 2, \ldots, N$, and possibly cause divergence of the accelerated iterative process.

The estimated limit $\mathbf{F}^{(n)*}$ computed using Eq. (8.65) can be related to the equilibrium distribution $\mathbf{F}^{(\infty)}$ by

$$\mathbf{F}^{(n)*} = \mathbf{F}^{(\infty)} + \mathbf{E}^{(n)} \tag{8.69}$$

where $\mathbf{E}^{(n)}$ is a sequence of numbers $\{e_i^{(n)}\}$, $i = 0, 1, 2, \ldots, N$, representing the errors in the estimation $\mathbf{F}^{(n)*}$ of the limiting vector $\mathbf{F}^{(\infty)}$. These values $e_i^{(n)}$ are different as $K_i^{(n)}$ varies for different positions in the regenerator, specified by the subscript i. Further, the values of $e_i^{(n)}$ represent in part the deformations to the "natural" packing temperature distribution induced by the acceleration process. The deformations will be large if $K_i^{(n)}$ is close to one. On the other hand, the natural convergence of the iteration (8.61) will eliminate the effect of these deformations after a number of unaccelerated cycles.

The essence of the Willmott-Kulakowski method is, first, to restrain any unacceptable deformations in the temperature profiles induced by the acceleration algorithm (8.68) by regarding $\mathbf{K}^{(n)}$ as a heuristic parameter and employing a more tightly controlled value for estimating the limit $\mathbf{F}^{(\infty)}$. Any element of the vector

$\mathbf{K}^{(n)}$ that indicates divergence is rejected, safeguarding the range 0 to +1 with an additional guard ϵ: Thus we keep only those elements $K_i^{(n)}$ for which

$$0 \leqslant K_i^{(n)} \leqslant 1 - \epsilon \tag{8.70}$$

Willmott and Kulakowski recommended that a value of $\epsilon = 0.02$ be used. The average of the acceptable elements of $K^{(n)}$ is taken; denoting this average value by $\bar{K}^{(n)}$, Eq. (8.68) is modified so that it becomes

$$f_i^{(n)*} = f_i^{(n)} + \frac{\bar{K}^{(n)}}{1 - \bar{K}^{(n)}} (f_i^{(n)} - f_i^{(n-1)}) \tag{8.71}$$

If none of the elements of $K^{(n)}$ lies in the range 0 to $1 - \epsilon$, no acceleration is applied and the normal iterative formula (8.61) is used.

Second, even employing a tightly controlled parameter $\bar{K}^{(n)}$, it is necessary to allow a "free running" of the simulation in order to smooth out the deforming effects of the acceleration. If the acceleration process is applied immediately after the next three cycles $n + 1$, $n + 2$, and $n + 3$ following an acceleration, and so using possibly unacceptably disturbed temperature distributions $\mathbf{F}^{(n+1)}$, $\mathbf{F}^{(n+2)}$, $\mathbf{F}^{(n+3)}$ as a basis for a further acceleration, the new estimated limit $\mathbf{F}^{(n+3)*}$ may be grossly deformed. Indeed direct application of the Aitken method every three cycles will cause the simulation process to become divergent.

The Steffensen algorithm is therefore further modified; following each acceleration, the simulation is run for six cycles before another acceleration is applied. Just prior to a possible acceleration, it is ascertained if cyclic equilibrium has been achieved, in which case the simulation process is terminated. The criteria by which a simulation is judged to have reached equilibrium are discussed in Chap. 10.

The temperature profiles $\mathbf{F}^{(n)}$ can be evaluated using either the Willmott [2] or the Iliffe [6] open schemes.

Inspection of Figs. 8.5 and 8.6 clearly indicates the effectiveness of the acceleration procedure. For a reduced period $\Pi = 2.5$, the relation between the number of cycles to equilibrium and reduced length Λ of the regenerator is shown graphically in Fig. 8.5. Note the N_{CO} is the number of cycles to equilibrium without acceleration, N_{CA} is the number when the acceleration algorithm has been applied.

Two measures of the effectiveness of the acceleration method are the ratios $N_{\mathrm{CA}}/N_{\mathrm{CO}}$ and $\tau_{\mathrm{A}}/\tau_{\mathrm{O}}$, where τ_{A} and τ_{O} are the computer times necessary for simulation to equilibrium with and without acceleration, respectively. The ratio $\tau_{\mathrm{A}}/\tau_{\mathrm{O}}$ is useful because it takes into account the computer time necessary for the intervention of the acceleration process into the regenerator simulation. Both these ratios are displayed graphically in Fig. 8.6 as a function of N_{CO}, the number of cycles to equilibrium without acceleration.

The general conclusion must be that the greater the reduced length, or, equally, the greater the thermal inertia, the greater the benefit of the acceleration. When $\Lambda/\Pi = 10$, the computer time simulation can be reduced by between 30% and 70% by employing the method of acceleration described here. Experience in using this acceleration procedure suggests that its basic advantages are its simplicity and reliability over a large range of model parameter values. On the other hand, this algorithm is heuristic and, as such, may not be regarded as an optional one. One must keep in mind, however, that any optimal (i.e., minimizing the number of cycles to equilibrium) algorithm of acceleration may involve a great deal of computation outside the regenerator simulation, thereby reducing the effectiveness of the method.

The significance of this acceleration method lies in its potential application to nonlinear models, such as those mentioned in Chap. 7. Computer simulations embodying realistic nonlinear features can be attempted with the knowledge that a means exists to cut down the calculation time, which in some instances may be a deterrent to

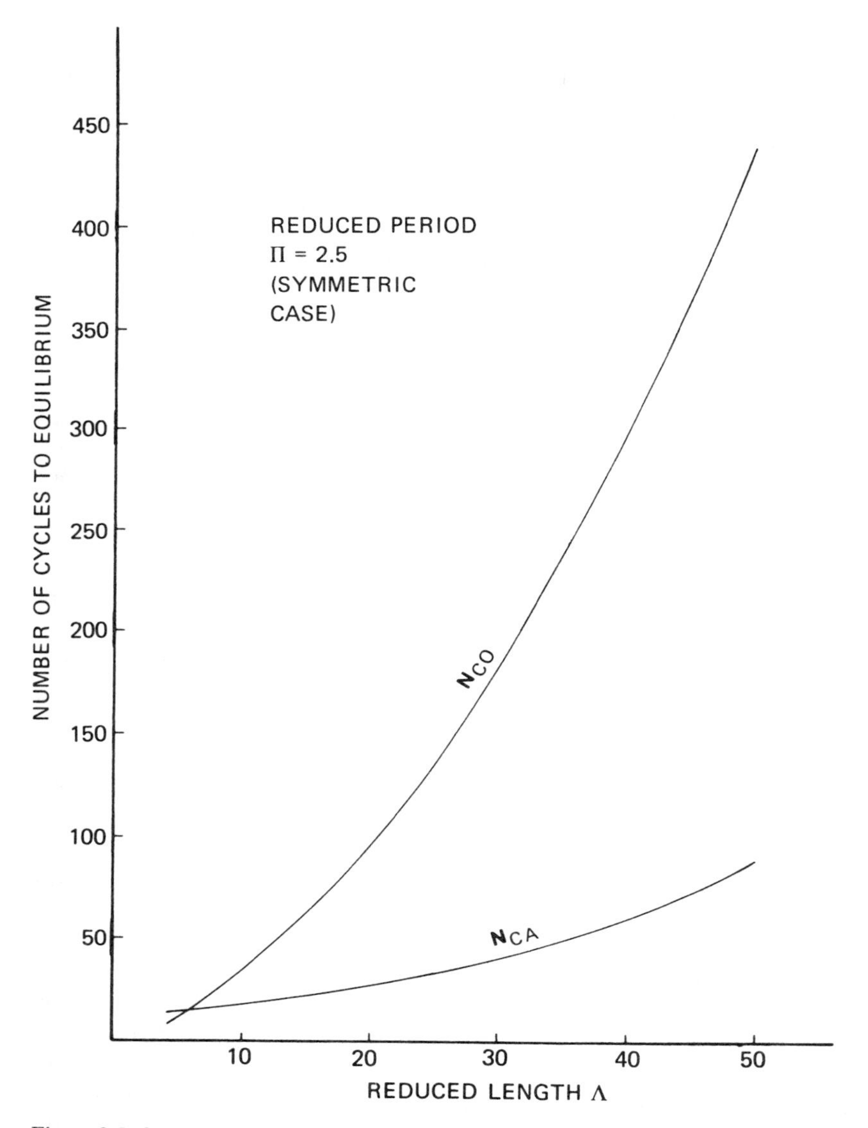

Figure 8.5 Comparison between accelerated and unaccelerated models of a regenerator.

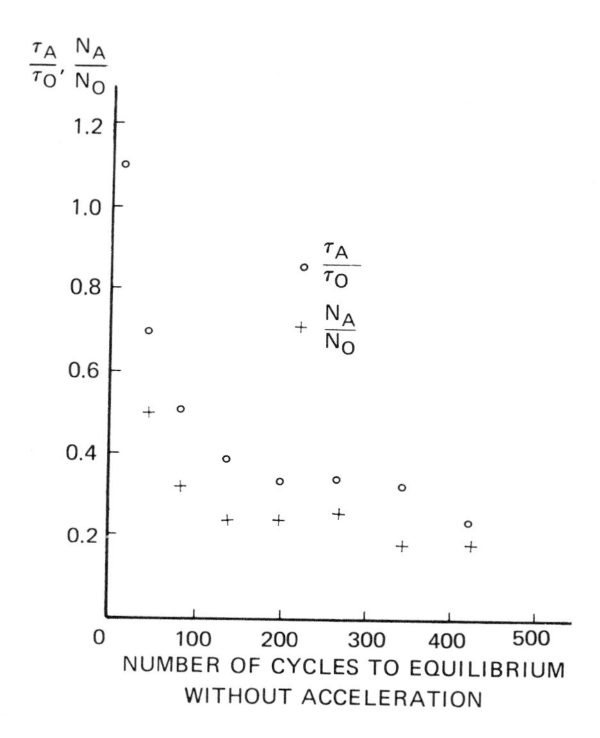

Figure 8.6 Effect of acceleration for linear model.

further research or to the solution of practical industrial problems. The method is particularly useful for the calculation of "long" regenerators, where $\Lambda > 10$ and $\Lambda/\Pi > 3$.

REFERENCES

1. W. Nusselt, "Die Theorie des Winderhitzers," *Z. Ver. Deutsch. Ing.*, vol. 71, 1927, pp. 85–91.
2. A. J. Willmott, "Digital Computer Simulation of a Thermal Regenerator," *Int. J. Heat Mass Transfer*, vol. 7, 1964, pp. 1291–1303.
3. H. Hausen, *Wärmeübertragung im Gegenstrom, Gleichstrom und Kreuzstrom*, Springer-Verlag, Berlin, 1950.
4. D. N. de G. Allen, "The Calculation of the Efficiency of Heat Exchangers," *Quart. J. Mech. Appl. Math.*, vol. 4, 1952, pp. 455–461.
5. A. N. Nahavandi and A. S. Weinstein, "A Solution to the Periodic-Flow Regenerative Heat Exchanger Problem," *Appl. Sci. Res.*, vol. A10, 1961, pp. 335–348.
6. C. E. Iliffe, 1948, "Thermal Analysis of Counterflow Regenerative Heat Exchanger," *J. Inst. Mech. Eng.*, vol. 159, 1948, pp. 363–372 (war emergency issue no. 44).
7. A. J. Willmott and R. J. Thomas, "Analysis of the Long Contra-Flow Regenerative Heat Exchanger." *J. Inst. Math. Appl.*, vol. 14, 1974, pp. 267–280.
8. A. J. Willmott and B. Kulakowski, "Numerical Acceleration of Thermal Regenerator Simulations," *Int. J. Numerical Methods Eng.*, vol. 11, 1977, pp. 533–551.
9. A. C. Aitken, "On Bernoulli's Numerical Solution of Algebraic Equations," *Proc. Roy. Soc. Edinburgh*, vol. 46, 1926, pp. 289–305.
10. J. F. Steffensen, "Remarks on Iteration," *Scand. Akt. Tidskr.*, vol. 16, 1933, pp. 64–72.

CHAPTER
NINE
PARALLEL-FLOW REGENERATORS

9.1 INTRODUCTION

The prediction of the transient response of storage units that are initially at a uniform temperature and are then suddenly exposed to a step change in inlet fluid temperature (single-blow operating mode) was presented in Chaps. 2 and 3. The use of these solutions to obtain the transient response of storage units operating under timewise variations in inlet fluid temperature and/or mass flow rate by using superposition was discussed in Chap. 4. The mathematical treatment of a storage unit subject to periodic, square wave, inlet fluid temperature (see Fig. 9.1) can be considered to be an extension of the modeling techniques described in Chaps. 2 and 3. Alternatively, the unit can be regarded as a regenerator in which the hot and cold fluids pass through the channels of the checkerwork in the same direction. It is thus possible to extend the methodology discussed in Chaps. 5, 6, 7, and 8 for counterflow regenerators to such parallel-flow regenerators, as they are called.

Although it may appear far more convenient to operate a regenerator in parallel-flow mode, it turns out that the counterflow regenerator is far more thermally effective than its parallel-flow counterpart. This can be concluded from the simple observation that at the end of the hot period of regenerator operation, the packing temperature steadily decreases as one moves down the length of the regenerator from the hot fluid entrance to the exit. In the subsequent cold period, if the regenerator runs in counterflow mode, the fluid encounters a continuously increasing temperature in the regenerator packing, and in this way the fluid is steadily raised in temperature. In the cold period of parallel-flow mode operation, however, the entrance gas is exposed initially to relatively high temperatures in the solid but then proceeds to meet decreasing solid temperatures. It is thus possible, especially in the early part of the cold period of parallel-flow operation, for the cold gas to pick up heat

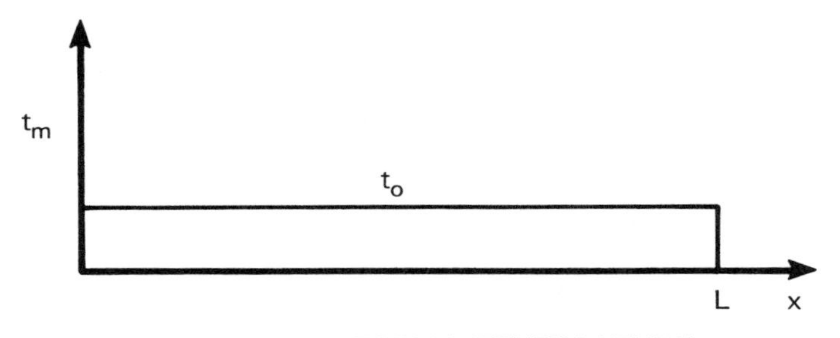

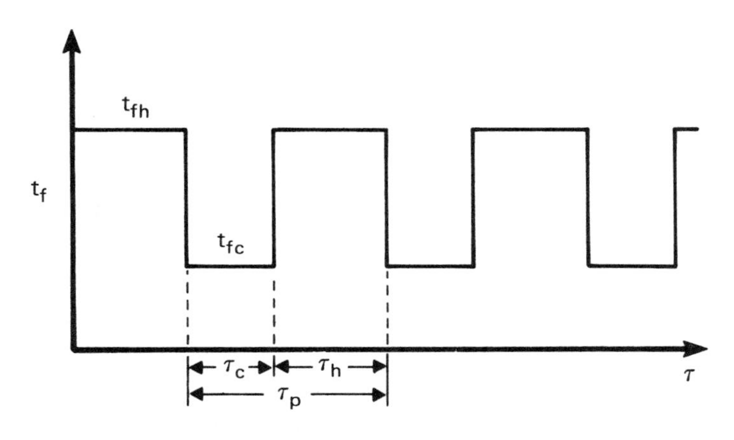

Figure 9.1 Fluid inlet temperature and initial material temperature for periodic operation.

from the solid in the region near the regenerator entrance, then to exchange no heat with the solid in approximately the middle of the regenerator, and finally to lose heat to the colder solid near the gas exit. In this way, the gas acts to transport heat from the hot to the cold end of the regenerator packing. This process is almost equivalent to the regenerator packing temporally having a high longitudinal thermal conductivity. Such a process will continue until, at every position in the regenerator, the solid is hotter than the gas passing through the channels. An analogous process may take place in the hot period of the regenerator operation.

Nevertheless it is sometimes impossible, or inconvenient, to run a regenerator in other than the parallel-flow mode. In some industrial applications, the condition of the fluid entering the storage unit has timewise variations that are repeated in a regular fashion. The fluid entering the unit need not involve only a single stream. As an example, during a complete cycle the fluid entering the heat storage unit could be composed of a hot stream, which supplies heat, followed by a cold stream of a different fluid, which removes heat from the storage unit. The initial performance of these units is transient in nature and depends to a great extent on the initial temperature distribution in the storage material, as well as the timewise-varying inlet fluid conditions. Under continuous operation, the transient component of the unit's response diminishes and the unit's performance becomes periodic in a fashion dependent only on the timewise-varying fluid inlet conditions. Cyclic equilibrium is said to have been reached when the heat stored by the warmer fluid as it passes through the unit is equal to the heat removed from the storage material by the colder fluid as it moves through the unit.

Rotary regenerators can also be operated in parallel-flow mode, but this is unusual. The idealization discussed here is equally applicable to fixed bed and rotary parallel-flow regenerators.

9.2 METHOD OF ANALYSIS

The linear model discussed in Chap. 5 for counterflow regenerators is equally applicable to parallel-flow regenerators. The equations in dimensionless form are employed:

$$\text{Fluid:} \qquad \frac{\partial t_f}{\partial \xi} = t_m - t_f \tag{5.7}$$

$$\text{Storage material:} \qquad \frac{\partial t_m}{\partial \eta} = t_f - t_m \tag{5.8}$$

with the dimensionless parameters, reduced length Λ', Λ'' and reduced period Π' and Π'' as defined previously. However, the reversal conditions for parallel-flow mode corresponding to Eqs. (5.9) and (5.10) for counterflow operation are different and are given by Eqs. (9.1) and (9.2):

$$t'_m(\xi', 0) = t''_m\left(\frac{\Lambda''\xi'}{\Lambda'}, \Pi''\right) \tag{9.1}$$

$$t''_m(\xi'', 0) = t'_m\left(\frac{\Lambda'\xi''}{\Lambda''}, \Pi'\right) \tag{9.2}$$

The same normalized [0, 1] temperature scale can be used here with the fluid temperature T_f and the solid temperature T_m, defined by Eqs. (5.11) and (5.12).

The model assumes that the solid thermal conductivity in the direction normal to fluid flow is infinite, although it is possible to use the overall heat transfer coefficient $\bar{h}$ for storage materials with low finite conductivity. The thermal conductivity in the direction parallel to fluid flow is assumed to be zero.

Hausen [1] predicted the performance of parallel-flow units employing this model, examining the cyclic equilibrium behavior using a closed method. Kardas [2] has investigated the effect of the finite conductivity of the material in the direction perpendicular to the fluid flow for the case where both the hot and cold fluid flow periods are equal. The results obtained from the use of his analytical techniques do not appear to converge at large values of the reduced length Λ.

A more complete model of the storage material embodies longitudinal conductivity and conductivity normal to fluid flow. The governing equation for the storage material is then

$$\frac{1}{\alpha}\frac{\partial t_m}{\partial \tau} = \frac{\partial^2 t_m}{\partial x^2} + \frac{\partial^2 t_m}{\partial y^2} \tag{9.3}$$

The boundary conditions are discussed in Chap. 4, with the exception that the inlet gas temperature varies in a square wave fashion, as shown in Fig. 9.1. Presented later are results obtained by Kumar [3] using this more extensive model.

The steady-state behavior of a parallel-flow regenerator may be obtained by an open or a closed method. In the former the computations begin by assuming a definite temperature distribution in the storage material and then cycling the model to equilibrium. In the latter method, the linear nature of the problem allows superposition techniques to be employed. This leads to a reasonably accurate calculation with considerably less computation time than is required for an open method.

The inlet conditions for the fluid shown in Fig. 9.1 can be written as

$$\text{Hot period:} \qquad n\tau_p \leqslant \tau \leqslant n\tau_p + \tau_n \qquad t_f = t_{fh} \tag{9.4}$$

$$\text{Cold period:} \qquad n\tau_p + \tau_h \leqslant n\tau_p + \tau_h + \tau_c = (n+1)\tau_p \qquad t_f = t_{fc} \tag{9.5}$$

where τ_h and τ_c are the time durations that the hot and cold fluids remain in contact with the storage material.

The superposition solution technique was proposed initially by Hausen [1] and called by him the heat pole method. It is equally applicable to both parallel-flow and counterflow regenerators. Because the differential equations (5.7) and (5.8), or (5.7) and (9.3), are linear, it is possible to add together arbitarily any set of particular solutions. The way in which these particular solutions are added together or superimposed is determined by the reversal conditions (9.1) and (9.2).

Before this matter is discussed further, it is necessary to consider briefly a simplification of the model possible for symmetric regenerators.

9.2.1 Symmetric Regenerators: Reversal Conditions

We showed in Chap. 8 that for the counterflow regenerator, the reversal conditions can be simplified; this is also the case for parallel-flow regenerators. Where $\Lambda = \Lambda' = \Lambda''$ and $\Pi = \Pi' = \Pi''$, the temperature distribution in the packing at the end of the hot period must be exactly symmetric with respect to the temperature distribution corresponding to the end of the cold period. This is shown in Fig. 9.2. It follows that the two temperature profiles on the [0, 1] scale are related by Eq. (9.6):

$$T''_m(\xi, \Pi) = 1.0 - T'_m(\xi, \Pi) \tag{9.6}$$

This considerably simplifies any discussion of the closed methods, since the examination of symmetric regenerators can be readily extended to nonsymmetric forms.

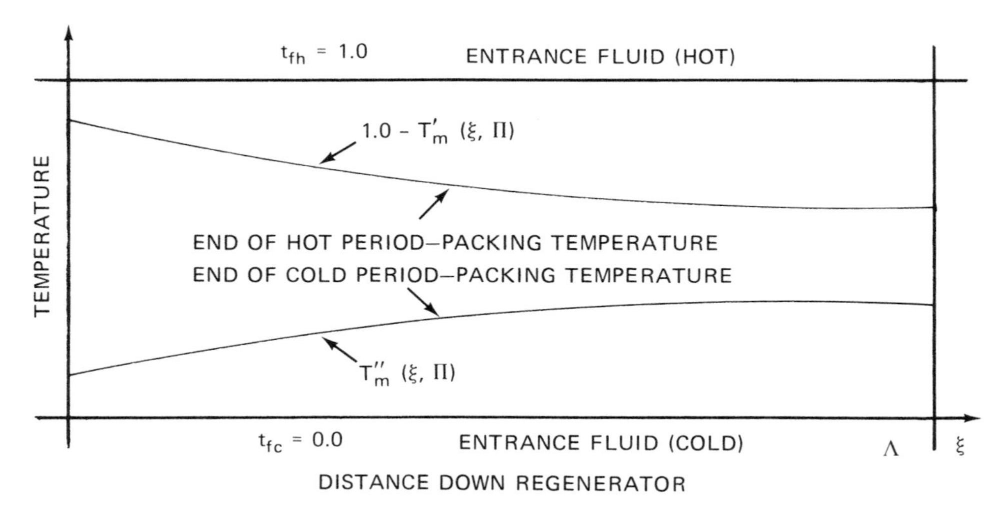

Figure 9.2 Boundary conditions for symmetric regenerators–parallel flow.

9.3 THE HEAT POLE METHOD

The regenerator is divided into strips $\Delta\xi$ wide, and the average solid temperature in each strip is denoted by $T'_{m,1}, T'_{m,2}, \ldots, T'_{m,N}$ for the hot period and by $T''_{m,1}, T''_{m,2}, \ldots, T''_{m,N}$ for the cold period, as shown in Fig. 9.3.

Hausen [1] defined a heat pole of width $\Delta\xi$ with temperature unity but with zero temperature at all other strips. This represents a concentration of heat within a narrow strip of checkerwork. If fluid of zero entrance temperature is considered to pass over the heat pole in the solid, from left to right as shown in Fig. 9.4, the strip in which the heat pole is located will be cooled and the fluid will be heated. As the now warm fluid passes over the remainder of the solid, it loses heat and the following strips experience an increase in temperature. The solid to the left of the heat pole is not affected.

The temperature in the heat pole drops to Δw_1. The temperature of the strips downstream rise by $\Delta w_2, \Delta w_3, \ldots, \Delta w_N$, as shown in Fig. 9.5. The increase in temperatures, $\Delta w_1, \Delta w_2, \Delta w_3, \ldots, \Delta w_N$, are called heat pole functions by Hausen. Because the equations describing regenerator behavior are linear, the temperature effect of a heat pole, not of unit height, but of height, for example, $F'_2(\Pi)$, the temperature in strip 2 at cyclic equilibrium at the end of the hot period, will be

$$F'_2(\Pi)\,\Delta w_1, F'_2(\Pi)\,\Delta w_2, \ldots, F'_2(\Pi)\,\Delta w_N$$

The effect of sequential heat poles is shown in Table 9.1. These heat pole functions, for example, Δw_2 and Δw_1, represent particular solutions to the differential equations. These equations are considered to be linear, and therefore the total temperature effect of heat poles to the right of and at the position considered can be

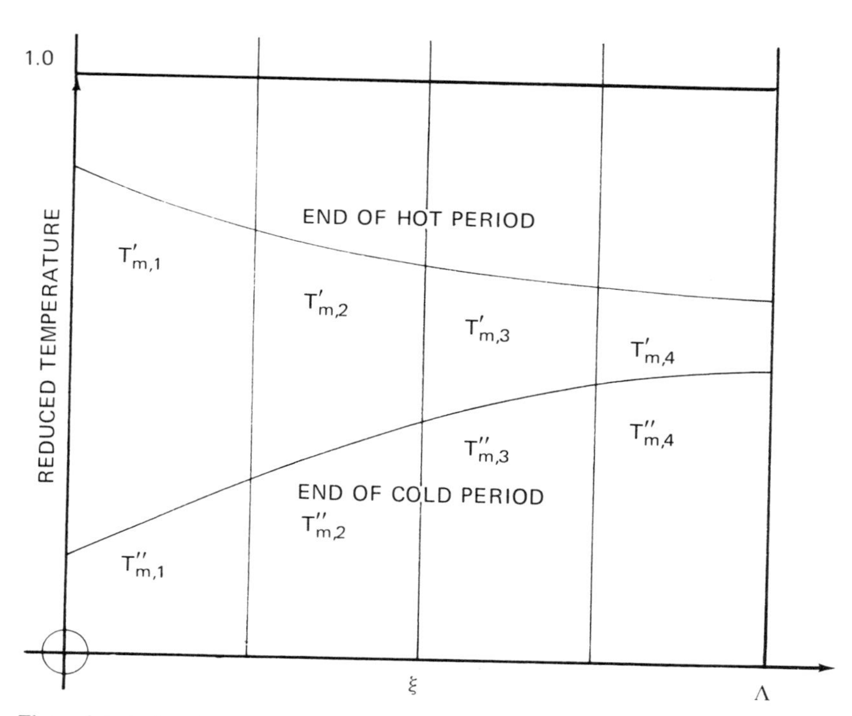

Figure 9.3 Reduced temperature distribution in parallel-flow regenerator.

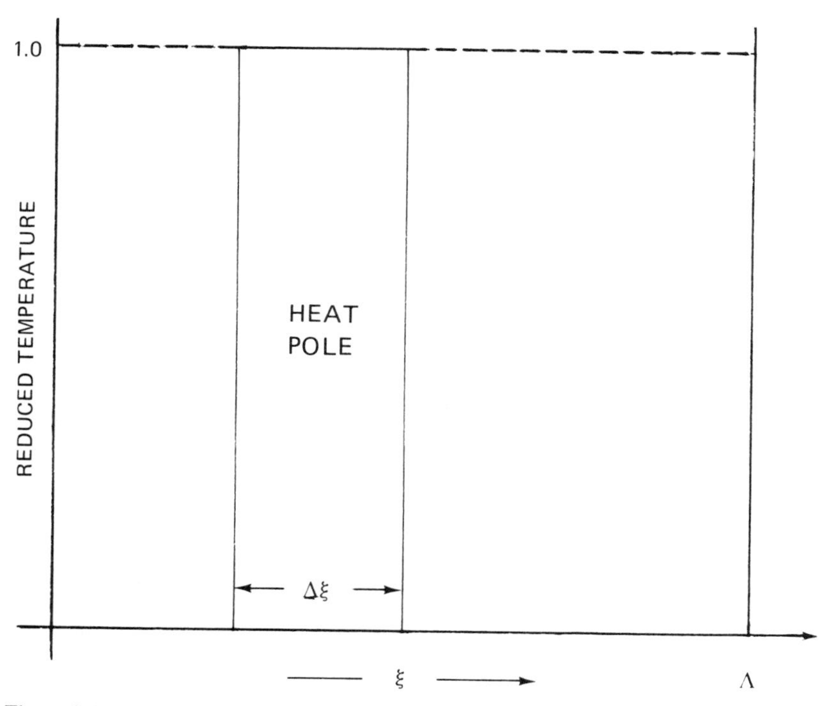

Figure 9.4 Hausen heat pole. From [1].

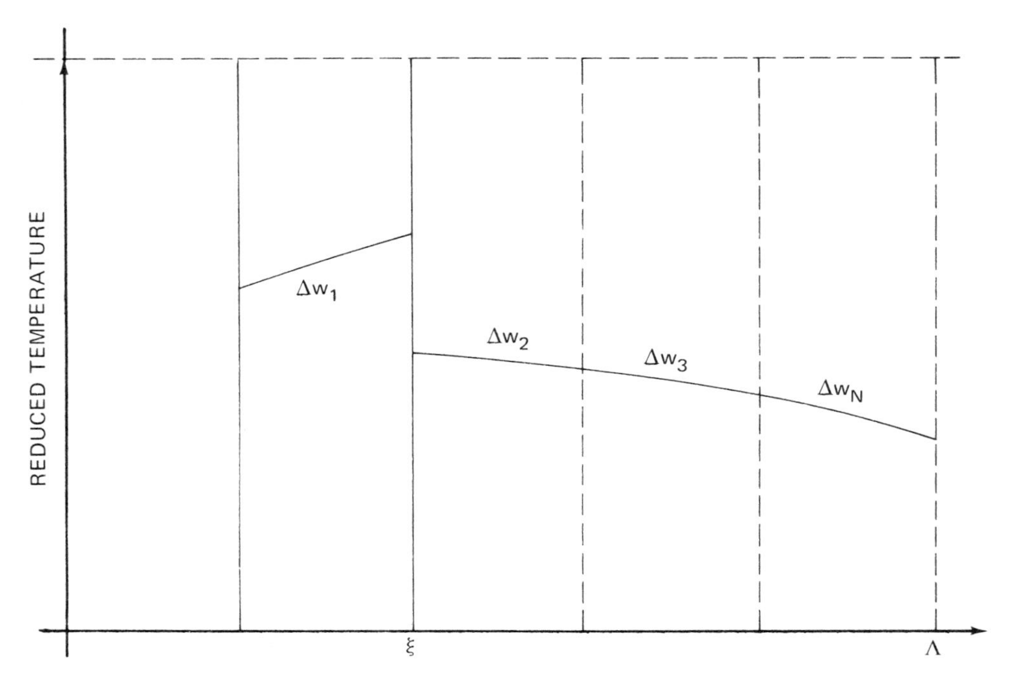

Figure 9.5 The effects of cooling a Hausen heat pole.

computed as the sum of the effects on that strip. This can be written in matrix form:

$$\begin{bmatrix} \Delta w_1 & & & & \\ \Delta w_2 & \Delta w_1 & & & \\ \Delta w_3 & \Delta w_2 & \Delta w_1 & & \\ \cdot & \cdot & \cdot & & \\ \cdot & \cdot & \cdot & & \\ \cdot & \cdot & \cdot & & \\ \Delta w_N & \Delta w_{N-1} & \Delta w_{N-2} & \cdots & \Delta w_1 \end{bmatrix} \begin{bmatrix} F'_1\{\Pi\} \\ F'_2\{\Pi\} \\ F'_3\{\Pi\} \\ \cdot \\ \cdot \\ \cdot \\ F'_N\{\Pi\} \end{bmatrix} = \begin{bmatrix} F''_1\{\Pi\} \\ F''_2\{\Pi\} \\ F''_3\{\Pi\} \\ \cdot \\ \cdot \\ \cdot \\ F''_N\{\Pi\} \end{bmatrix} \tag{9.7}$$

where the heat pole functions $\Delta w_1, \Delta w_2, \ldots, \Delta w_N$, are computed for the length of a whole period Π, and where $F''_k\{\Pi\}$ is the temperature in strip k ($k = 1, 2, \ldots, N$) at the end of the cold period. This corresponds to the method of superposition discussed in Chap. 4.

The strip heights $F'_k\{\Pi\}$ also represent the relative weights given to the heat pole functions $\Delta w_1, \Delta w_2, \ldots, \Delta w_N$, in summing the particular solutions of the differential equations in order to obtain the required solid temperature distributions at cyclic equilibrium. Use is now made of the reversal condition Eq. (9.6), which can be rewritten for each strip k ($k = 1, 2, \ldots, N$) as

$$F''_k\{\Pi\} = 1 - F'_k\{\Pi\} \tag{9.8}$$

The matrix equation becomes

$$\begin{bmatrix} 1+\Delta w_1 & & & & \\ \Delta w_2 & 1+\Delta w_1 & & & \\ \Delta w_3 & \Delta w_2 & 1+\Delta w_1 & & \\ \cdot & \cdot & \cdot & & \\ \cdot & \cdot & \cdot & & \\ \cdot & \cdot & \cdot & & \\ \Delta w_N & \Delta w_{N-1} & \Delta w_{N-2} & \cdots & 1+\Delta w_1 \end{bmatrix} \begin{bmatrix} F'_1\{\Pi\} \\ F'_2\{\Pi\} \\ F'_3\{\Pi\} \\ \cdot \\ \cdot \\ \cdot \\ F'_N\{\Pi\} \end{bmatrix} = \begin{bmatrix} 1 \\ 1 \\ 1 \\ \cdot \\ \cdot \\ \cdot \\ 1 \end{bmatrix} \tag{9.9}$$

The elegance of this approach is twofold. First, no particular method of computing Δw_k ($k = 1, 2, \ldots, N$) is specified; the pair of equations (5.7) and (5.8) or the pair (5.7) and (9.3) can be used. Second, the matrix in Eq. (9.9) is triangular and $F'_1\{\Pi\}$ can be computed directly, followed by $F'_2\{\Pi\}$, $F'_3\{\Pi\}$, and so on up to $F'_N\{\Pi\}$.

The fluid temperature curve can also be calculated in the same manner using as a basis another heat pole function corresponding to fluid temperature. It is simpler, however, to determine only the solid temperature $T_m\{\xi\}$ using the heat pole method. The fluid temperature $T_f\{\xi\}$ can then be computed using Eq. (5.7) in the finite difference form used in Chap. 5.

The method described so far deals only with the cold period, that is, a period of duration Π when cold fluid (entrance temperature zero, on the reduced scale) passes over a heat pole. This is all that is necessary for the symmetric case, since the reversal condition Eq. (9.8) can be used. In order to extend the method to unsymmetric regenerators, one first denotes the heat pole functions for the cold period (relating to reduced length Λ'' and reduced period Π'') by $\Delta'' w_k$ ($k = 1, 2, \ldots, N$).

Next the solid is considered to be isothermally equal to one at the start of the hot period, except that there is a "cold pole" as shown in Fig. 9.6. A hot period of

Table 9.1 Effect of heat poles at sequential strips

	Effect at strip		
Strip	1	2	N
1	$F'_1\{\Pi\}\,\Delta w_1$	$F'_1\{\Pi\}\,\Delta w_2$	$F'_1\{\Pi\}\,\Delta w_N$
2	0	$F'_2\{\Pi\}\,\Delta w_1$	$F'_2\{\Pi\}\,\Delta w_{N-1}$
3	0	0	$F'_3\{\Pi\}\,\Delta w_{N-2}$
.	.	.	.
.	.	.	.
.	.	.	.
N	0	0	$F'_N\{\Pi\}\,\Delta w_1$

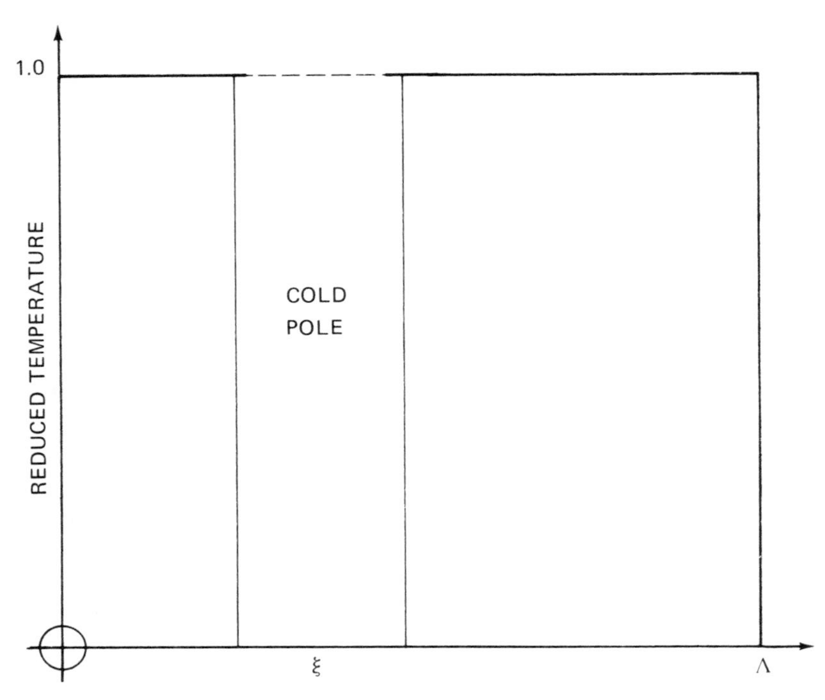

Figure 9.6 Cold pole.

duration Π' is then studied in which hot gas of entrance temperature unity passes over the cold pole. It is, of course, quite possible to calculate the "cold pole" functions $\Delta w_1'$, $\Delta w_2'$, . . . , $\Delta w_N'$. It is easier, however, to compute them from heat pole considerations:

$$\Delta' w_k = 1 - \Delta w_k^* \tag{9.10}$$

where Δw_k^* is the "heat pole" function for a regenerator of reduced length Λ' and reduced period Π'. See Fig. 9.7. Again, if the cold pole has temperature $F_k''(\Pi'')$ $(k = 1, 2, \ldots, N)$ instead of zero, the equivalent heat pole has temperature $1 - F_k''(\Pi'')$.

The matrix equation equivalent to Eq. (9.7) is now

$$\begin{bmatrix} \Delta w_1^* & & & & \\ \Delta w_2^* & \Delta w_1^* & & & \\ \Delta w_3^* & \Delta w_2^* & \Delta w_1^* & & \\ \cdot & \cdot & \cdot & & \\ \cdot & \cdot & \cdot & & \\ \cdot & \cdot & \cdot & & \\ \Delta w_N^* & \Delta w_{N-1}^* & \Delta w_{N-2}^* & \ldots & \Delta w_1^* \end{bmatrix} \begin{bmatrix} 1 - F_1''(\Pi) \\ 1 - F_2''(\Pi) \\ \\ \cdot \\ \cdot \\ \cdot \\ 1 - F_N''(\Pi) \end{bmatrix} = \begin{bmatrix} 1 - F_1'(\Pi) \\ 1 - F_1'(\Pi) \\ \\ \cdot \\ \cdot \\ \cdot \\ 1 - F_N'(\Pi) \end{bmatrix} \tag{9.11}$$

Matrix equations (9.7) and (9.11) can now be brought together to form one set of linear equations, which can be solved for

$$F_1'\{\Pi\}, F_1''\{\Pi\}, F_2'\{\Pi\}, F_2''\{\Pi\}, \ldots, F_N'\{\Pi\}, F_N''\{\Pi\}$$

9.4 REFINEMENT OF THE HEAT POLE METHOD

The calculation of the temperatures $F_1'\{\Pi\}, F_2'\{\Pi\}, \ldots, F_N'\{\Pi\}$ by solving the simultaneous linear equation (9.9) for the symmetric case [or eqs. (9.7) and (9.11) for the unsymmetric case] can be regarded as a possible approximation method for solving an integral equation equivalent to Eq. (8.47) as set out in Chap. 8 in our discussion of counterflow regenerators. The relevant equations (8.45) and (8.46) apply to the non-symmetric counterflow case.

The relationship between the integral equation and the heat pole method is obtained by considering what happens when a transition is made to infinitely narrow heat poles. Hausen [1] demonstrates that for the model of the regenerator embodying Eqs. (5.13) and (5.14), with the gas and solid temperatures on the [0, 1] scale, the limiting form of the matrix equation (9.7) with infinitely narrow heat poles is Eq. (8.22), namely,

$$T_m''(\xi'', \Pi'') = e^{-\Pi''}F''(\xi'') + \int_0^{\xi''} K''(\xi'' - \epsilon)F''(\epsilon)\, d\epsilon \tag{8.22}$$

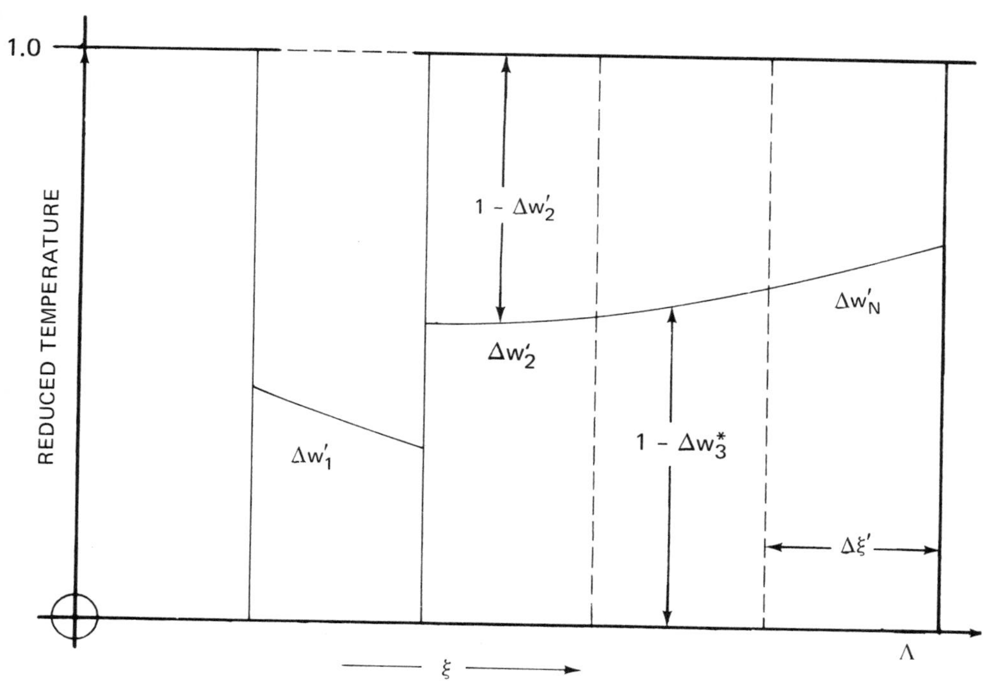

Figure 9.7 The effect of heating a cold pole.

In other words, by an appropriate choice of heat pole function, the heat pole method can be regarded as an approximate method for solving (8.22) [and (8.23) if necessary], and can be treated as one that embodies the principles of superposition included in other methods.

9.5 APPLICATION OF THE NAHAVANDI AND WEINSTEIN METHOD TO PARALLEL-FLOW REGENERATORS

Without loss of generality, we describe the solution of the symmetric parallel-flow regenerator problem. The reversal condition Eq. (9.6) can be written in the form

$$T''_m(\xi, \Pi) = 1.0 - F''(\xi) \tag{9.12}$$

This condition is inserted into Eq. (8.22), which becomes

$$(1 + e^{-\Pi})F''(\xi) + \int_0^{\xi} K(\xi - \epsilon)F''(\epsilon)\,d\epsilon = 1 \qquad \text{parallel flow} \tag{9.13}$$

This corresponds to the term $F''(\Lambda - \xi)$ being replaced by $F''(\xi)$ in the integral equation (8.47) for counterflow operation.

In the Nahavandi and Weinstein method [4], Eq. (9.13) takes the form for parallel flow:

$$(1 + e^{-\Pi}) \sum_{j=0}^{n} a_j \xi^j + \sum_{j=0}^{n} a_j \int_0^{\xi} \epsilon^j K(\xi - \epsilon)\,d\epsilon = 1 \tag{9.14}$$

equivalent to Eq. (8.48) for counterflow. This equation is applied at $n + 1$ distinct values ξ_i $(i = 0, 1, 2, \ldots, n)$ of ξ on the range $0 \leqslant \xi \leqslant \Lambda$. The set of linear equations thus generated are very similar to the equation for counterflow

$$R\mathbf{a} = \mathbf{1} \tag{8.49}$$

except that the (i, j)th element of the $(n + 1) \times (n + 1)$ matrix R takes the form

$$(1 + e^{-\Pi})\,\xi_i^j + \int_0^{\xi_i} \epsilon^j K(\xi_i - \epsilon)\,d\epsilon$$

for $0 \leqslant i, j \leqslant n$. The solution to the equations consists of the vector $\mathbf{a} = \{a_0, a_1, a_2, \ldots, a_n\}$ of coefficients of the approximating polynomial

$$F(\xi) = \sum_{j=0}^{n} a_j \xi^j$$

The chief advantage of this method lies in the fact that a computer program developed for the counterflow case using the Nahavandi and Weinstein method can be readily modified to deal with the parallel-flow case.

The regenerator effectiveness can be computed using Eq. (8.50):

$$\eta_{\text{REG}} = \frac{1}{\Pi} \int_0^{\Lambda} (F''(\xi) - F'(\xi))\,d\xi \qquad \text{symmetric case} \tag{8.50}$$

Upon applying the reversal condition (9.6), Eq. (8.50) takes the form

$$\eta_{\text{REG}} = \frac{1}{\Pi} \int_0^{\Lambda} (2F''(\xi) - 1)\, d\xi \tag{9.15}$$

9.6 PARALLEL-FLOW REGENERATOR PERFORMANCE

Hausen [1] published a graph displaying the thermal effectiveness (thermal ratio) η_{REG} as a function of reduced length for different values of reduced period for a symmetric parallel-flow regenerator. This is shown here as Fig. 9.8. The figure also includes two corresponding curves for symmetric counterflow regenerators, which show clearly that these types of regenerators are more efficient than parallel-flow devices.

We have computed the values of the thermal effectiveness for a range of dimensionless parameters, first for $\Lambda = \Pi$ and then $2.5 \leqslant \Lambda \leqslant 30.0$ for different values of Π in the range $2.5 \leqslant \Pi \leqslant 15.0$ for symmetric parallel-flow regenerators, using the Nahavandi and Weinstein method discussed previously in this chapter. These values are set out in Tables 9.2 and 9.3.

It will be seen that for low values of reduced period Π, the effectiveness approaches the value of 0.5 asymptotically as Λ increases. For large values of reduced period, however, the curves have an alternating oscillation about 0.5, and the larger the reduced period, the more pronounced is the peak value of the effectiveness. It must be pointed out that a counterflow regenerator with reduced length $\Lambda = 10$ can achieve

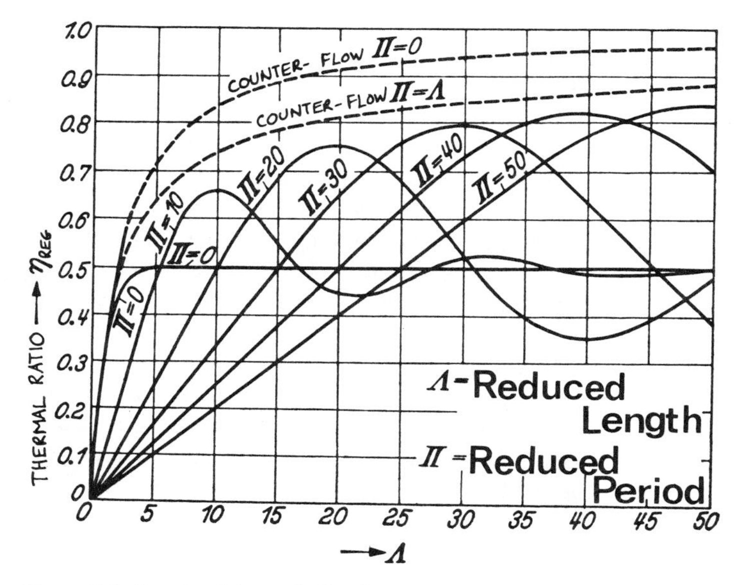

Figure 9.8 Graph of thermal effectiveness (ratio) in relation to reduced length and reduced period for symmetric parallel-flow regenerators [1].

Table 9.2 Comparison between parallel-flow and counterflow symmetric regenerators, thermal effectiveness η_{REG}

$\Lambda = \Pi$	Parallel-flow η_{REG}	Counterflow η_{REG}
2.5	0.462	0.512
5.0	0.562	0.638
7.5	0.618	0.700
10.0	0.658	0.738
12.5	0.689	0.764
15.0	0.713	0.784

82% effectiveness by a suitable choice of reduced period $\Pi < 3.0$, whereas a parallel-flow regenerator must be at least four times as large, $\Lambda = 40.0$, for the same value of η_{REG} to be achieved. Further, for a parallel-flow symmetric regenerator of this size, the effectiveness is very sensitive to reduced period; for the cycle time to be halved so that Π drops from 40.0 to 20.0, the effectiveness collapses from 82% to 35%, a phenomenon that has no equivalent in counterflow regenerators.

Although it is clearly advantageous to operate regenerators in counterflow mode, when parallel flow is preferred from constructional or operational considerations, great care must be exercised in matching the regenerator size and flow rates with the cycle time in order to maximize the effectiveness.

9.7 THE APPROACH OF KUMAR

Kumar recently studied [3] the performance of parallel-flow regenerators, using the method of superposition discussed earlier. He used the finite conductivity model, which takes into account both longitudinal and latitudinal conductivity in the heat storing mass. In terms of the heat pole method, he computed the equivalent of the heat pole functions using this finite conductivity model rather than the model used by Hausen and ourselves (in the tables presented earlier in this chapter). Kumar's method of analysis, however, is equally applicable for either the simplified or the finite conductivity model.

An open method can be used in which the computations begin by assuming a particular temperature distribution in the storage material and the model run through a number of cycles until cyclic equilibrium is reached. Alternatively, a second, closed method that can be used exploits the linear nature of the problem by employing superposition techniques. Kumar begins by defining the inlet conditions for the fluid, which are shown in Fig. 5.1 and which can be written as

$$n\tau_P \leqslant \tau < n\tau_P + \tau_h \qquad t_f = t_{fh}$$

$$n\tau_P + \tau_h \leqslant \tau < (n+1)\tau_P \qquad t_f = t_{fc}$$

Table 9.3 Symmetric parallel-flow regenerators, thermal effectiveness η_{REG}

Reduced length, Λ	Reduced period, Π	Thermal effectiveness, η_{REG}	Reduced period, Π	Thermal effectiveness, η_{REG}	Reduced period, Π	Thermal effectiveness, η_{REG}	Reduced period, Π	Thermal effectiveness, η_{REG}	Reduced period, Π	Thermal effectiveness, η_{REG}	Reduced period, Π	Thermal effectiveness, η_{REG}
2.5	1.0	0.461	2.5	0.462	5.0	0.407	7.5	0.137	10.0	0.247	15.0	0.167
5.0	1.0	0.499	2.5	0.514	5.0	0.562	7.5	0.542	10.0	0.467	15.0	0.331
7.5	1.0	0.5	2.5	0.504	5.0	0.547	7.5	0.618	10.0	0.613	15.0	0.486
10.0	1.0	0.5	2.5	0.499	5.0	0.505	7.5	0.583	10.0	0.658	15.0	0.613
12.5	1.0	0.5	2.5	0.5	5.0	0.491	7.5	0.516	10.0	0.620	15.0	0.692
15.0	1.0	0.5	2.5	0.5	5.0	0.495	7.5	0.474	10.0	0.543	15.0	0.713
17.5	1.0	0.5	2.5	0.5	5.0	0.5	7.5	0.470	10.0	0.476	15.0	0.681
20.0	1.0	0.5	2.5	0.5	5.0	0.501	7.5	–	10.0	–	15.0	0.613
22.5	1.0	0.5	2.5	0.5	5.0	0.5	7.5	0.502	10.0	0.447	15.0	0.531
25.0	1.0	0.5	2.5	0.5	5.0	0.5	7.5	–	10.0	–	15.0	0.458

where τ_h and τ_c are the time durations that the hot and cold fluids remain in contact with the storage material.

The superposition solution technique is based on the summation of the response to two subproblems. The first, in which the initial material temperature is t_o and the fluid enters at a uniform temperature t_{fc}, and the second, in which the initial material temperature is zero and the fluid temperature is a periodic function of time, are shown in Fig. 9.9.

When cyclic equilibrium is achieved, the contribution of the first subproblem to the total heat stored or removed will be equal to zero, and the subproblem's solution will indicate that both the material and the fluid will have a uniform temperature t_{fc}.

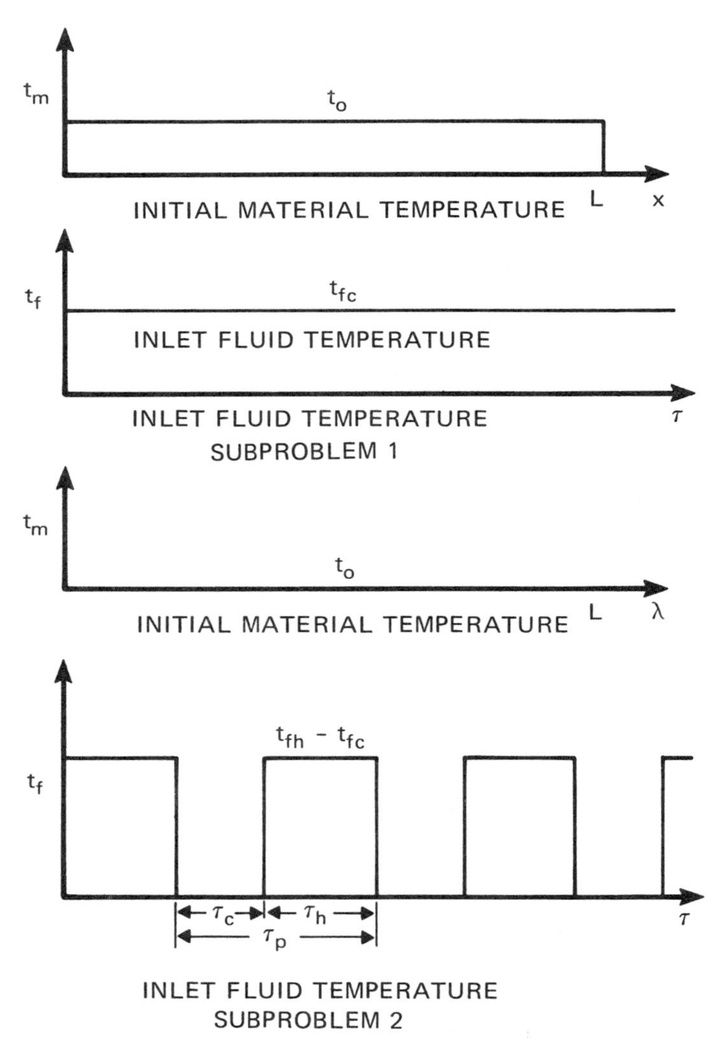

Figure 9.9 Periodic operation subproblems.

This is very similar to the way in which the effect of the heat poles in Hausen's method are put together in order that the reversal conditions at cyclic equilibrium are satisfied.

The fluid temperature in the parallel-flow regenerator can be found by applying the techniques described in Chap. 4, which yields

$$t_f = t_{fc} + \Delta t \left(\sum_{j=0}^{J} T_f(\eta - j\eta_p, \xi) - \sum_{j=1}^{J} T_f(\eta - (j-1)\eta_p - \eta_h, \xi) \right) \tag{9.16}$$

where $\Delta t = t_{fh} - t_{fc}$

J is an integer obtained by dividing τ by τ_p with the quotient rounded down. The second term on the right-hand side is the contribution of the second subproblem. When the heating and cooling periods are equal, the expression reduces to

$$t_f = t_{fc} + \Delta t \left(\sum_{j=0}^{J} T_f(\eta - j\eta_p, \xi) - \sum_{j=1}^{J} T_f(\eta - (j-0.5)\eta_p, \xi) \right) \tag{9.17}$$

The amount of heat stored or retrieved during the hot and cold periods, respectively, after the equilibrium has been achieved, is associated only with the solution of subproblem 2; that is, the heat stored or retrieved is independent of the initial material temperature and is a function of the magnitude of the temperature step of the fluid and the time durations only, if all the other parameters remain the same. In this way a closed solution to the problem is obtained; the results presented here were generated in this manner.

The heat stored, when cyclic equilibrium has been reached, at any time τ is given by

$$\begin{aligned} H(\tau) &= Q(\tau) - Q(\tau - \tau_h) + Q(\tau - \tau_p) - Q(\tau - \tau_p - \tau_h) + \cdots \\ &= \sum_{j=0}^{J} Q(\tau - j\tau_p) - \sum_{j=1}^{J} Q(\tau - (j-1)\tau_p - \tau_h) \end{aligned} \tag{9.18}$$

or

$$H(\tau) = MLc_m \, \Delta t \left(\sum_{j=0}^{J} Q^+(\eta - j\eta_p, \lambda) - \sum_{j=1}^{J} Q^+(\eta - (j-1)\eta_p - \eta_h, \lambda) \right) \tag{9.19}$$

where the Q's are the contributions of various steps, under single-blow operation, into which the second subproblem has been broken up.

When the heating and cooling periods are equal, this expression reduces to

$$H(\tau) = MLc_m \, \Delta t \left(\sum_{j=0}^{J} Q^+(\eta - j\eta_p, \lambda) - \sum_{j=1}^{J} Q^+(\eta - (j-0.5)\eta_p, \lambda) \right) \tag{9.20}$$

After equilibrium has been achieved, the heat stored during the hot period is equal to the heat retrieved during the cold period, and is given by

$$Q_{\text{eq}} = H\{\tau_e + \tau_h\} - H\{\tau_e\}$$

$$= MLc_m\,\Delta t\left(\sum_{j=0}^{J} Q^+\{\eta_e + \eta_h - j\eta_p\} - \sum_{j=1}^{J} Q^+\{\eta_e - (j-1)\eta_p\}\right.$$

$$\left. - \sum_{j=0}^{J} Q^+\{\eta_e - j\eta_p\} + \sum_{j=1}^{J} Q^+\{\eta_e - (j-1)\eta_p - \eta_h\}\right) \tag{9.21}$$

where τ_e is greater than the time to reach equilibrium and is an integral multiple of the time period for the cycle, τ_p.

The equilibrium heat stored or retrieved can be expressed as

$$Q^+ = \frac{Q_{\text{eq}}}{MLc_m\,\Delta t} \tag{9.22}$$

The effectiveness, A^+, can be expressed as

$$A^+ = \frac{Q_{\text{eq}}}{\dot{m}_f c_f\,\Delta t\,\tau^*} \tag{9.23}$$

where τ^* is the harmonic mean of hot and cold periods, $N = \tau_h/\tau_c = 1$,

$$\frac{1}{\tau^*} = \frac{1}{2}\left(\frac{1}{\tau_h} + \frac{1}{\tau_c}\right) \tag{9.24}$$

When the hot and cold periods are equal,

$$\tau^* = \tau_h = \tau_c = \frac{\tau_p}{2}$$

The functional relationship between the effectiveness A^+ and the fraction of maximum heat stored Q^+ is given by

$$A^+ = Q^+\,\frac{\lambda}{\eta^*}$$

where η^* is the nondimensional time, based on the harmonic mean of the time periods.

Kumar presented results first from symmetric parallel-flow regenerators. He related his thermal conductivity model to the finite conductivity model using the relations

$$\lambda = \text{Bi}\,\frac{G^+}{V^+} \tag{9.25}$$

$$\pi = \text{Bi}\,F_o^* \tag{9.26}$$

This is in contrast to the relationships employed by Willmott [5] for the symmetric case, namely,

$$\Lambda_{2D} = \Lambda_{3D}\left(1 + \text{Bi}\frac{\phi}{3}\right)^{-1} \tag{9.27}$$

$$\Pi_{2D} = F_o\left(\frac{1}{\text{Bi}} + \frac{\phi}{3}\right)^{-1} \tag{9.28}$$

Clearly, the Kumar π and λ are equivalent to the Π_{2D} and λ_{2d} parameters if the $\phi/3$ term is neglected and τ_c equals τ_h (N = 1).

The ranges of the various independent nondimensional parameters used in Kumar's work were

$$0 \leqslant \text{Bi} \leqslant 10$$

$$0 \leqslant \frac{\lambda}{\pi} \leqslant 10$$

$$0 \leqslant \lambda \leqslant 10$$

A comparison of Kumar's results for a symmetric parallel-flow regenerator with those obtained by Hausen (Fig. 9.8) is possible if τ_h equals τ_c. The results for a unit having a reduced time of 10 (π = 10) are shown in Fig. 9.10. As the Biot number

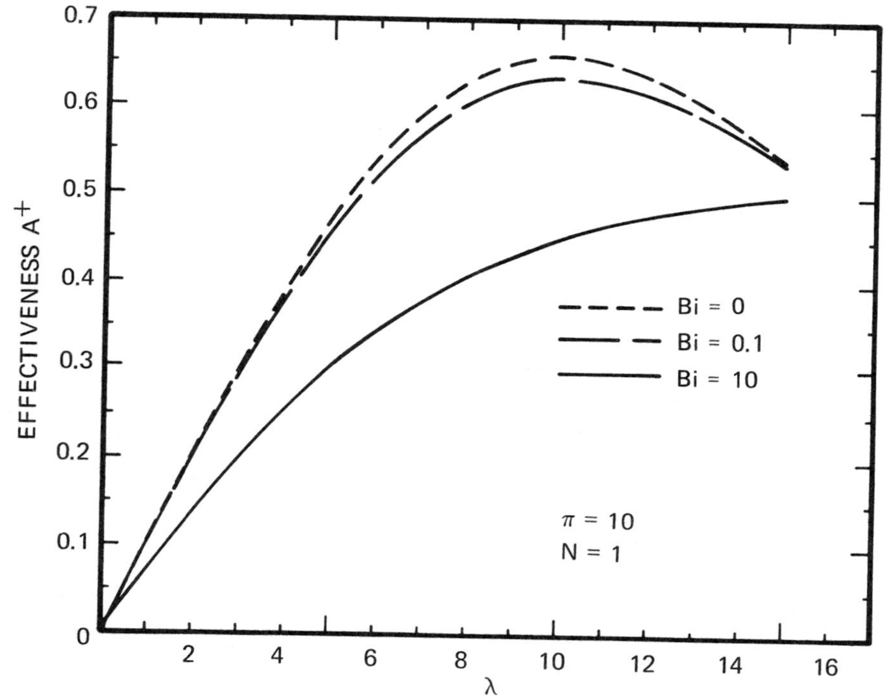

Figure 9.10 Effectiveness of periodic regenerator, $N = 1$.

increases, the effectiveness decreases. For a Biot number of 10, the maximum in the effectiveness curve observed at lower Biot number has disappeared.

His results for four different Biot numbers are shown in Figs. 9.11 through 9.14. A comparison of the results indicates that the effectiveness A^+ is very dependent on reduced length λ, as has been shown previously.

However, whereas Hausen displayed effectiveness as a function of reduced length Λ for different values of Π, Kumar presented his results in the form of effectiveness as a function of λ/π, for different values of λ. Kumar observed that if the Biot number is small, Bi $\leqslant 1$, and λ is held constant while λ/π is varied, the regenerator's effectiveness will reach a maximum value and then decrease to its asymptotic value. For small λ, $\lambda \leqslant 2$, as λ/π is increased, the effectiveness increases monotonically and reaches an asymptotic value. In other words, for small time periods ($\pi < \lambda/3.5$), the effectiveness is independent of the time period with everything else remaining the same. However, as the time period is increased the effectiveness decreases, and eventually reaches a value of zero for infinitely long time periods.

For large λ, $\lambda \geqslant 5$, and small Biot numbers (Bi = 0, 0.1, and 1.0), as λ/π is increased, the effectiveness increases and shows maxima near $\lambda/\pi = 1$, after which it reduces and reaches an asymptote. This means that for large λ, the effectiveness again becomes independent of the π when the nondimensional time is small, $\pi < \lambda/3.5$.

The asymptotic value of the effectiveness increases as λ is increased, but becomes constant (approximately equal to 0.5) for $\lambda \geqslant 5$. The asymptotic value of the effectiveness, observed at large λ/π–that is, small time periods–will be equal to that of a parallel-flow recuperator.

For the simplified model (Bi = 0),

$$U = \frac{h}{2}$$

$$\text{NTU} = \frac{\Pi A}{C_{\min}} = \frac{hA}{2C} = \frac{\lambda}{2}$$

and for a recuperator,

$$\epsilon = f\left(\text{NTU}, \frac{C_{\min}}{C_{\max}}\right) = f\left(\frac{\lambda}{2}, 1\right) = A^+$$

Table 9.4 gives the values of effectiveness for a parallel-flow recuperator [4], and the asymptotic values obtained in this study for Bi = 0. Once again, the values compare extremely well.

As the Biot number is increased and the other parameters are held constant, the effectiveness decreases. For large λ, $\lambda \geqslant 5$, this decrease is much more significant at or near the maxima. For example, with $\lambda = 10$, as the Biot number is increased from 0 to 1.0, the effectiveness at the maximum decreases from 0.66 to 0.60, or about 9%, whereas the asymptotic value decreases from 0.5 to 495 for the same increase in Biot number. For λ = 5, as the Biot number is increased from 0 to 1.0, the maximum decreases from 0.562 to 0.516, or about 8%.

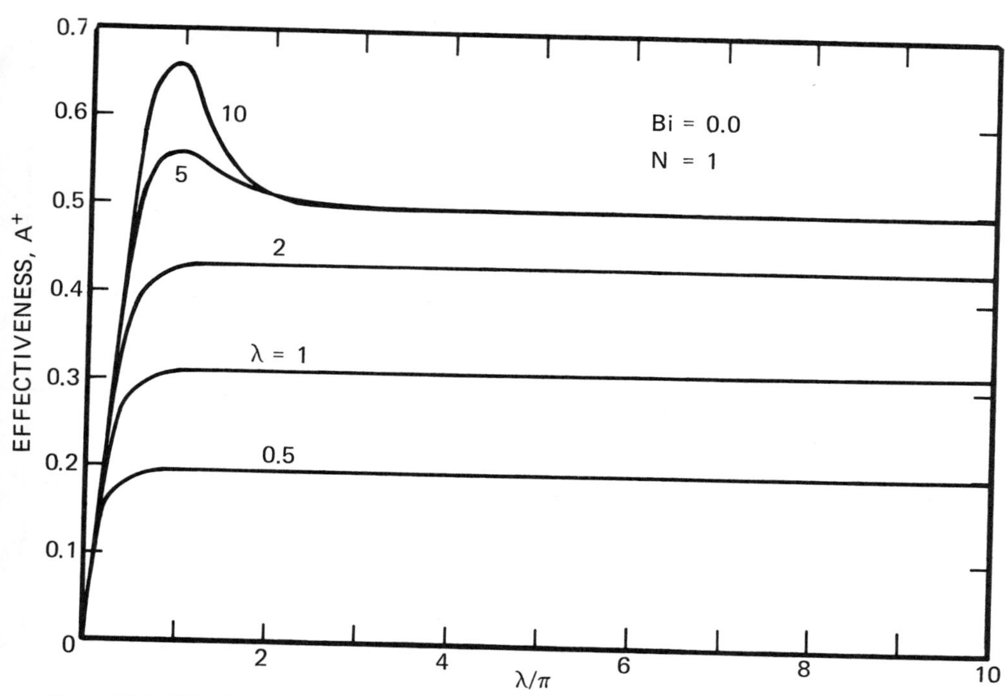

Figure 9.11 Effectiveness of periodic regenerator, $N = 1$.

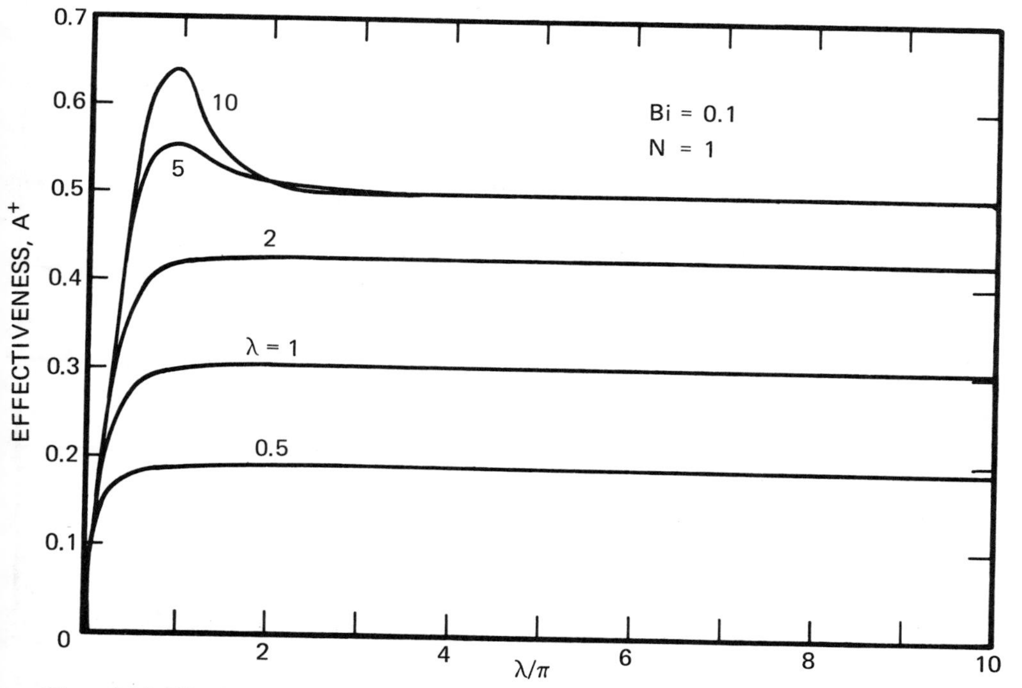

Figure 9.12 Effectiveness of periodic regenerator, $N = 1$.

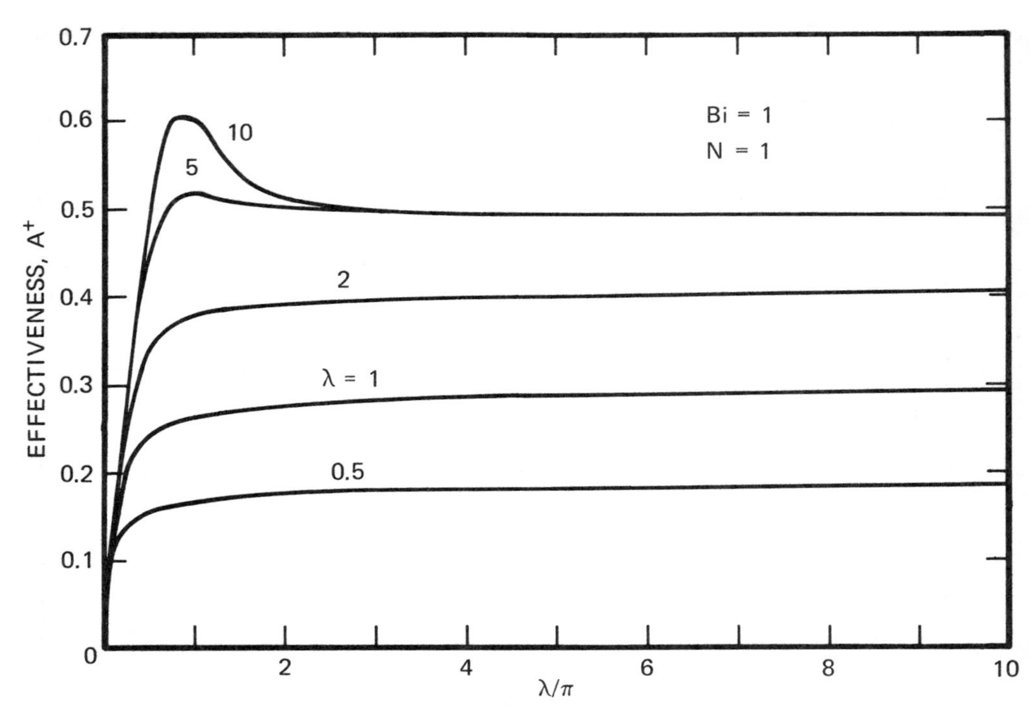

Figure 9.13 Effectiveness of periodic regenerator, $N = 1$.

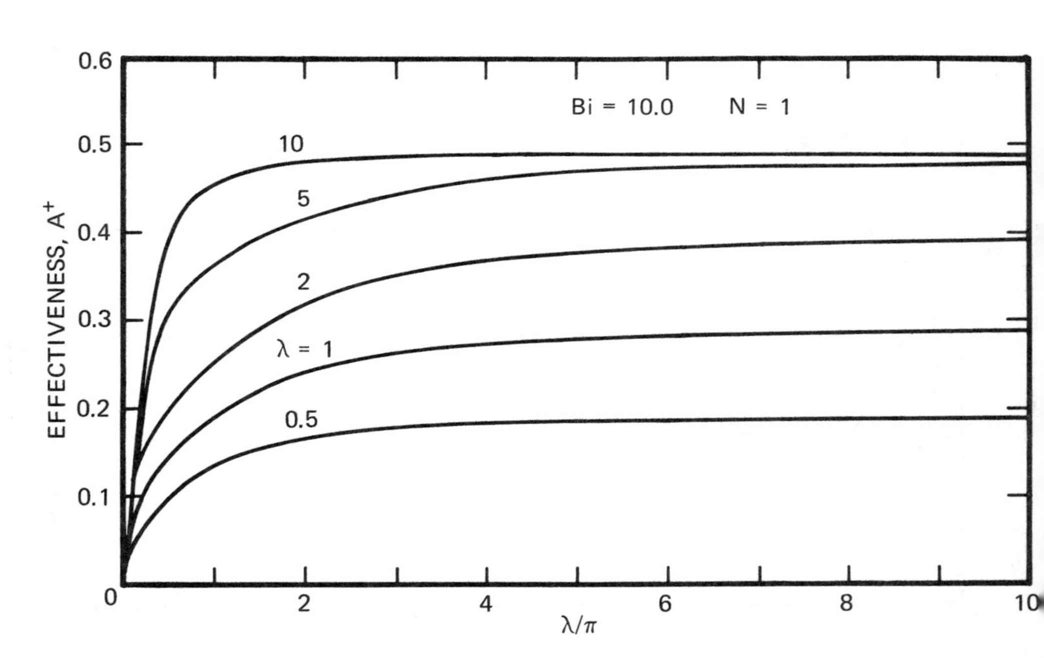

Figure 9.14 Effectiveness of periodic regenerator, $N = 1$.

Table 9.4 Effectiveness for parallel-flow regenerator

λ	ϵ^a	A^{+b}
0.05	0.197	0.197
1.0	0.316	0.316
2.0	0.432	0.432
5.0	0.497	0.4975
10.0	0.500	0.500

[a] Efficiency of a parallel-flow recuperator obtained from [4].
[b] Asymptotic value of the effectiveness for (Bi = 0, $N = 1$) obtained in this study.

When the Biot number is increased, the constant λ curves tend to keep their general shape up to Bi = 1.0. At Bi = 10, the nature of the constant λ curves for $\lambda \geqslant 5.0$ changes; the maxima disappear and the response becomes analogous to that shown by curves for $\lambda \leqslant 2.0$. By comparing the constant λ curve for $\lambda = 5$ and $\lambda = 10$ of Fig. 9.14 (Bi = 10) with the corresponding curves of Figs. 9.11 through 9.13, the disappearance of the maxima can be observed. The increase of Biot number from 0.0 to 10.0 results in a 6% to 10% decrease in the asymptotic value of the effectiveness, but, the effectiveness near $\lambda/\pi = 1$ may decrease by as much as 30%.

Another effect of increase in the Biot number is that the asymptotic value is obtained at a smaller time period, or at larger λ/π. For Bi = 0.0, the asymptotic value of the effectiveness is reached near $\lambda/\pi = 1.5$; for Bi = 0.1, it is near 2.0; and for Bi = 10.0, it is near 4.0.

The effect of a change in the mass flow rate can be seen if the convective film coefficient is evaluated from the following expression for fully developed turbulent flow:

$$Re = 4.0 \frac{\dot{m}_f}{P\mu}$$

$$h = \text{constant} \frac{Re^{0.8}}{d} = \frac{\text{constant}}{d} \left(\frac{\dot{m}_f}{P\mu} \right)^{0.8}$$

If the mass flow rate is increased and the heat transfer coefficient h is kept constant with the other parameters, both λ and λ/π will decrease, and in most cases the effectiveness will decrease. If the operating conditions are near the right-hand side of the maxima, however, the effectiveness may go up. The effect of decreasing the mass flow rate will be the opposite to that described above.

If the length of the unit L is increased, λ will increase and so will λ/π. The effect will be an increase in effectiveness for most cases, unless the operating conditions are near and to the left of the maxima, in which case it will go down.

The effect of increase in the wetted perimeter P while keeping the mass flow rate constant can be obtained using the following expression:

$$\lambda = \frac{hPL}{\dot{m}_f c_f} = K_1 P^{0.2} \qquad \text{(substituting for } h\text{)}$$

and $$\frac{\lambda}{\pi} = K_2 P$$

This means that λ will increase at a much slower rate than λ/π as P is increased, and once again the effectiveness will increase unless the operating condition is near and to the left of the maxima.

If only the storage material is changed, such that $\rho_m c_m$ increases, λ will remain constant while λ/π will increase. The effectiveness in this case will increase unless the operating conditions are near the maxima.

9.8 UNBALANCED REGENERATOR

Kumar next considered the cases where the heating and cooling periods are not equal and when the nondimensional lengths are equal, $\lambda_c = \lambda_h$; the unit in this case is unbalanced. Results for N, the ratio of the heating and the cooling periods, equal to 2 and 3 are presented in Figs. 9.15 through 9.22. The general shape of all the curves remains the same for the corresponding equal heating and cooling periods, $N = 1$. It can be observed that the effectiveness of a unit remains unchanged when the heating and cooling periods for any given unit are interchanged. This happens because of the

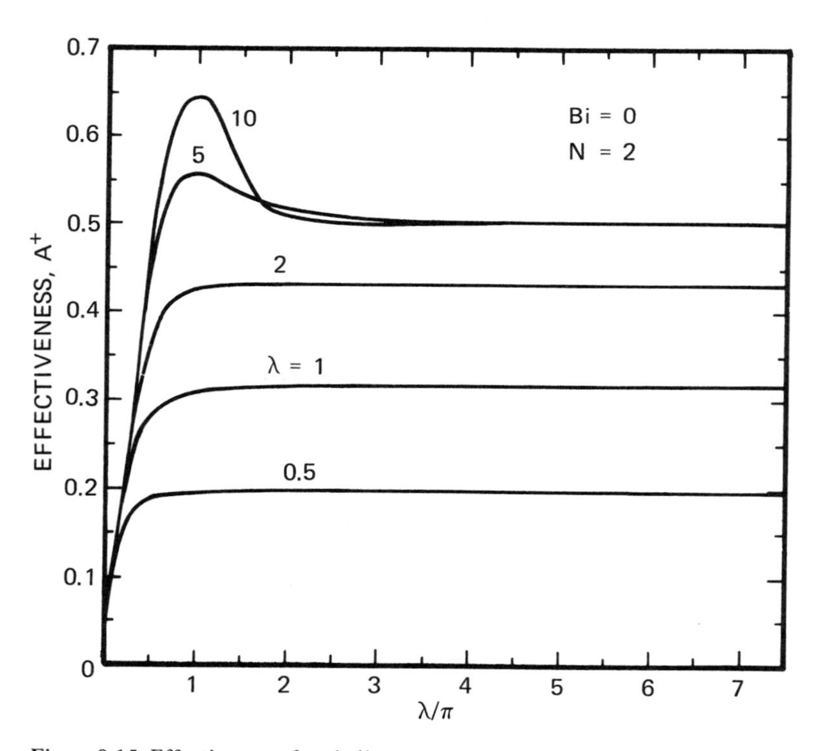

Figure 9.15 Effectiveness of periodic regenerator, $N = 2$.

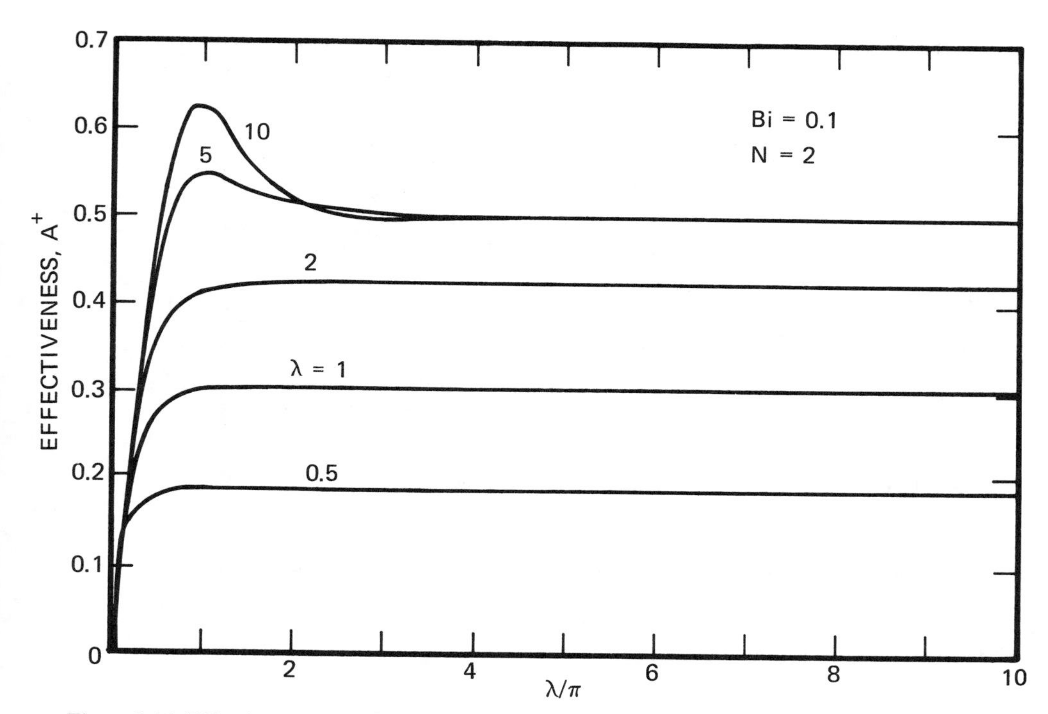

Figure 9.16 Effectiveness of periodic regenerator, $N = 2$.

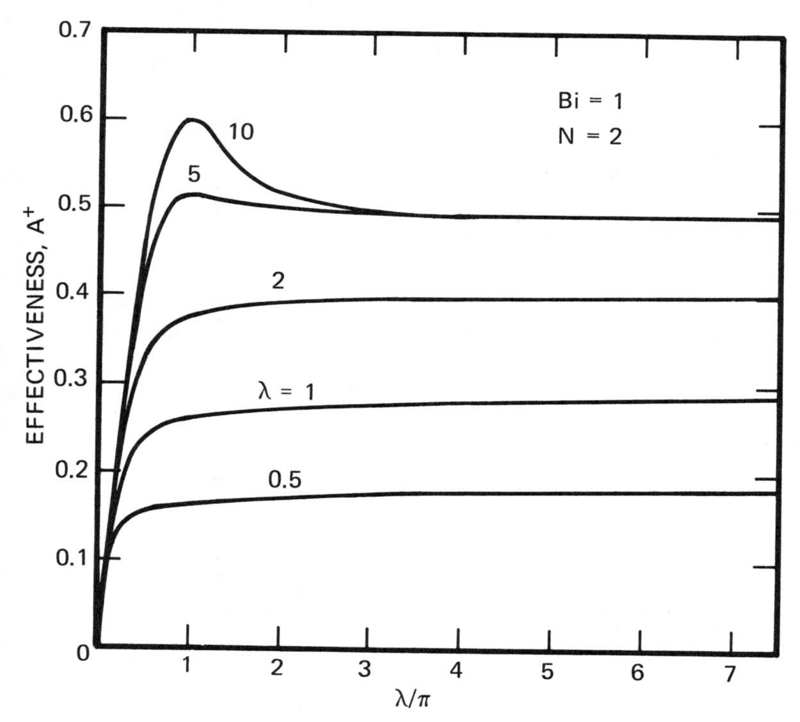

Figure 9.17 Effectiveness of periodic regenerator, $N = 2$.

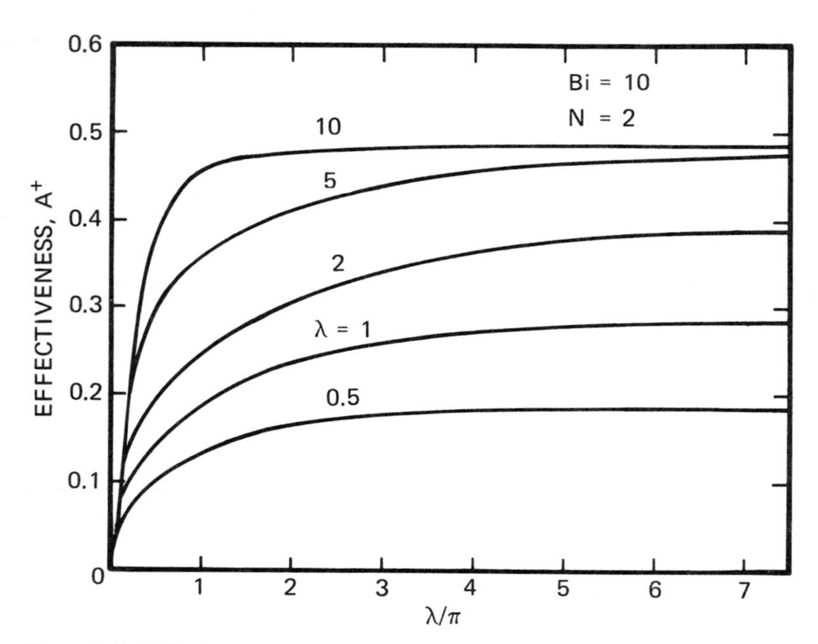

Figure 9.18 Effectiveness of periodic regenerator, $N = 2$.

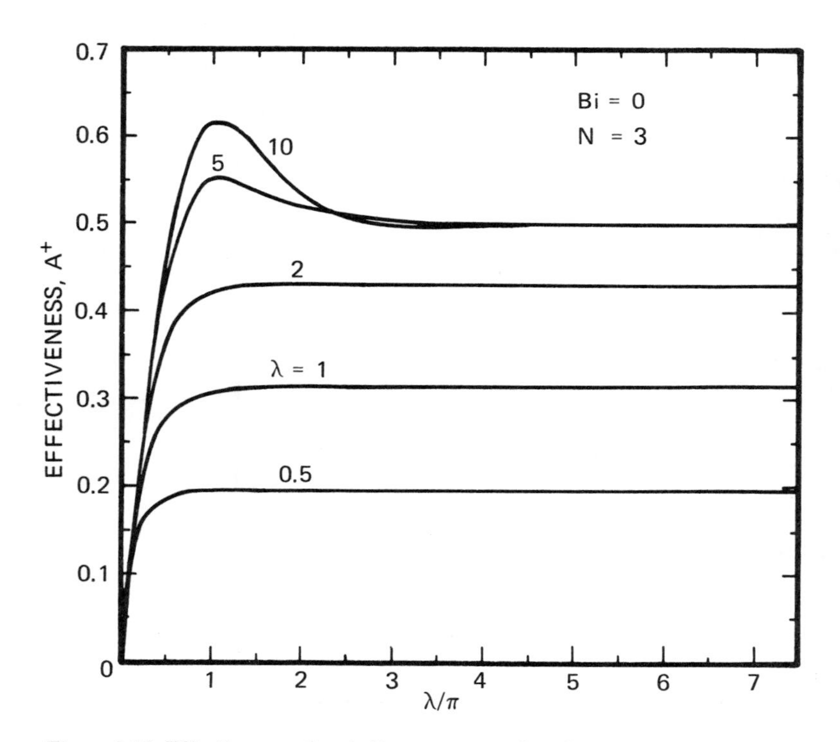

Figure 9.19 Effectiveness of periodic regenerator, $N = 3$.

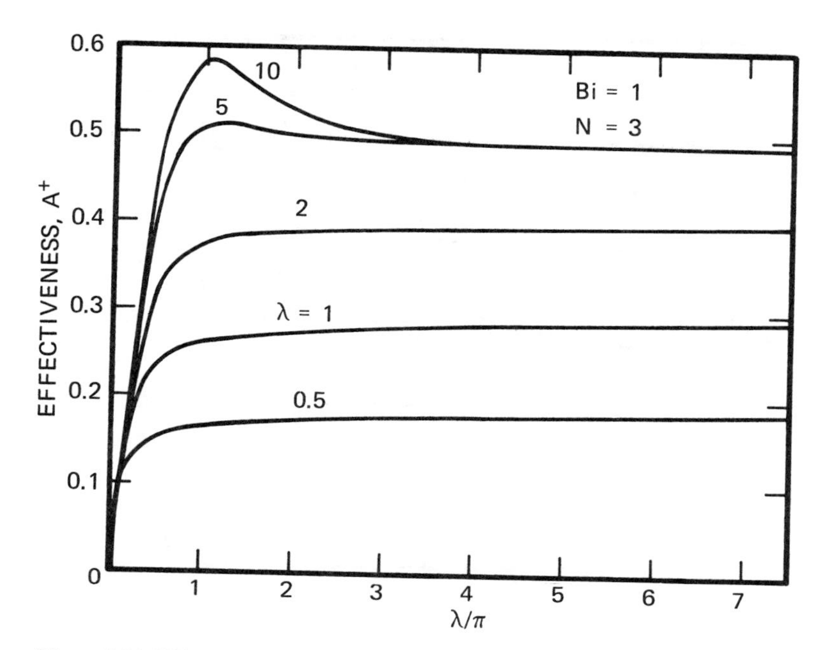

Figure 9.20 Effectiveness of periodic regenerator, $N = 3$.

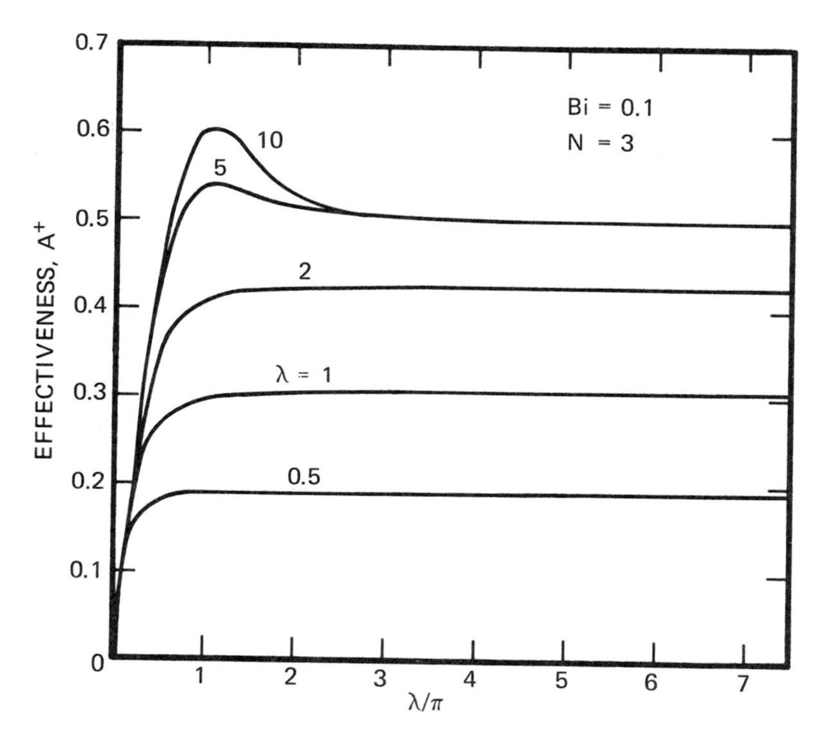

Figure 9.21 Effectiveness of periodic regenerator, $N = 3$.

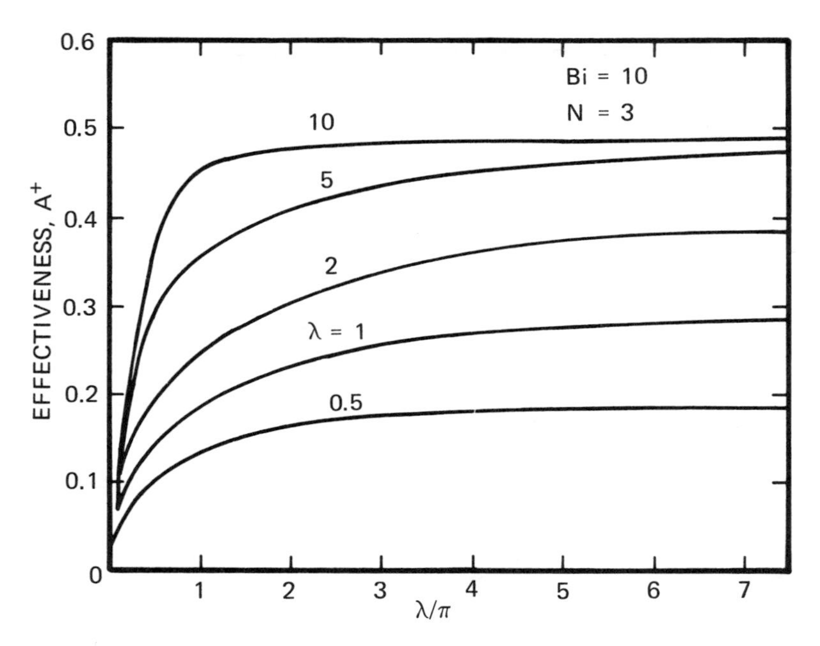

Figure 9.22 Effectiveness of periodic regenerator, $N = 3$.

symmetry of the expressions used to evaluate the cyclic heat storage with respect to τ_h and τ_c [4].

For small time periods, $\pi < \lambda/4$, the effectiveness is observed to be independent of N, as predicted by Hausen. For large time periods, however, that is, for $\pi \geqslant \lambda$, the effectiveness decreases as the ratio of heating to cooling periods, N, is increased. This effect can be readily observed near the maxima. For $\lambda = 10$ and Bi $= 0.1$, by comparing Figs. 5.11, 5.16, and 5.20, it is observed that as N is increased from 1 to 3, the maximum decreases from 0.635 to 0.60, or about 5.5%.

For large λ and $1 > \lambda/\pi < 4$, the effectiveness increases by a small amount as N is increased from 1 to 3, since the curves near the maxima smooth out and the decrease from the maxima to the asymptotic value is less steep.

REFERENCES

1. H. Hausen, *Wärmeübertragung im Gegenstrom, Gleichstrom und Kreuzstrom*, Springer-Verlag, Berlin, 1950.
2. A. Kardas, "On a Problem in the Theory of the Unidirectional Regenerator," *Int. J. Heat Mass Transfer*, vol. 19, 1966, p. 567.
3. M. Kumar, "Periodic Response of a Parallel Flow, Solid Sensible Heat Thermal Storage Unit," M.S. thesis, Pennsylvania State University, University Park, Pa., 1978.
4. A. N. Nahavandi and A. S. Weinstein, "A Solution to the Periodic Flow Regenerative Heat Exchanger Problem," *Appl. Sci. Res.*, vol. A10, 1961, pp. 335–348.
5. A. J. Willmott, "The Regenerative Heat Exchanger Computer Representation," *Int. J. Heat Mass Transfer*, vol. 12, 1969, pp. 997–1014.

CHAPTER

TEN

TRANSIENT PERFORMANCE OF COUNTERFLOW REGENERATORS

10.1 INITIAL CONSIDERATIONS

The regenerator can be regarded as a forced oscillation device, where oscillations in the temperature of the heat storing packing are imposed by the alternate washing of the packing surface by the hot and the cold gases. In counterflow regenerators, the hot and cold gases pass alternately through the same passages in this packing but in opposite directions. The temporary storage of heat in the regenerator packing or checkerwork implies necessarily that the regenerator possesses an inertia whereby changes in the operating conditions do not result immediately in changes in the level and amplitude of the oscillations in gas and solid temperature. Instead, following such a change in operating conditions—for example, an increase in hot gas inlet temperature and/or a decrease in cold gas flow rate—the inertia of the regenerator restrains the response of the system and cyclic equilibrium is reestablished only gradually. In this chapter theoretical work on this transient performance is presented and attempts to parameterize the nature and scale of the behavior of a regenerator under non-equilibrium conditions are discussed. For the most part, considerations are restricted to the linear model presented in Chap. 5, embodied in Eqs. (5.1) and (5.2) or in dimensionless form in Eqs. (5.7) and (5.8).

Fluid: $$\frac{\partial t_f}{\partial \xi} = t_m - t_f \tag{5.7}$$

Storage material: $$\frac{\partial t_m}{\partial \eta} = t_f - t_m \tag{5.8}$$

The dimensionless parameters reduced length Λ', Λ'' and reduced period Π', Π'' turn out to be as significant in describing transient performance as they are for cyclic equilibrium behavior.

There have been two different approaches to this problem, which in the current state of the art remain to be synthesized. The first approach, applied by London and co-workers [1-3] to rotary regenerators and by Willmott and Burns [4-6] to a whole range of fixed bed and rotary configurations, has sought to parameterize the response of a regenerator initially at cycle equilibrium to one or more step changes in operation. "What is the nature and scale of the transient response to step changes in operation?" and "How long does a regenerator take to regain cyclic equilibrium following a step change?" are typical questions this first approach seeks to answer. It is this treatment that is presented in this book.

The other approach, typified by papers by Beets and Elshout [7] and by Strausz [8], has been to devise feedback control strategies to enable Cowper stove systems, typically operating in staggered parallel mode, to deliver time-varying thermal loads or to maintain constant thermal loads under circumstances where there are chronological variations in the operating conditions.

One possible method of control is to operate a Cowper stove system with a target cycle time. An increase in thermal load (that is, cold period air flow rate and exit temperature) will require an increase, known for cyclic equilibrium conditions, in the heat input to the system (that is, hot period gas flow rate and inlet temperature). Despite the application of such a necessary increase in the thermal input, the inertia of the regenerator system precludes the immediate satisfaction of the new thermal demand. Provided that the step increase in load is not too large, the new load can be accommodated with a smaller cycle time. A control strategy for a Cowper stove system in such circumstances will regulate the hot period gas flow rate and the cycle time in such a way that, on the one hand, during transient performance, the target demand is always satisfied, and on the other, at cyclic equilibrium the target cycle time is restored. The hot period gas flow rate is regulated in such a way that any shortfall in the heat input to the system caused by the decreased hot periods is hopefully eliminated. Needless to say, large step increases in load cannot be accommodated in this way, and this has been discussed by Zuidema [9]. The computer simulation of such systems is discussed in Chap. 7.

This latter approach is applicable directly to industrial regenerators and requires no information about the theoretical transient response of the system, as set out by London and Cima [1] and by Willmott and Burns [4]. It requires only a record of measurements of present and past operation and performance of the regenerator system; from these measurements, simple heat balance calculations can be undertaken and the required hot period flow rate estimated for the next cycle. Any error associated with this estimate can be at least partially corrected at the next stove reversal when the control strategy is again applied.

10.2 RESPONSE TO A STEP CHANGE IN OPERATION CONDITIONS

London et al. [3] illustrated the responses to changes in operating conditions by defining the following dimensionless parameters $\epsilon f_1\{\tau\}$ and $\epsilon f_2\{\tau\}$, given by

$$\epsilon f_1\{\tau\} = \frac{t'_{fo}\{\tau\} - t'_{fo}\{0\}}{t'_{fo}\{\infty\} - t'_{fo}\{0\}} \tag{10.1}$$

$$\epsilon f_2\{\tau\} = \frac{t''_{fo}\{\tau\} - t''_{fo}\{0\}}{t''_{fo}\{\infty\} - t''_{fo}\{0\}} \tag{10.2}$$

where $0 \leqslant \tau \leqslant \infty$.

At equilibrium, immediately prior to a step change ($\tau = 0$), $\epsilon f_1\{0\} = \epsilon f_2\{0\} = 0$; once cyclic equilibrium has been restored ($\tau = \infty$), $\epsilon f_1\{\infty\} = \epsilon f_2\{\infty\} = 1$.

London restricted his considerations to rotary regenerators where the exit gas temperature t_{fo} used in Eqs. (10.1) and (10.2) is an average over the whole face of the rotating heat storing mass currently exposed to the gas stream under consideration. Since this type of regenerator operates continuously, London was able to express $t_{fo}\{\tau\}$ and thus $\epsilon f_1\{\tau\}$ and $\epsilon f_2\{\tau\}$ as continuous functions of time.

This approach ignores two factors, namely: (1) the thermal state of each sector of the checkerwork will be different at the moment when a step change is effected, and (2) the response on the cold side to changes on the hot side (and vice versa) will be delayed by a time determined by the speed of rotation of the checkerwork. London was able to ignore these problems by considering only cases where $\Lambda'/\Pi' = \Lambda''/\Pi'' > 100$, that is, where the angular velocity of the rotor is great relative to the gas flow rate.

This treatment, however, can be extended to fixed bed regenerators. Step changes here are regarded as taking place only at the reversals. The responses ϵf_1 and ϵf_2 are measured as chronological averages for each successive period. The current values of t'_{fo} and t''_{fo} for the nth cycle following a step change are computed using (10.3) and (10.4):

Hot period, nth cycle:

$$t'_{fo}\{n(P' + P'') - P''\} = \frac{1}{P'} \int_{(n-1)(P'+P'')}^{n(P'+P'')-P''} t'_{fo}\{\tau\}\, d\tau \tag{10.3}$$

Cold period, nth cycle:

$$t''_{fo}\{n(P' + P'')\} = \frac{1}{P''} \int_{n(P'+P'')-P''}^{n(P'+P'')} t''_{fo}\{\tau\}\, d\tau \tag{10.4}$$

These values of t'_{fo} and t''_{fo} are inserted in Eqs. (10.1) and (10.2) to yield $\epsilon f_1\{n(P' + P'') - P''\}$ and $\epsilon f_2\{n(P' + P'')\}$, respectively.

Notice that $\tau = n(P' + P'') - P''$ corresponds to a time at the end of the hot period and $\tau = n(P' + P'')$ to the end of the cold period of the nth cycle. These times are employed for fixed bed systems so that ϵf_1 and ϵf_2 can represent the fact that the effect of a step change becomes completely manifest only at the end of a period under consideration. At the corresponding times for rotary regenerator operation, only a gas temperature averaged over the whole face of the rotating checkerwork is normally available. This spatial average has as its equivalent, the chronological average of Eqs. (10.3) and (10.4) for fixed bed systems. As close a correspondence as possible between the rotary and fixed checkerwork systems is thereby achieved.

When a step change is made in the gas inlet temperature, the dimensionless parameters Λ', Λ'', Π', and Π'' are unaffected and the thermal ratios at cyclic equilibrium η'_{REG} and η''_{REG} remain unchanged. This enables $t'_{fo}(\infty)$ and $t''_{fo}(\infty)$ to be calculated prior to the step change for a step in hot inlet temperature from t'_{fi} to t'^{*}_{fi}, for example,

$$\frac{t'^{*}_{fi} - t'_{fo}(\infty)}{t'^{*}_{fi} - t''_{fi}} = \eta'_{\mathrm{REG}} = \frac{t'_{fi} - t'_{fo}(0)}{t'_{fi} - t''_{fi}} \tag{10.5}$$

$$\frac{t''_{fo}(\infty) - t''_{fi}}{t'^{*}_{fi} - t''_{fi}} = \eta''_{\mathrm{REG}} = \frac{t''_{fo}(0) - t''_{fi}}{t'^{*}_{fi} - t''_{fi}} \tag{10.6}$$

It follows that in any regenerator calculations, $\epsilon f_1(\tau)$ and $\epsilon f_2(\tau)$ can be calculated immediately as $t'_{fo}(\tau)$ and $t''_{fo}(\tau)$ are computed. However, when step changes are made in gas flow rate, some or all of the parameters Λ', Λ'', Π', and Π'' are altered. Unless the thermal ratios for periodic behavior for the revised dimensionless parameters are known, $\epsilon f_1(\tau)$ and $\epsilon f_2(\tau)$ can be calculated only once cyclic equilibrium is re-established and $t'_{fo}(\infty)$ and $t''_{fo}(\infty)$ have been computed.

London [3] exhibited the response of a regenerator to changes in operation by displaying graphically the variation of ϵf_1 and ϵf_2 with time. Green [10], on the other hand, compared the effect of different step changes in operation in terms of the total time taken to reestablish cyclic equilibrium. Here, both approaches are employed and we investigate the dependence of ϵf_1 and ϵf_2 on the parameters of the system, which are the reduced lengths and reduced periods for the hot and cold periods since they can completely describe the regenerator both before and after the step change.

A further parameter is the magnitude of the step change, although as London pointed out, the linear nature of the model implies that ϵf_1 and ϵf_2 will not be dependent on the size of any step change made in inlet gas temperature. This is not the case for step changes in flow rate, however.

10.3 STEP CHANGES IN INLET GAS TEMPERATURE

10.3.1 Symmetric Case

All earlier work was restricted to symmetric regenerators ($\Lambda' = \Lambda''$, $\Pi' = \Pi''$). In Fig. 10.1 are displayed the responses to a step change in hot period inlet gas temperature. These responses ϵf_1 and ϵf_2 are represented as functions of the dimensionless time parameter η for a reduced length $\Lambda = 10$. It will be seen that the response is

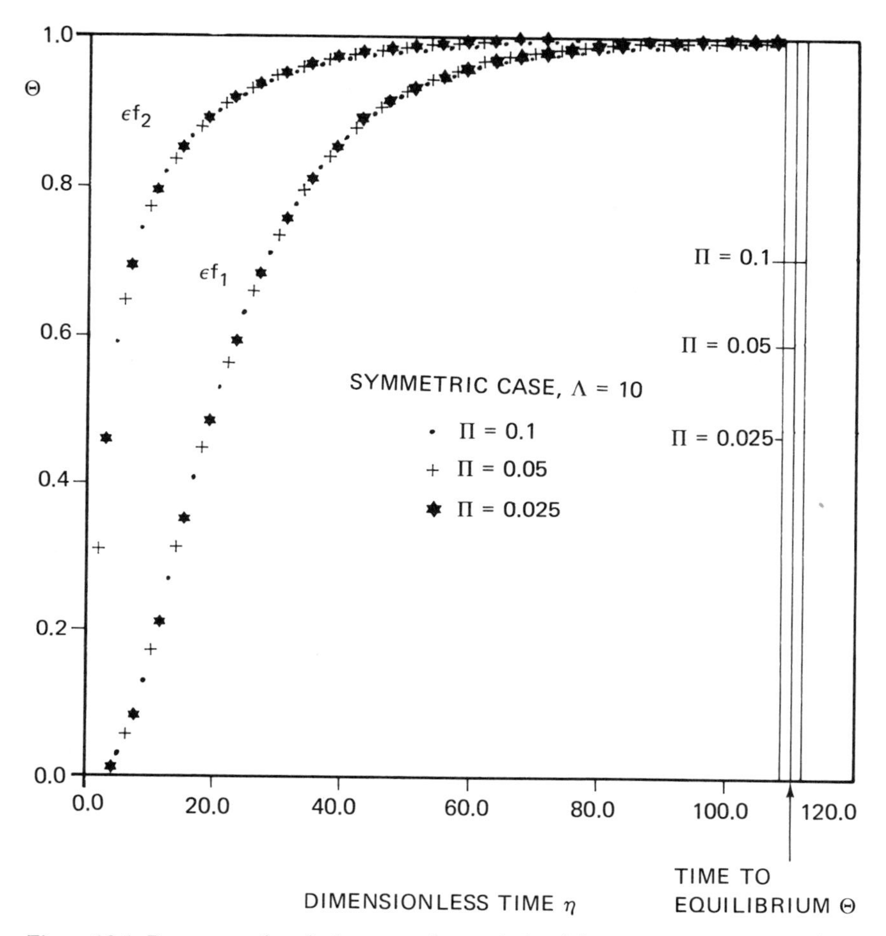

Figure 10.1 Responses ϵf_1 and ϵf_2 to step changes in hot inlet gas temperature for $\Lambda/\Pi > 100$.

independent of reduced period Π for $\Lambda/\Pi = 100$, 200, and 400, thereby confirming the observations of London et al. [3] for $\Lambda/\Pi > 100$. Willmott and Burns investigated cases outside the ranges examined by London and have shown (see Fig. 10.2) that, for example, with $\Lambda = 28$, ϵf_1 and ϵf_2 do exhibit some dependence on reduced period Π for $\Pi = 16$, 8, 4, and 2. Nevertheless, the total dimensionless time Θ needed to re-establish cyclic equilibrium remains independent of reduced period.

It will be seen that ϵf_1 represents the exit gas temperature response on the same side as that on which the step change inlet gas temperature is made; ϵf_2 represents the response on the opposite side to that of the step change. The linear nature of the model permits consideration of step changes on the hot side alone. The results are equally applicable to step changes on the cold side if ϵf_1 is interpreted as ϵf_2 and vice versa; see Eqs. (10.1) and (10.2).

The response displayed in Figs. 10.1 and 10.2 together with London's results indicate the response ϵf_1 on the hot side to a step change on the hot side converges

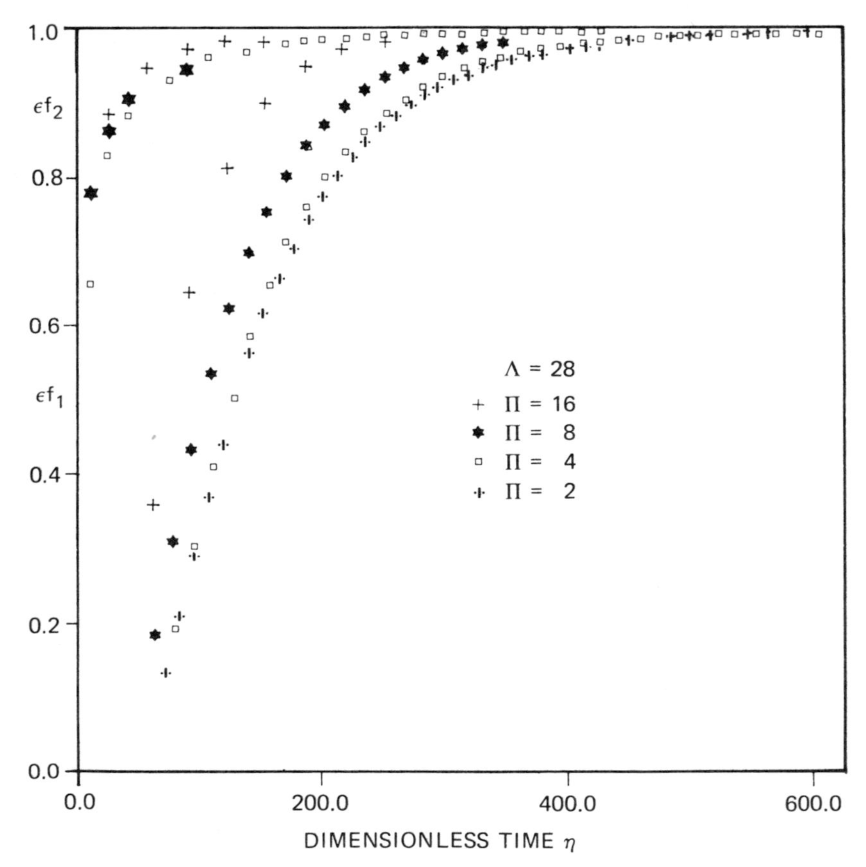

Figure 10.2 Dependence of responses ϵf_1 and ϵf_2 on Π for step changes in hot inlet gas temperature.

toward equilibrium more slowly than ϵf_2, the corresponding response on the cold side. Without loss of generality, it is possible to examine the response ϵf_1 alone to facilitate clearer graphical presentation.

That the reduced length Λ is an effective measure of the inertia of a symmetric regenerator is illustrated in Fig. 10.3, where ϵf_1 and Θ are displayed for reduced period fixed at $\Pi = 4$ for the different reduced lengths $\Lambda = 32$, 24, 16, and 8. The initial lag in the response, which increases with reduced length, is typical of a distributed parameter system of which a thermal regenerator system is an example.

This reflection of the inertia of a regenerator in total dimensionless time Θ needed to reestablish cyclic equilibrium following a step change must be examined further. In a computer simulation of a regenerator, it is important that a suitable measure of convergence (to cyclic equilibrium) be adopted, which takes note of the different rates of convergence between the hot and cold period responses, and between different regenerator configurations.

Willmott [11] proposed pseudo-thermal ratios for the nth cycle,

$$Z'_n = \frac{t'_{fi} - t'_{fo}(P')}{t'_{fi} - t''_{fi}} \tag{10.7}$$

$$Z''_n = \frac{t''_{fo}(P'') - t''_{fi}}{t'_{fi} - t''_{fi}} \tag{10.8}$$

where $t'_{fo}(P')$ and $t''_{fo}(P'')$ are the fluid outlet temperatures at the end of the hot and

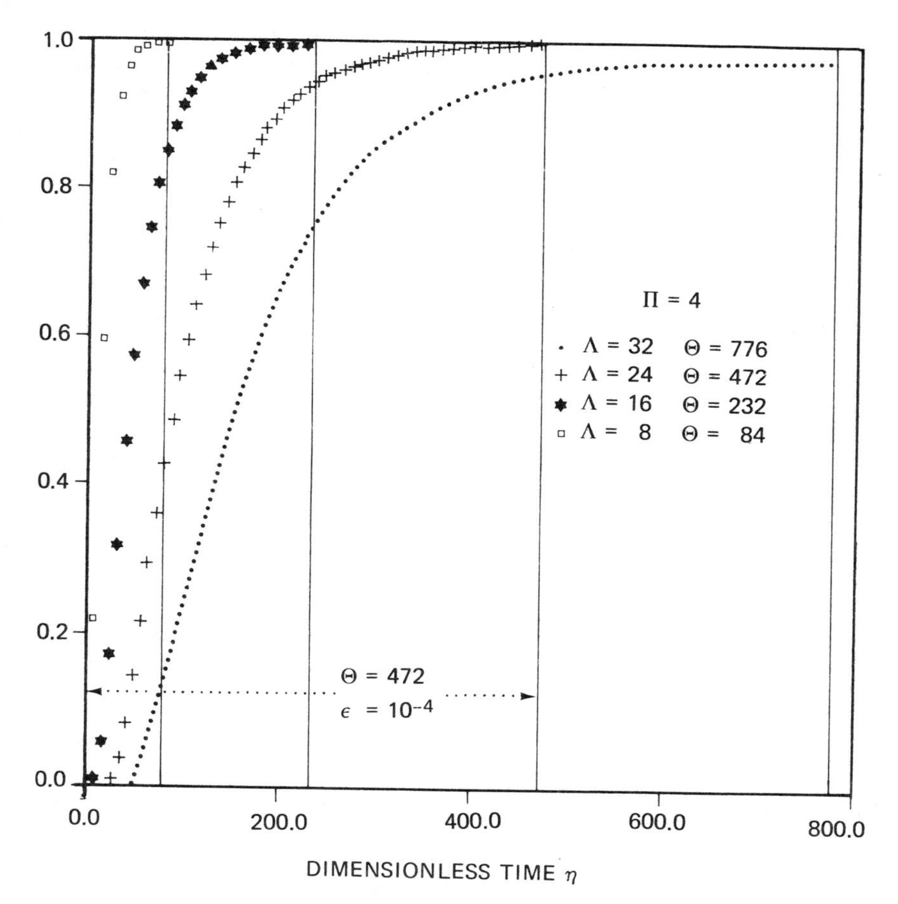

Figure 10.3 The effect of reduced length Λ on the response ϵf_1 and the time to equilibrium Θ. Vertical lines denote time when cyclic equilibrium is restored.

cold periods, respectively, of the nth cycle. He suggested that, provided that

$$|Z''_n - Z''_{n-1}| < \epsilon \tag{10.9}$$

$$(\text{or } |Z'_n - Z'_{n-1}| < \epsilon)$$

where ϵ is a small number, say, 10^{-4} or 10^{-5}, a computer simulation of a regenerator could be considered to have achieved cyclic equilibrium in the nth cycle. If the true equilibrium values of Z'' is α'', that is,

$$\lim_{n\to\infty} Z''_n = \alpha''$$

then this criterion presumes that

$$\left.\begin{array}{ll} \text{if} & |Z''_n - Z''_{n-1}| < \epsilon \\ \text{then} & |Z''_n - \alpha''| < \epsilon \quad \text{also} \end{array}\right\} \tag{10.10}$$

However, if the regenerator possesses a large thermal inertia and convergence of Z''_n toward α'' is very slow, although the values of Z''_n and Z''_{n-1} are close together, then the assumption (10.10) may not be valid and the simulation may be still some way from equilibrium.

An estimate α''^* of the value of α'' can be extrapolated [12] using Aitken's formula:

$$\alpha''^* = Z''_n - \frac{(Z''_n - Z''_{n-1})^2}{Z''_n - 2Z''_{n-1} + Z''_{n-2}} \tag{10.11}$$

Willmott and Burns [4] have found that for regenerators with large thermal inertia, provided that

$$|Z''_n - \alpha''| \approx \left| \frac{(Z''_n - Z''_{n-1})^2}{Z''_n - 2Z''_{n-1} + Z''_{n-2}} \right| < \epsilon \tag{10.12}$$

the simulation can be safely regarded as having reached cyclic equilibrium. The integration of the differential equations (5.7) and (5.8) over a complete cycle can be regarded as a function f acting on Z such that

$$Z''_n = f(Z''_{n-1}) \tag{10.13}$$

If this function could be expressed explicitly, then the iterative process would converge to α'' provided that a value of ξ existed on the range between Z''_{n-1} and α'' such that

$$|f'(\xi)| < 1 \tag{10.14}$$

However, the function f cannot be expressed explicitly. Instead, only local estimates $K_Z(n)$ of the derivative can be obtained:

$$f'(Z''_n) \approx K_Z(n) = \frac{f(Z''_n) - f(Z''_{n-1})}{Z''_n - Z''_{n-1}} = \frac{Z''_{n+1} - Z''_n}{Z''_n - Z''_{n-1}} \tag{10.15}$$

It can be shown that [12]

$$\alpha'' - Z''_{n+1} \approx -\frac{K_Z\{n\}}{1-K_Z\{n\}}(Z''_{n+1} - Z''_n) \tag{10.16}$$

It follows that, provided that $K_Z\{n\} < 0.5$,

$$|\alpha'' - Z''_{n+1}| \leqslant |Z''_{n+1} - Z''_n| \tag{10.17}$$

in which case the convergence criteria (10.9) can be used. Where the convergence is slow and

$$0.5 < K_Z\{n\} < 1 \tag{10.18}$$

an estimate of $|\alpha'' - Z''_n|$ can be used by expanding Eq. (10.16). This yields the Aitken convergence criterion (10.12).

Employing this revised convergence criterion as necessary, the times Θ to equilibrium have been determined and are displayed as part of Fig. 10.3. A value of $\epsilon = 10^{-4}$ was employed in the convergence criterion. Willmott and Burns developed a general formula for symmetric regenerators,

$$\Theta = 0.622\Lambda^2 + 4.144\Lambda + 6.464 \tag{10.19}$$

relating dimensionless time to equilibrium and reduced length for $\Lambda < 40$. Equation (10.19) assumes that $\epsilon = 10^{-4}$. For another value of $\epsilon = E$, Θ_E may be defined as the time to equilibrium; this can be estimated using

$$\Theta_E = \frac{\Theta \log_e E}{\log_e \epsilon} \tag{10.20}$$

For example, if $E = 10^{-2}$,

$$\Theta_E = \frac{\Theta \log_e (10^{-2})}{\log_e (10^{-4})}$$

and

$$\Theta_E = \frac{\Theta}{2} \tag{10.21}$$

Example 10.1 Employing the convergence criterion $\epsilon = 10^{-4}$, estimate how many cycles are required to restore equilibrium following a step change in inlet temperature for the symmetric regenerator configuration $\Lambda' = \Lambda'' = 10$, $\Pi' = \Pi'' = 0.5$.

Using Eq. (10.19), the time to equilibrium is Θ, where

$$\begin{aligned}\Theta &= (0.622)(10)^2 + (4.144)(10) + 6.464 \\ &= 62.2 + 41.4 + 6.446 \\ &= 110.046\end{aligned}$$

The total cycle time is $\Pi' + \Pi'' = 1.0$, and therefore approximately 110 cycles are required before equilibrium is reestablished.

Example 10.2 Employing the convergence criterion $\epsilon = 10^{-2}$, again estimate the number of cycles to equilibrium following a step change for $\Lambda' = \Lambda'' = 10$, $\Pi' = \Pi'' = 2$.

The total time to equilibrium is $110.046/2 = 55.023$ (see Eq. 10.21). The cycle time is $\Pi' + \Pi'' = 4$, and therefore to reestablish equilibrium $55.023/4 \approx 14$ cycles are required.

10.3.2 Unsymmetric-Balanced Regenerators

Can the results developed in Sec. 10.3.1 for symmetric regenerators be extended to the unsymmetric case, where $\Lambda' \neq \Lambda''$ and/or $\Pi' \neq \Pi''$? Considered first is the balanced case where $\Lambda'/\Pi' = \Lambda''/\Pi''$, where the hot period thermal ratio η'_{REG} is equal to that for the cold period, η''_{REG}. In discussing cyclic equilibrium performance, Hausen [13] proposed that the performance of an unsymmetric but balanced regenerator could be replicated by the performance of a symmetric regenerator whose reduced length and reduced period was Λ_H and Π_H, respectively, for both periods of the cycle. The Λ_H and Π_H are "harmonic means" of the reduced lengths Λ' and Λ'' and the reduced periods Π' and Π'' that define the unsymmetric-balanced system:

$$\frac{1}{\Lambda_H} = \frac{1}{2}\left(\frac{1}{\Lambda'} + \frac{1}{\Lambda''}\right) \qquad (10.22)$$

$$\frac{1}{\Pi_H} = \frac{1}{2}\left(\frac{1}{\Pi'} + \frac{1}{\Pi''}\right) \qquad (10.23)$$

Iliffe [14] first confirmed the acceptability of this parameterization using Λ_H and Π_H for the calculation of thermal ratios at cyclic equilibrium. The degree of unbalance can be specified by a parameter p, where $p = \Lambda'/\Lambda'' = \Pi'/\Pi''$, from which it follows that

$$\Lambda' = (p + 1)\frac{\Lambda_H}{2} \qquad (10.24)$$

and
$$\Pi' = (p + 1)\frac{\Pi_H}{2} \qquad (10.25)$$

and that $\Lambda'/\Pi' = \Lambda''/\Pi'' = \Lambda_H/\Pi_H$. Iliffe examined cases over the range $0 \leqslant \Pi'' \leqslant 24$ and $0 \leqslant \Lambda'' \leqslant 24$ for $p = 2$ and $0 \leqslant \Pi'' \leqslant 18$ and $0 \leqslant \Lambda'' \leqslant 18$ for $p = 3$.

Willmott and Burns [4] verified that the transient behavior of balanced but unsymmetric regenerators can also be adequately parameterized by the use of harmonic means. Figure 10.4 illustrates that the responses ϵf_1 and ϵf_2 to a step change on the hot side inlet gas temperature for the symmetric regenerator with $\Lambda = \Lambda_H = 24$ and $\Pi = \Pi_H = 8$ follow very closely the responses ϵf_1 and ϵf_2, respectively, for the corresponding unsymmetric-balanced cases for

1. $\Lambda' = 36$, $\Lambda'' = 18$, $\Pi' = 12$, and $\Pi'' = 6$ $(p = 2)$
2. $\Lambda' = 18$, $\Lambda'' = 36$, $\Pi' = 6$, and $\Pi'' = 12$ $(p = \frac{1}{2})$

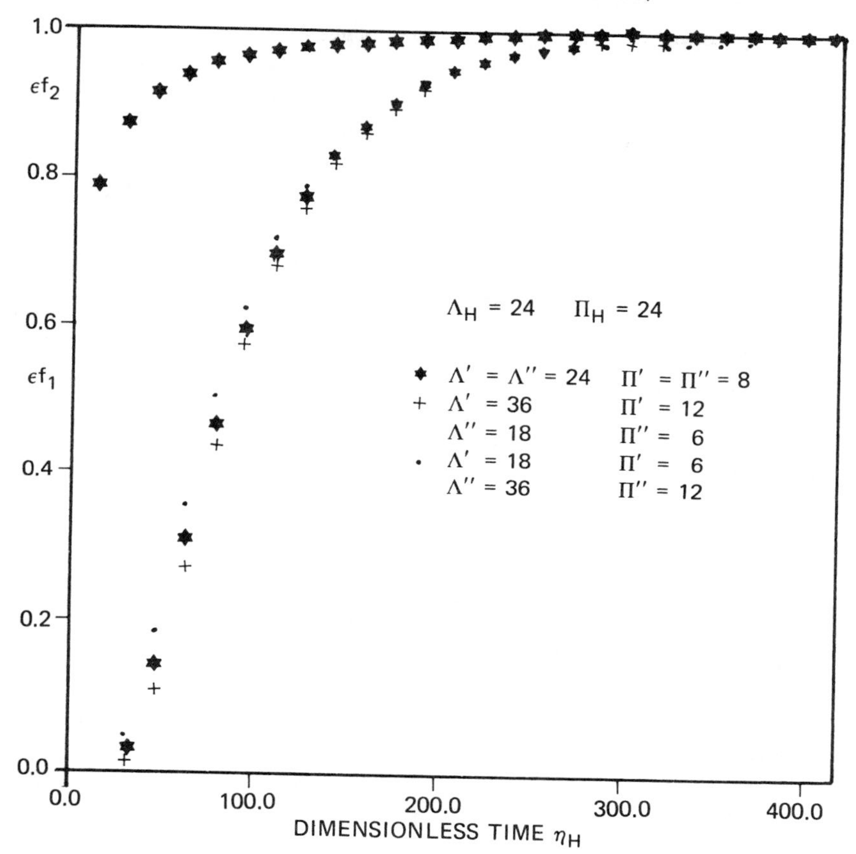

Figure 10.4 Responses ϵf_1 and ϵf_2 to step changes in hot inlet gas temperature.

Note that because the step change considered is applied to the hot inlet gas temperature, it is necessary to consider both cases 1 and 2, which correspond to $p = 2$ and $p = \frac{1}{2}$. Further, the response ϵf_1 for $p = 2$ and that for $p = \frac{1}{2}$ lie on opposite sides of the response ϵf_1 for the corresponding symmetric case ($p = 1$).

The time to equilibrium can be computed approximately using a modified form of Eq. (10.19) for the equivalent symmetric case

$$\Theta_H = 0.622\Lambda_H^2 + 4.144\Lambda_H + 6.464 \tag{10.26}$$

for $\Lambda_H \leqslant 40$, $\epsilon = 10^{-4}$. The actual dimensionless time Θ for the unsymmetric-balanced regenerator to regain cyclic equilibrium can then be found using Eq. (10.27):

$$\Theta = \frac{\Theta_H}{4}(1+p)\left(1+\frac{1}{p}\right) \tag{10.27}$$

The approximation (10.26) provides a relative accuracy of 10% or better.

10.3.3 Unbalanced Regenerators

In the balanced case, the forces imposing the results of the step change in operation in the hot period, in this case an increase in inlet gas temperature, might be regarded as being partly counterbalanced by opposing forces in the cold period. For example, the propagation of the heat from down the length of the checkerwork resulting from a step increase in the rate of heat input to the regenerator system will be delayed to a greater or lesser extent by the action of the cold gas passing through the packing in successive cold periods.

One might imagine the regenerator to be a reservoir of heat from which energy is drawn in the cold period of operation, and topped up again in the hot period. If the hot period inlet gas temperature is increased, the rate of energy supply to the reservoir is increased and the level of heat in the reservoir (for example, the average temperature of the checkerwork) begins to rise; consequently, the rate of heat extraction also begins to increase. However, the rise in level in the reservoir achieved in the hot period, over and above the normal topping up of the system, following this step change in operation is counteracted by the drawing-off of energy from the reservoir in the subsequent cold period.

When the regenerator is unbalanced, then $\Lambda'/\Pi' \neq \Lambda''/\Pi''$ and the two thermal ratios η'_{REG} and η''_{REG} are unequal. Further, the regenerator is unbalanced in the sense that the forces imposing temperature changes in the packing in the hot period are not matched by the counterforces operating in the cold period. In terms of our reservoir analogy, if the amount of heat supplied during the hot period at the start of the transient phase is significantly greater or smaller than the quantity of heat drawn off during the subsequent cold period, the system will move toward its equilibrium more rapidly than if the system were balanced. Therefore, when the regenerator is unbalanced, the packing responds more quickly to changes imposed upon it and it is necessary to introduce an extra parameter β, defined in Eq. (10.28), to account for the effect of the unbalance on the transient performance:

$$\beta = \frac{\Lambda'\Pi''}{\Lambda''\Pi'} \tag{10.28}$$

The use of the harmonic mean reduced length Λ_H is extended to this case to parameterize the factors governing transient performance for unbalanced regenerators. Displayed in Fig. 10.5 is the dependence of the time to regain cyclic equilibrium on Λ_H and the degree of unbalance β. It is quite clear that the regenerator exhibits greatest inertia when $\beta = 1$; the time to reestablish equilibrium is reduced significantly for $\beta = \frac{1}{2}, 2, \frac{1}{4}, 4$.

10.4. INTERPRETATION OF THE RELATION BETWEEN THE TRANSIENT PERFORMANCE OF A REGENERATOR AND ITS DIMENSIONLESS PARAMETERS

The reduced length (Λ' or Λ'') is a measure of the size of the regenerator relative to the thermal load imposed on it by the mass flow rate of the heating or cooling gas. On

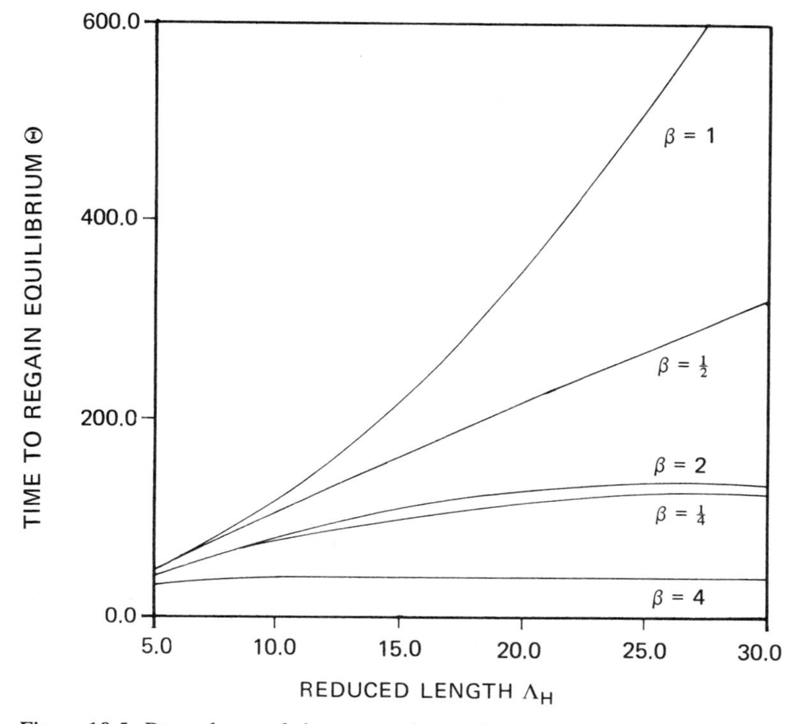

Figure 10.5 Dependence of time to regain equilibrium, Θ, on reduced length Λ_H and the measure of unbalance β, where $\epsilon = 10^{-4}$.

the other hand, the reduced period (Π' or Π'') is a measure of the hot or cold period length relative to heat capacity of the checkerwork and the available heating surface area. The reduced period embodies the idea that thin wall packing of small heat capacity but large heating surface area enforces short periods for efficient regenerator operation. On the other hand, a regenerator with a packing of smaller heating surface area but greater capacity to store heat in thicker walls may be operated with longer periods of operation.

The force imposing the oscillations in the packing temperature, and then any change in these oscillations resulting from step changes in operation, is derived from the alternate passage of hot and cold gas through the channels of the packing. The "size" of the regenerator can be measured by the product hA in either period of operation. However, the larger the product $m_f c_f$, the smaller the "relative size" of the regenerator. What we might expect is that the smaller the relative size of $\dot{m}_f c_f$—that is, relative to hA—the smaller the size of the force imposing oscillations of temperature, or changes in the oscillations of temperature of the packing. In other words, as we have observed in Sec. 10.3, the smaller the reduced length, the sooner the packing will respond to oscillations imposed on it. On the other hand, the larger the reduced length, the more resistant is the regenerator to the consequences of changes made to the operating conditions.

The reduced period does not reflect the "size" of the regenerator as does the reduced length. The reduced period represents in contrast the heat capacity of the system relative to the "force" hA imposing oscillations, and the period length. A change in operating conditions results in a change in the average temperature of the packing over a complete cycle, that is, a change in the average "level of heat" held in the reservoir. The total dimensionless time Θ taken for the change to be completed is determined by the heat capacity of the packing or the "size of the reservoir," and not the cycle time. It follows that this time Θ can be made up of a few long cycles or many short cycles.

However, in the same way that the actual cycle times are necessarily short for regenerators with a low packing heat capacity to surface area ratio, so the actual time for a regenerator to reach equilibrium following a step change in operation will be correspondingly small for such packings and vice versa.

10.5 STEP CHANGES IN GAS FLOW RATE

The first work on this problem was undertaken by Green [10] in 1967 when he examined, using a computer simulation of a regenerator, the effect of step changes in gas flow rate on a regenerator with a symmetric configuration *prior* to the step change. He observed that the total time required to reestablish equilibrium reached a maximum for changes of approximately 10–20% and then was reduced for larger step changes in flow rate (see Fig. 10.6).

Willmott and Burns extended these results for step changes in inlet gas temperature to step changes in gas flow rate by considering the state of the regenerator (symmetric or unsymmetric, balanced or unbalanced) *after* the step change, because the final condition toward which the regenerator will "converge" is largely determined by this final state. It will be shown that these results can be used to predict the transient response observed by Green previously.

The definition of reduced length and reduced period (see Chap. 5) indicate the dependence of Λ and Π on $\dot{m}_f$ and $\bar{h}$. If we consider the case where $\bar{h}$ is approximately linearly proportional to $\dot{m}_f$, Λ is independent of flow rate and the response of the regenerator following a step change in the hot period gas flow rate can be investigated by considering step changes in Π' alone. This simplifying assumption is examined in detail by Burns [15].

Whereas the responses ϵf_1 and ϵf_2 for step changes in inlet gas temperature are independent of the size of the step change, the magnitude of the step increase in gas flow rate cannot be neglected. Up to a certain threshold value of the percentage change made to the reduced period Π' and implicitly to the hot gas flow rate, the time Θ taken to reestablish cyclic equilibrium increases with the size of the step change. Beyond this threshold value, the transient performance is dependent only on the final operating conditions. The threshold value increases with reduced length. These features are displayed in Fig. 10.7 for the case where the final state of the regenerator is symmetric with $\Pi = 4$. Reduced lengths $\Lambda = 5, 10$, and 20 are considered.

If step changes beyond this threshold are considered alone, then the transient response is almost the same as that for step changes in inlet gas temperature. This

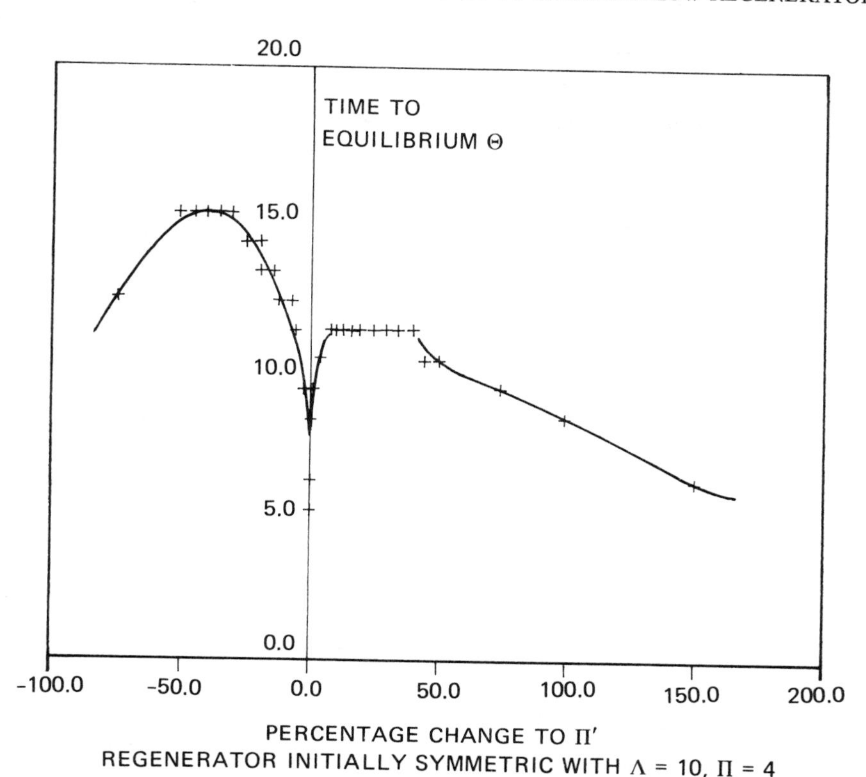

Figure 10.6 Green's representation of the effect of step changes in gas flow rate (symmetric case).

response can be parameterized in terms of the harmonic mean reduced length Λ_H and the unbalance factor β, discussed previously. It turns out one additional factor is required to relate the transient response to step changes in gas flow rate below the threshold value to the response to changes beyond the threshold. This factor is $\gamma\ (0<\gamma<1)$:

$$\gamma = \frac{\text{magnitude of step change}}{\text{magnitude of step change at threshold}} \tag{10.29}$$

However, this parameter γ has relatively little effect on the dimensionless time Θ to regain cyclic equilibrium ($\epsilon = 10^{-4}$ has been considered). The effect is displayed in Fig. 10.8 for $2 \leqslant \Lambda_H \leqslant 30$ and $\gamma = 1, \frac{1}{2}, \frac{1}{4}$. The important feature to notice in Figs. 10.7 and 10.8 is that the greater the reduced length Λ_H, the more sensitive is the transient response to the value of the factor γ.

When the step change in Π' (hot gas flow rate) is greater than the threshold value ($\gamma > 1$), the transient response of the regenerator is parameterized only by Λ_H and β. The threshold value is usually less than 50%; for $\Lambda = 10$, $\Pi = 4$ it is 10%, and for $\Lambda = 20$, $\Pi = 4$ it is 35%. The dependence of the transient response on reduced length, reduced period, and β for step changes of 50%, well beyond the threshold limit, is

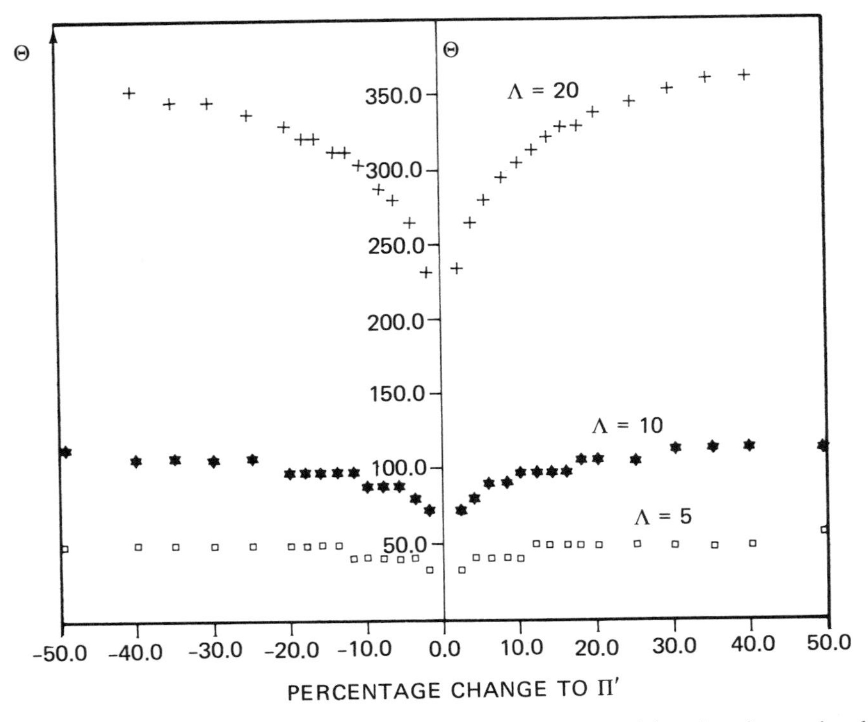

Figure 10.7 Dependence of time to equilibrium Θ on reduced length Λ for varying degrees of percentage change to Π′. (The final state of the regenerator is symmetric with reduced period $\Pi = 4$.) $\epsilon = 10^{-4}$.

therefore now presented. The results of Willmott and Burns are presented in Figs. 10.9 and 10.10.

For such 50% changes to Π′, for the symmetric case, ϵf_1 and Θ are observed in Fig. 10.9 to be independent of reduced period. The total reduced time to equilibrium can thus be expressed as a function of Λ only (as $\beta = 1$), Eq. (10.30), which is of similar type to Eq. (10.19) relating Θ to Λ for step changes in inlet gas temperature. Equations (10.19) and (10.30) are compared in Fig. 10.11.

$$\Theta = 0.72\Lambda^2 + 4.2\Lambda + 8 \qquad \text{for} \qquad \Lambda \leqslant 30 \tag{10.30}$$

The use of harmonic mean reduced lengths (and reduced periods) and the unbalance factor β for the transient response of unsymmetric regenerators to step changes in inlet gas temperature can be extended to the parallel treatment for gas flow rate step changes. Although small differences arise between the actual transient response computed for an unsymmetric-balanced regenerator and that of the corresponding "harmonic equivalent" symmetric regenerator, it is doubtful if these differences could be detected experimentally. In any event, the total time Θ computed to reestablish equilibrium for $\epsilon = 10^{-4}$ remains accurately represented by the equivalent symmetric case.

The effect of unbalance on the system, parameterized by β, is shown in Fig. 10.10, where the transient response is displayed for Λ_H fixed at 10.0 and $\beta = 1, 2, 0.5, 4, 0.25$. The same maximal inertia exhibited by a balanced regenerator ($\beta = 1$) is found for step changes both in gas flow rate and gas inlet temperature.

Green's initial observations, Fig. 10.6, can now be interpreted in terms of these results. Step changes are made to a regenerator that is initially symmetric in Green's work. For small step changes, the regenerator's final state will still be approximately symmetric and the time to reestablish cyclic equilibrium will increase with the size of the step change γ. For larger step changes, the regenerator becomes increasingly unbalanced, and the dimensionless time Θ to regain equilibrium decreases. There is therefore a value of the step change in gas flow rate for a particular regenerator configuration at which the time to equilibrium is a maximum beyond which the effect of the unbalance β becomes increasingly predominant.

10.6 FURTHER PARAMETERIZATION OF TRANSIENT RESPONSE

Although a number of parameters have been collected, namely, harmonic mean reduced length Λ_H, unbalance factor β, and relative step size γ, which almost fully

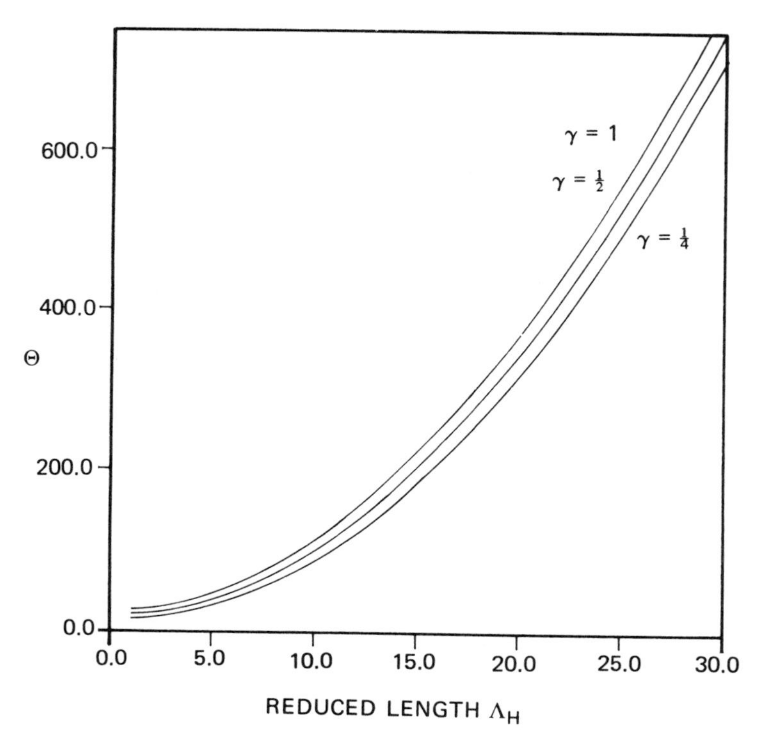

Figure 10.8 Dependence of time to equilibrium Θ on reduced length Λ_H and γ, the step size parameter.

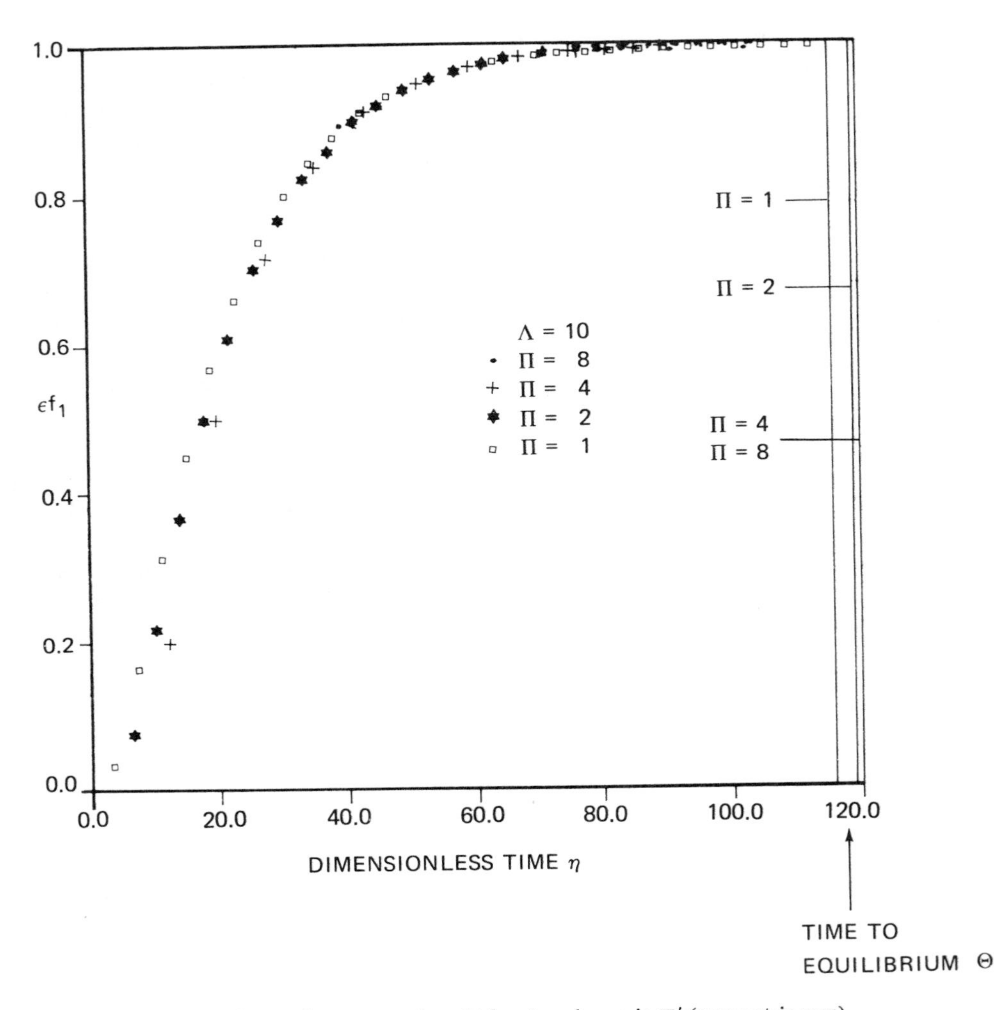

Figure 10.9 Dependence of response ϵf_1 on Π for step change in Π' (symmetric case).

describe the transient response of a regenerator to both step changes in inlet gas temperature and gas flow rate, it is most useful to be able to measure the rates of convergence of the responses ϵf_1 and ϵf_2. In Fig. 10.12 are plotted $\log_e(1 - \epsilon f_1)$ and $\log_e(1 - \epsilon f_2)$ against dimensionless time η, illustrating clearly the experimental nature of the transient responses [6].

The data points $n(\Pi' + \Pi'')$ have been joined by straight lines to emphasize the almost parallel gradients of $\log(1 - \epsilon f_1)$ and $\log(1 - \epsilon f_2)$. Apart from the initial few cycles after the step change, the responses can be represented approximately by ϵf_1^* and ϵf_2^*.

$$\epsilon f_1^* = 1 - \exp\left(-\frac{\eta_H - C_1}{\eta_c}\right) \tag{10.31}$$

$$\epsilon f_2^* = 1 - \exp\left(-\frac{\eta_H + C_2}{\eta_c}\right) \tag{10.32}$$

Equations (10.31) and (10.32) effectively map the overall asymptotic approach of the thermal regenerator toward a new equilibrium after a step change. These equations point toward a single "time constant" η_c as a measure of the single rate of convergence of both responses, ϵf_1 and ϵf_2, and hence of the thermal inertia of the regenerator. The time lag between the responses ϵf_1 and ϵf_2 is given by lag_H, where

$$\text{lag}_H = C_2 + C_1 \tag{10.33}$$

Willmott and Burns [6] examined the properties of these parameters η_c, C_1, C_2, and lag_H over a whole range of operating parameters for both step changes in inlet gas temperature and gas flow rate. Their conclusions can be summarized:

1. For any regenerator, the responses ϵf_1 and ϵf_2 generate the same time constants η_c.

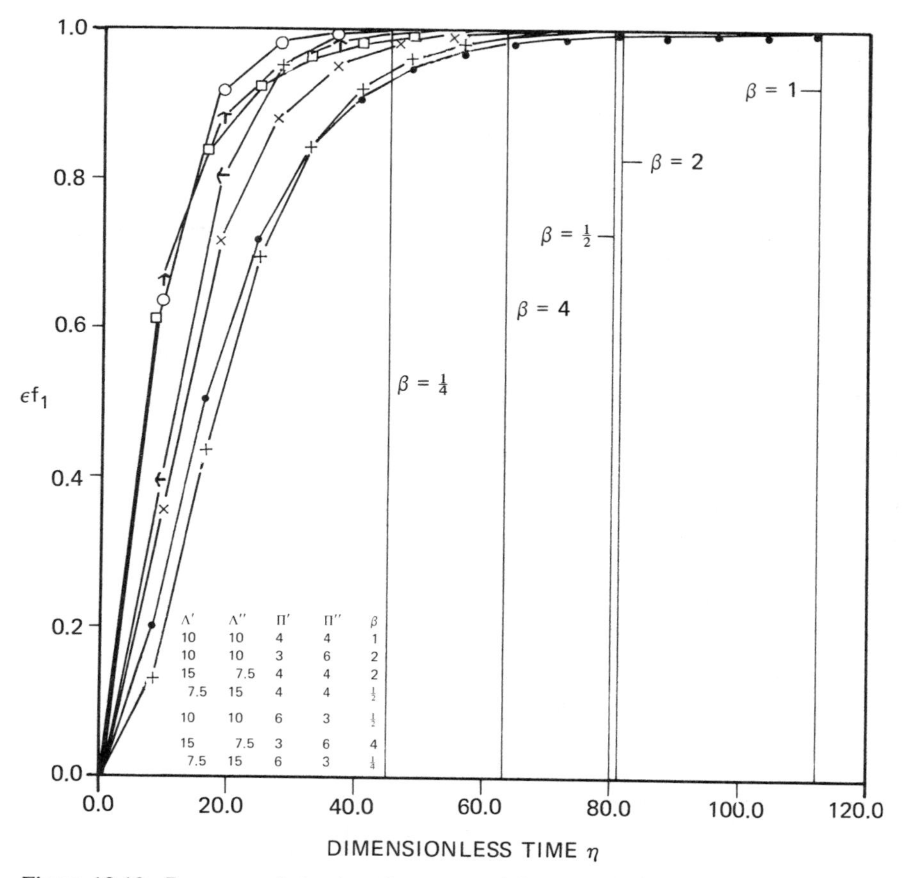

Figure 10.10 Response ϵf_1 to step change to Π' for unbalanced regenerator ($\Lambda'/\Pi' \neq \Lambda''/\Pi''$). Vertical lines denote time when equilibrium is restored.

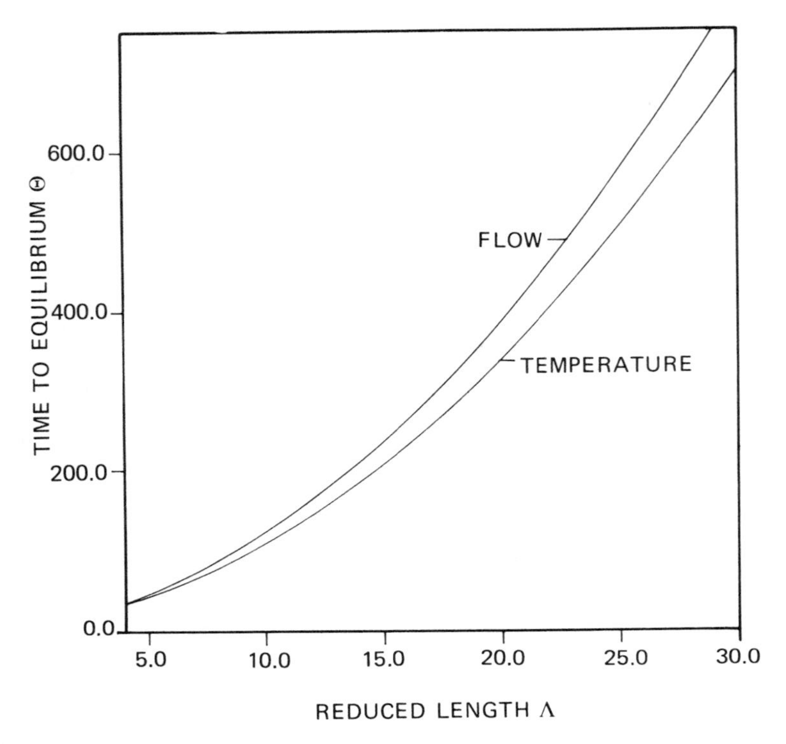

Figure 10.11 Comparison between the equations for time and equilibrium for step changes to inlet gas temperature and gas flow rate. $\epsilon = 10^{-4}$.

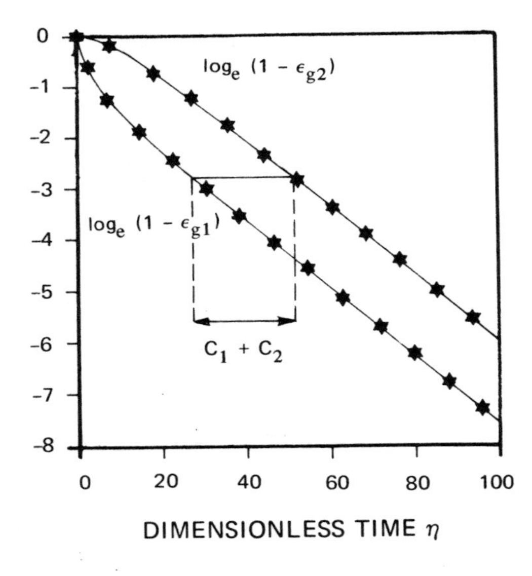

Figure 10.12 Responses $1 - \epsilon_{g1}$ and $1 - \epsilon_{g2}$ ($\log_e$ scale) to step change in inlet gas temperature.

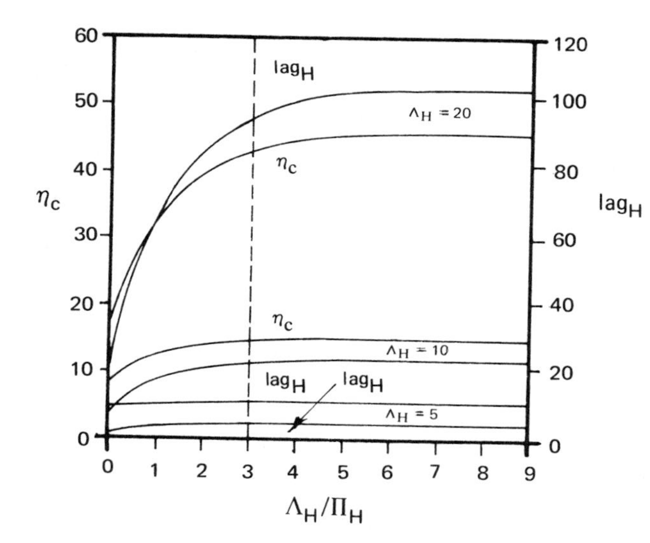

Figure 10.13 The dependence of lag_H and η_C on the ratio of reduced length to reduced period.

2. The parameters η_C, C_1, and C_2 and hence the time lag all exhibit independence of reduced period Π_H for $\Lambda_H/\Pi_H > 3$. This is illustrated in Fig. 10.13.
3. The time constant is determined by the final operating parameters of the regenerator in terms of the harmonic mean reduced length Λ_H and the unbalance of the regenerator, β. It is independent of the magnitude of the step change in either inlet gas temperature or gas flow rate.
4. The transient performance is determined by the final state of the regenerator; this determines the time constant for the response to a step change in either inlet gas temperature or gas flow rate.
5. The constants C_1 and C_2 both increase with reduced length for step changes in inlet gas temperature. This is shown in Fig. 10.14. For step changes in gas flow

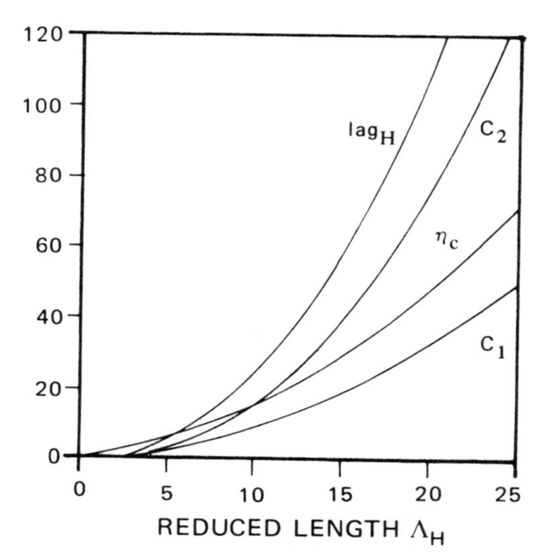

Figure 10.14 Dependence of lag_H, η_C, C_1, and C_2 on reduced length.

rate, the situation is complicated by the dependence of the time lag (lag_H) on the magnitude of the step change. However, for increasingly large step changes in gas flow rate, these changes cross a threshold value beyond which the value of the time lag (and C_1, C_2) approaches that computed for step changes in inlet gas temperature. The time lag is independent of the size of the inlet gas temperature step change.

The relationship between the time constant η_c and the harmonic mean reduced length Λ_H is depicted in Fig. 10.14; Willmott and Burns [6] fitted a quadratic function in Λ_H by the method of least squares to the values of η_c obtained by computer simulation, and this is shown as Eq. (10.34):

$$\eta_c \approx 0.0922\Lambda_H^2 + 0.489\Lambda_H + 0.928 \tag{10.34}$$

For a typical hot blast stove installation with $\Lambda_H = 12$, the time constant $\eta_c = 20$.

In Sec. 10.3.3, it was discussed that a regenerator exhibits greatest inertia when it is balanced, $\beta = 1$. This feature can be displayed in terms of the variation of the time constant η_c with regenerator imbalance β, as shown in Fig. 10.15. It will be observed that the sensitivity of the inertia of a regenerator system to changes in β becomes increasingly marked for larger values of reduced length Λ_H.

10.7 THERMAL INERTIA OF VARIABLE GAS FLOW REGENERATOR SYSTEMS

In Chap. 7, the operation of Cowper stoves in bypass main mode and in staggered parallel mode was discussed. Figures 7.2 and 7.3 illustrated how such systems operated. Willmott and Burns [6] made an initial examination of the thermal inertia of such systems by examining bypass main operation and considered a step increase in gas flow rate or inlet gas temperature in the hot period to a hot blast stove system

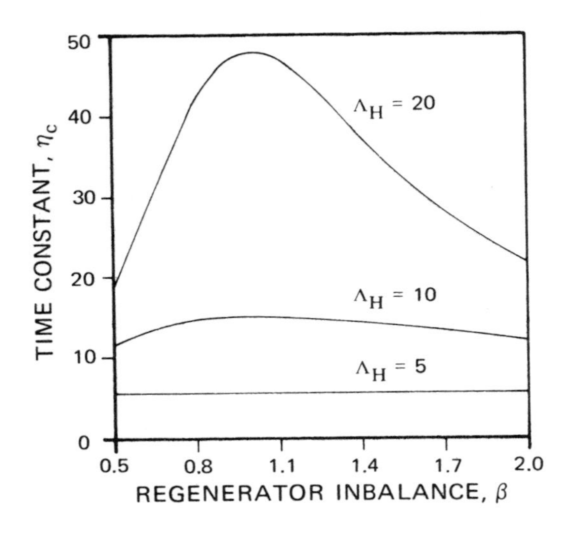

Figure 10.15 Dependence of time constant η_c on regenerator imbalance β.

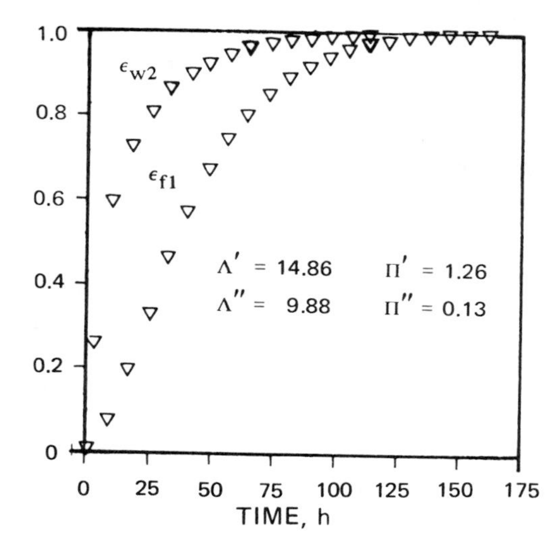

Figure 10.16 Responses ϵ_{f1} and ϵ_{w2} of a bypass main regenerator to step change in inlet gas temperature.

initially at cyclic equilibrium. During the transient phase while equilibrium is being reestablished, the target blast temperature t_B'' is left unaltered. The thermal load on the regenerator system remains unchanged (see Eq. 7.30), and the result of the step changes is therefore forced to be manifest in the hot side exit gas temperature and in the way $\dot{m}_f''(\tau)$ varies with time in successive cycles during transcience. A response ϵw_2 may be defined using the cold period average value of $\dot{m}_f''$ instead of t_{fo}'' in Eq. (10.2). Such a response together with ϵf_1 is shown in Fig. 10.16. The responses that relate to a regenerator of typical Cowper stove dimensions are identical to the form of those extracted for constant-flow regenerator systems, and approximations of the form of (10.30) and (10.31) may be employed.

Now, changes in flow rate, certainly over the range of change of $\dot{m}_f''$ within a period of bypass main operation, have little influence on reduced length Λ''. It is also known that for Cowper stoves, where typically $\Lambda_H/\Pi_H > 3$, variations in reduced period caused by changes in $\dot{m}_f''$ have no influence on the time constants and lags parameterizing the inertia of such a regenerator system. The cold period dimensionless parameters can thus be computed using flow rates and convective heat transfer coefficients averaged over each period under consideration, and the common basis of the inertia regenerator systems, namely, reduced length Λ_H and imbalance factor β, utilized for both constant flow and variable flow schemes of regenerator operation.

10.8 CONCLUSIONS

Whereas there has been much published work on the steady-state performance of regenerators, and to this reference has been made in earlier chapters, this work presented here is relatively new. Although London et al. undertook some preliminary work in this area between 1958 and 1964, it is only very recently that further computer simulations have been undertaken by Willmott and Burns. It is evident that these theoretical studies should be complemented now by experimental work. At present, we know only of the work of Chao in 1955 [16]; our theoretical results compare favorably with his experimental values of ϵf_2, obtained in work for the Ford Motor Company on an experimental rotary regenerator.

Mention has already been made of Cowper (hot blast) stoves that are required to operate under varying conditions. Improved control strategies can be evolved if notice is taken of the possible parameterizations of the thermal inertia of regenerator systems discussed in this chapter. It seems likely that some form of regenerative apparatus will be embodied in, for example, solar energy or geothermal energy collection systems. Solar systems must be able to accommodate a minimal thermal load and to smooth out the effect of inevitable short-term variations in the weather. An understanding of the transient performance of regenerators within such systems must be important if the full potential of the regenerator is to be exploited. It is hoped that the work presented in this chapter will contribute to any efforts being made in this area.

REFERENCES

1. R. M. Cima and A. L. London, "The Transient Response of a Two Fluid Counterflow Heat Exchanger–The Gas Turbine Regenerator," *J. Eng. Prog.*, vol. 80A, 1958, pp. 1169–1179.
2. A. L. London, F. R. Biancardi, and J. W. Mitchell, "The Transient Response of Gas Turbine Plant Heat Exchangers–Regenerators, Intercoolers, Precoolers and Ducting," *J. Eng. Prog.*, vol. 81A, 1959, pp. 433–448.
3. A. L. London, P. F. Sampsell, and J. G. McGowan, "The Transient Response of Gas Turbine Plant Heat Exchangers–Additional Solutions for Regenerators of the Periodic-Flow and Direct-Transfer Type," *J. Eng. Prog.*, vol. 86A, 1964, pp. 127–135.
4. A. J. Willmott and A. Burns, "Transient Response of Periodic-Flow Regenerators," *Int. J. Heat Mass Transfer*, vol. 20, 1977, pp. 753–761.
5. A. Burns and A. J. Willmott, "Transient Performance of Periodic-Flow Regenerators," *Int. J. Heat Mass Transfer,* vol. 21, 1978, pp. 623–627.
6. A. J. Willmott and A. Burns, "Periodic-Flow Regenerators: Parameter Identification for Transient Performance," *Proc. VI Int. Heat Transfer Conf.*, HX-19, 1978, pp. 297–302.
7. J. Beets and J. Elshout, "Control Model for a Hot Blast Stove System," *Int. Meeting on Iron and Steel*, Brussels, 1976.
8. I. Strausz, "Automatic Control of a Hot Stove System at a Blast Furnace, by Use of a Digital Computer," *Quart. J. Automatic Control (Belgium)*, vol. 1, 1970, pp. 15–20.
9. P. Zuidema, "Non-Stationary Operation of a Staggered Parallel System of Blast Furnace Stoves," *Int. J. Heat Mass Transfer*, vol. 15, 1972, pp. 433–442.
10. D. R. Green, "Some Aspects of Hot Blast Stove Development and Operation," M.Sc. thesis, UMIST, Manchester, 1967.
11. A. J. Willmott, "Digital Computer Simulation of a Thermal Regenerator," *Int. J. Heat Mass Transfer*, vol. 7, 1964, 1291–1302.
12. P. Henrici, *Elements of Numerical Analysis*, Wiley, New York, 1974, pp. 70–74.
13. H. Hausen, "Vervollstandigte Berechnung des Wärmeaüstausches in Regeneratoren," *Z. Ver. Deutsch. Ing.*, vol. 2, 1942, pp. 31–43.
14. C. E. Iliffe, "Thermal Analysis of the Counterflow Regenerative Heat Exchanger," *J. Inst. Mech. Eng.*, vol. 159, 1948, pp. 363–372 (war emergency issue no. 44).
15. A. Burns, "Heat Transfer Coefficient Correlations for Thermal Regenerator Calculations–Transient Response," *Int. J. Heat Mass Transfer*, vol. 22, 1979, pp. 969–973.
16. W. W. Chao, "Research and Development of an Experimental Rotary Regenerator for Automotive Gas Turbines," *Proc. Amer. Power Conference*, vol. 17, 1955, pp. 358–374.

CHAPTER

ELEVEN

HEAT STORAGE EXCHANGERS

11.1 INTRODUCTION

In the previous chapters the analyses have been concerned with heat storage units in which the heat transfer surfaces of the individual elements of the storage unit are in contact with identical fluid streams. There are many applications where two different streams are in contact with the storage material or where a heat flux is applied to one surface while the other surface is in contact with a fluid. Units operating under these conditions will be termed heat storage exchangers. The transient response of these units is discussed in this chapter.

In energy management systems there will be many occasions when it is necessary to work with several fluid streams which may, on an intermittent basis, result in heat being added and withdrawn from storage at a given instant of time. If the fluids are different and cannot be mixed, the design of the storage system becomes more complicated. An example of such a system is the use of the energy contained in the hot exhaust gases of an industrial process, waste heat recovery, to provide domestic hot water and space heating. The space heating could involve either a hot air, hot water, or steam unit. There would be three subsystems involved, one supplying energy while the other two withdraw energy. If the waste heat was available only during the working hours whereas the domestic hot water and space heating requirements were spread in a time-varying manner over the complete 24-h period, the system would fall into the category noted above.

Since the thermal energy is usually transported from one location to another by streams of liquids or gases, a heat storage exchanger can be used to store or release energy as well as to enable the thermal energy to be transferred directly between the flowing streams. The rate of heat transfer for steady-state operation is easily calculated by using conventional techniques described in basic heat transfer textbooks. If it is desired, at certain periods, to remove a larger amount of the energy from the hot stream than is required by the thermal load, the thermal energy demand is less than the energy available at that particular period of time, some of the energy will be stored and a transient analysis of the performance of the heat storage exchanger is necessary. The essential difference between a conventional heat exchanger and the heat storage exchanger is that the thin metallic walls in the heat exchanger have been replaced with a thick layer of material that has good heat storage characteristics. A sketch of a typical sensible heat storage exchanger is given in Fig. 11.1. The outlet temperature of both fluid streams will be time dependent, and in many applications the inlet temperature of one or both of the fluids may be time dependent.

There are many applications where the second type of heat storage exchanger, with a constant heat flux, is encountered. In England and several other European countries, off-peak electrical energy is stored at night for utilization as needed during the daytime hours to meet residential demands. The heat supplied by the electrical resistance heaters is stored in a solid. When space heating is required, air is heated by passing it through the storage unit and then circulating it through the home.

Another application that falls into the category of the second type is solar energy striking a wall that has an appreciable heat capacity. If a fluid passes over the other surface of the wall, a portion of the solar energy is stored in the wall whereas the remaining energy is transferred to the fluid stream. The response of this system is needed in the prediction of the cooling load of the structure during the hours of solar incidence and the heating load during the other periods.

11.2 TWO-FLUID HEAT STORAGE EXCHANGERS

11.2.1 Single-Blow Operation: Mathematical Model

The transient response of heat exchangers has been the topic of a number of papers. A summary of these studies has been presented by Schmidt [1]. Most authors have been concerned mainly with the response of the energy transporting fluids to step changes in either the mass flow rate or the inlet temperature of one of the fluids. The heat capacity of the separating wall has either been neglected or serious restrictions have been imposed when its contribution to the response of the heat exchanger is considered.

Studies that have taken the heat capacity of the wall into consideration are described by Cima and London [2] and London, Lampsell, and McGowan [3]. They have presented several solutions for the transient response of a direct transfer counterflow regenerator for a gas turbine application. These results have been obtained from an electrical analog. A study by Schmidt and Szego [4] used a finite-difference method to predict the transient response of a two-fluid heat storage exchanger for a solar energy system, but no general analysis of the response characteristics of these units was reported. Recently, Szego and Schmidt [5] presented a more general discussion of the transient response of thermal heat storage exchangers.

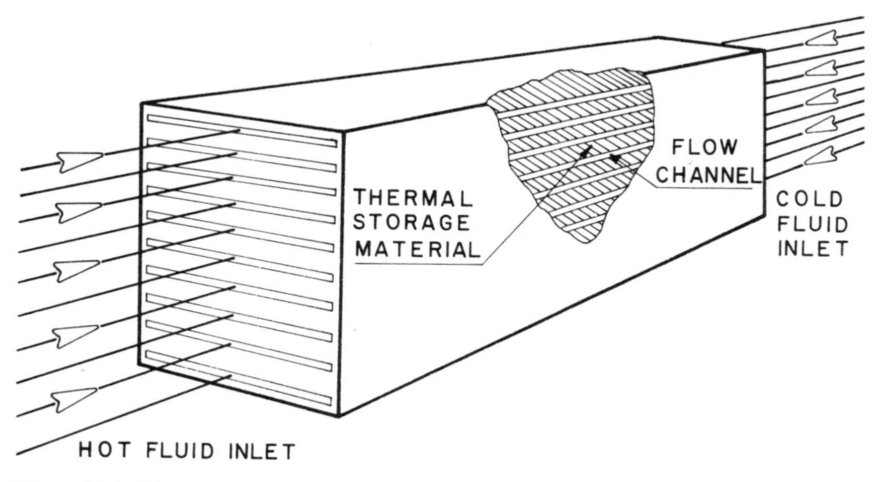

Figure 11.1 Schematic diagram of typical heat storage exchanger.

The flat slab type of heat storage exchanger, as shown in Fig. 11.1, is composed of a series of large aspect ratio, rectangular cross-sectional channels for the energy transporting fluids. The hot and cold fluids flow in a countercurrent fashion, in alternate channels, so that each slab of the storage material is in contact with both fluids. Planes of symmetry, surfaces across which no heat is transferred, are assumed to exist at the midlocation of the channels; this allows the analysis to be restricted to the section shown in Fig. 11.2.

The following assumptions will be used in establishing the mathematical model describing the physical systems under consideration:

1. Constant fluid and material properties
2. Uniform convective film coefficient
3. Two-dimensional conduction within the storage material

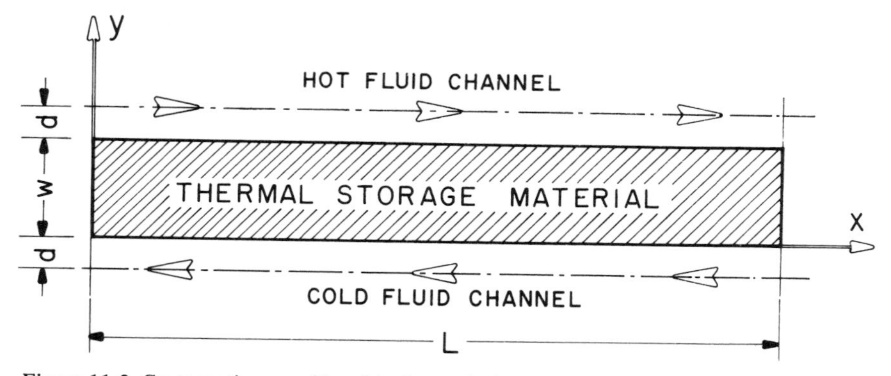

Figure 11.2 Cross section considered in the analysis.

4. Constant fluid mean velocity
5. Uniform initial temperature distribution in the storage material
6. Initiation of the process by a step change in the inlet temperature of the hot fluid

Based on these assumptions, the differential equations relating the temperature distribution in the storage material and the two fluid streams are the two-dimensional transient heat conduction equation for the storage material coupled to the one-dimensional conservation of energy equations for the energy transporting fluids. The resulting equations express the temperatures in the fluids and the storage material as a function of the two spacial coordinates, x and y, and the time, τ, and are

Cold fluid:
$$\rho_c c_c S_c \left(\frac{\partial t_c}{\partial \tau} - v_c \frac{\partial t_c}{\partial x}\right) = h_c P_c (t_{wc} - t_c) \tag{11.1}$$

Storage material:
$$\frac{1}{\alpha}\frac{\partial t_m}{\partial \tau} = \frac{\partial^2 t_m}{\partial x^2} + \frac{\partial^2 t_m}{\partial y^2} \tag{11.2}$$

Hot fluid:
$$\rho_h c_h S_h \left(\frac{\partial t_h}{\partial \tau} + v_h \frac{\partial t_h}{\partial x}\right) = h_h P_h (t_{wh} - t_h) \tag{11.3}$$

The associated initial and boundary conditions are

$\tau = 0 \qquad t_m = t_c = t_h = t_o$

$\tau > 0 \qquad x = 0 \qquad t_h = t_{hi} \qquad \dfrac{\partial t_m}{\partial x} = 0 \text{ for } 0 < y < w$

$\qquad\qquad x = L \qquad t_c = t_{ci} \qquad \dfrac{\partial t_m}{\partial x} = 0 \text{ for } 0 < y < w$

$\qquad\qquad y = 0 \qquad k_m \dfrac{\partial t_m}{\partial y} = h_c (t_{wc} - t_c) \text{ for } 0 \leqslant x \leqslant L$

$\qquad\qquad y = w \qquad k_m \dfrac{\partial t_m}{\partial y} = -h_h (t_{wh} - t_h) \text{ for } 0 \leqslant x \leqslant L$

For greater generality, the equations are nondimensionalized by introducing the following dimensionless parameters:

Biot numbers: $\quad \mathrm{Bi}_h \equiv \dfrac{h_h w}{k_m} \qquad \mathrm{Bi}_c \equiv \dfrac{h_c w}{k_m}$

Fourier number: $\quad \mathrm{Fo} \equiv \dfrac{\alpha \tau}{w^2}$

Dimensionless temperature:
$$T \equiv \frac{t - t_{ci}}{t_{hi} - t_{ci}} \tag{11.4}$$

Dimensionless flow length: $X \equiv \dfrac{x}{L}$

Dimensionless transverse coordinate: $Y \equiv \dfrac{y}{w}$

Thickness/length ratio: $V^+ \equiv \dfrac{w}{L}$

Capacity rate ratio for fluids: $C^+ \equiv \dfrac{\rho_c v_c S_c c_c}{\rho_h v_h S_h c_h} = \dfrac{\dot{m}_c c_c}{\dot{m}_h c_h} = \dfrac{E_c}{E_h}$

Convective resistance ratio: $R^+ \equiv \dfrac{1/P_c L h_c}{1/P_h L h_h} = \dfrac{\text{Bi}_h P_h}{\text{Bi}_c P_c} = \dfrac{\text{Bi}_h}{\text{Bi}_c}$

and $G_c^+ \equiv \dfrac{P_c k_m}{E_c} \qquad G_h^+ = \dfrac{P_h k_m}{E_h}$

The nondimensional equations for the fluids become

$$-\frac{1}{(V^+)^2}\left(\frac{\alpha}{w v_c}\right)\frac{\partial T_c}{\partial \text{Fo}} + \frac{\partial T_c}{\partial X} - \frac{G_c^+ \text{Bi}_c}{V^+}(T_c - T_{wc}) = 0 \tag{11.5}$$

and

$$\frac{1}{(V^+)^2}\left(\frac{\alpha}{w v_h}\right)\frac{\partial T_h}{\partial \text{Fo}} + \frac{\partial T_h}{\partial X} + \frac{G_c^+ \text{Bi}_c R^+ C^+}{V^+}(T_h - T_{wh}) = 0 \tag{11.6}$$

An order-of-magnitude analysis was used to evaluate the importance of each term in Eqs. (11.5) and (11.6). It was concluded that a negligible error was introduced when the transient terms in the fluid equations were abandoned. This assumption, which implies neglecting the heat capacities of the fluids, was previously discussed in [5] and can be further substantiated by the findings of Cima and London [2].

The complete set of governing equations in nondimensional form becomes

Cold fluid:
$$\frac{\partial T_c}{\partial X} - \frac{G_c^+ \text{Bi}_c}{V^+}(T_c - T_{wc}) = 0 \tag{11.7}$$

Storage material:
$$(V^+)^2 \frac{\partial^2 T_m}{\partial X^2} + \frac{\partial^2 T_m}{\partial Y^2} = \frac{\partial T_m}{\partial \text{Fo}} \tag{11.8}$$

Hot fluid:
$$\frac{\partial T_h}{\partial X} + \frac{G_c^+ \text{Bi}_c R^+ C^+}{V^+}(T_h - T_{wh}) = 0 \tag{11.9}$$

with the initial condition

$$\text{Fo} = 0 \qquad T_m = T_c = T_h = T_o$$

and the boundary conditions

$$X = 0 \qquad T_h = T_{hi} = 1 \qquad \frac{\partial T_m}{\partial X} = 0 \text{ for } 0 < Y < 1$$

$$X = 1 \qquad T_c = T_{ci} = 0 \qquad \frac{\partial T_m}{\partial X} = 0 \text{ for } 0 < Y < 1$$

$$Y = 0 \qquad \frac{\partial T_m}{\partial Y} = \mathrm{Bi}_c(T_{wc} - T_c) \text{ for } 0 \leqslant X \leqslant 1$$

$$Y = 1 \qquad -\frac{\partial T_m}{\partial Y} = \mathrm{Bi}_h(T_{wh} - T_h) \text{ for } 0 \leqslant X \leqslant 1$$

Equations (11.7) through (11.9) form a coupled set of equations and must therefore be solved simultaneously. This has been accomplished by a finite difference technique discussed by Szego and Schmidt [5]. The thermal storage material is subdivided in the x and y directions, which establishes a grid pattern. Equation (11.8) is rewritten in finite-difference form by using a backward difference for the time derivative and central differences for the spatial derivatives. Third-order orthogonal polynomials are fitted by least squares to T_{wh} and T_{wc} along the flow channels. These polynomials are then substituted into Eqs. (11.7) and (11.9), and an exact solution of the resulting equations for the fluids can be obtained at any instant of time. The conduction equation is solved by finite difference techniques to determine T_{wh} and T_{wc}. Since the fluid temperatures T_c and T_h appear in the boundary condition used to obtain a solution for the conduction equation, an iterative procedure is needed to satisfy the three partial differential equations and the appropriate boundary conditions. The procedure to be followed is very similar to that used in Chap. 3 and presented by Schmidt and Szego [6] for the determination of the transient response of a single-fluid heat storage unit.

11.2.2 Transient Response Parameters

Three new time-dependent dimensionless parameters are introduced as follows:

$$\epsilon_h = \frac{\text{heat transfer rate from hot fluid at time } \tau}{\text{heat transfer rate from hot fluid at steady state}} = \frac{E_h(t_{hi} - t_{ho})}{E_h(t_{hi} - t_{ho}^{ss})}$$

$$\epsilon_c = \frac{\text{heat transfer rate from cold fluid at time } \tau}{\text{heat transfer rate from cold fluid at steady state}} = \frac{E_c(t_{ci} - t_{co})}{E_c(t_{ci} - t_{co}^{ss})} \tag{11.10}$$

and

$$Q^+ = \frac{\text{amount of heat stored at time } \tau}{\text{amount of heat stored at steady state}} = \frac{Q}{Q^{ss}} = \frac{\overline{T}_m}{\overline{T}_m^{ss}} \tag{11.11}$$

These sets of dependent variables have been formulated because they have simple physical significance and are convenient for graphical presentation of the transient behavior of the heat storage exchanger. The basic analysis was carried out with the complete set of governing equations, Eqs. (11.7) through (11.9), and the associated boundary conditions. The results obtained indicated that for the case of practical interest, the effects of longitudinal heat conduction are negligible. Thus, it is possible to combine G^+ and V^+ into a single parameter, G^+/V^+.

The steady-state quantities used in the definition of ϵ_h, ϵ_c, and Q^+ can be obtained exactly. Under the assumption of steady state and negligible longitudinal heat conduction, the temperature distribution in the storage material in the y direction becomes linear at each location x along the flow channels. Hence, a mean wall temperature $\bar{t}_w$ is used and effective heat coefficients $1/R_h$ and $1/R_c$ are specified by combining one-half of the conductive resistance with each of the convective resistances of the fluid streams. The energy conservation equations thus become

$$\text{Hot fluid:} \qquad E_h \frac{dt_h}{dx} + R_h P_h (t_h - \bar{t}_w) = 0 \tag{11.12}$$

$$\text{Cold fluid:} \qquad E_c \frac{dt_c}{dx} - R_c P_c (t_c - \bar{t}_w) = 0 \tag{11.13}$$

$$\text{Storage material:} \qquad R_h P_h (t_h - \bar{t}_w) + R_c P_c (t_c - \bar{t}_w) = 0 \tag{11.14}$$

and for the entire system,

$$-R_c P_c L (t_c - \bar{t}_w) = R_h P_h L (t_h - \bar{t}_w) = UA(t_h - t_c) \tag{11.15}$$

Elimination of $\bar{t}_w$ from the above equations gives

$$E_h \frac{dt_h}{dx} + \frac{UA}{L}(t_h - t_c) = 0 \tag{11.16}$$

$$E_c \frac{dt_c}{dx} + \frac{UA}{L}(t_h - t_c) = 0 \tag{11.17}$$

Introducing the dimensionless parameters defined in the text and solving the resulting equations yields the steady-state temperature distributions in the fluids and the storage material. The temperature must be integrated over the volume of the storage material to obtain $\bar{T}_m^{ss}$. The final results are

$$T_{ho}^{ss} = \frac{C^+ e^M - e^M}{C^+ - e^M} \text{ for } C^+ \neq 1 \qquad T_{ho}^{ss} = \frac{1}{\text{NTU} + 1} \text{ for } C^+ = 1 \tag{11.18}$$

$$T_{co}^{ss} = \frac{1 - e^M}{C^+ + e^M} \text{ for } C^+ \neq 1 \qquad T_{co}^{ss} = \frac{\text{NTU}}{\text{NTU} + 1} \text{ for } C^+ = 1 \tag{11.19}$$

$$\bar{T}_m^{ss} = \frac{1}{2} \frac{[e^M (N/M - 2) - N/M]}{C^+ - e^M} \text{ for } C^+ \neq 1 \qquad \bar{T}_m^{ss} = \frac{1}{2}\left(1 + \frac{\text{N*}}{\text{NTU} + 1}\right) \text{ for } C^+ = 1 \tag{11.20}$$

where $M = \text{NTU}(1 - C^+)$ for $C^+ < 1$

$$M = \text{NTU}\left(\frac{1 - C^+}{C^+}\right) \text{ for } C^+ > 1$$

$$N = (C^+ + 1) + (C^+ - 1)N^*$$

$$N^* = \frac{(1/\text{Bi}_c) - (1/\text{Bi}_h)}{(1/\text{Bi}_c) + (1/\text{Bi}_h) + 1}$$

For all the cases under consideration in this study, the initial uniform temperature of the storage material is equal to the incoming temperature of the cold fluid, $T_o = T_{ci} = 0$, since the step change is imposed on the inlet temperature of the hot stream alone.

Performing an overall energy balance for the entire system without heat losses gives

$$\frac{dQ}{d\tau} = E_h(t_{hi} - t_{ho}) + E_c(t_{ci} - t_{co}) \tag{11.21}$$

Rearranging this equation and introducing the previously defined parameters yields a simple expression relating the three dependent variables, ϵ_h, ϵ_c, and Q^+:

$$K\frac{dQ^+}{d\text{Fo}} = \epsilon_h - \epsilon_c$$

where $K = \left(\dfrac{G^+}{V^+}\right)_c C^+ \left(\dfrac{\overline{T}_m^{ss}}{1 - T_{ho}^{ss}}\right)$

$\overline{T}_m^{ss}$ and T_{ho}^{ss} are given by Eqs. (11.20) and (11.18).

11.2.3 Results: Step Change in One Inlet Fluid Temperature

The temperature distribution in the storage material is a function of eight variables (X, Y, G_c^+, V^+, Bi_c, C^+, R^+, and Fo), whereas the temperature distribution in the fluids is a function of seven variables, with Y being omitted. The outlet fluid temperature and the total heat storage are functions of six variables (G_c^+, V^+, Bi_c, C^+, R^+, and Fo). The dependent variables ϵ_h, ϵ_c, and Q^+ are each functions of the five independent variables Bi_c, $(G^+/V^+)_c$, C^+, R^+, and Fo. It becomes apparent that it is impossible to generate a set of tables or an interpolation program that will be general enough to be used to predict the transient behavior of a large number of heat storage unit configurations. General trends in the transient response of the heat storage exchangers, however, will be presented for a unit where both the cold stream and the hot stream have the same physical properties (same fluid) and both channels have the same physical dimensions.

For turbulent flow, the convective resistance ratio can be reduced to the ratio of the mean velocity of the two streams raised to some exponent. The exponent commonly employed by most turbulent flow correlations is -0.8, and the relationship obtained between R^+ and C^+ for turbulent flows is

$$R^+ = (C^+)^{-0.8} \tag{11.22}$$

ϵ_h, ϵ_c, and Q^+ are functions of only four independent dimensionless parameters, Bi_c, $(G^+/V^+)_c$, C^+, and Fo.

A more convenient way of presenting results is obtained by plotting ϵ_h, ϵ_c, and Q^+ against $\mathrm{Fo}/(G^+/V^+)_{\max}$. In addition, the commonly used steady-state parameter NTU proved to be a very useful parameter. The final results for ϵ_h, ϵ_c, and Q^+ will be presented as functions of Bi_c, C^+, NTU, $\mathrm{Fo}/(G^+/V^+)_{\max}$. The following relationships between the variables hold:

$$\mathrm{Bi}_h = \frac{\mathrm{Bi}_c}{(C^+)^{0.8}} \tag{11.23}$$

$$\mathrm{NTU} = \frac{AU}{E_{\min}} = \left(\frac{G^+}{V^+}\right)_{\max} \left(\frac{1}{\mathrm{Bi}_c} + \frac{1}{\mathrm{Bi}_h} + 1\right)^{-1} \tag{11.24}$$

and

$$\frac{\mathrm{Fo}}{(G^+/V^+)_{\max}} = \frac{E_{\min}}{(\rho c V)_m}\tau \tag{11.25}$$

From the definitions of ϵ_c and Q^+, it is evident that both quantities range from a magnitude of zero at the beginning of the process, $\tau = 0$, to a value of unity as steady-state operation of the storage unit is approached. Moreover, ϵ_h, starting from some value larger than one, will also approach unity at steady-state conditions. The mathematical model described previously in the analysis does not account for the events that take place in the time interval between the introduction of the hot fluid when the step change in inlet temperature occurs and the moment when those particles leave the storage unit. However, this time interval, usually referred to as the dwell time τ_d, is typically very small compared to the characteristic time for the overall process. It is apparent that the hot fluid outlet temperature will remain at its initial value for a period of time less than the dwell time for the hot fluid. During this time interval the first fluid particles of the hot stream will have viewed a constant wall temperature as they move through the heat storage exchanger. The solution of Eq. (11.9) with a constant wall temperature $T_{wh} = T_o = 0$ is

$$T_{ho} = \exp\left[-\mathrm{Bi}_c \left(\frac{G^+}{V^+}\right)_c (C^+)^{0.2}\right] \qquad \text{for} \quad \tau = \tau_d \tag{11.26}$$

whereas $T_{ho} = T_o = 0$ for $0 < \tau < \tau_d$. It is apparent that an abrupt change in the hot fluid temperature leaving the heat storage unit will occur, and thus ϵ_h will experience an abrupt drop. A listing of T_{ho} at $\tau = \tau_d$ is presented in Table 11.1. Figure 11.3 shows a typical plot of the results to illustrate the foregoing conclusions. The area bounded by the curves for ϵ_h and ϵ_c, with $\mathrm{Fo}/(G^+/V^+)_{\max}$ varying from zero to some specified steady-state value, is given by $KQ^+/(G^+/V^+)_{\max}$, as can be inferred by inte-

Table 11.1 Operating characteristics of heat storage exchangers

C^+	NTU	Bi_c	T_{co}^{ss}	T_{ho}^{ss}	$\bar{T}_m^{ss}$	$T_{ho}(\theta_d)$
0.1	0.1	0.1	0.095	0.990	0.838	0.924
		1.0			0.706	0.872
		10.0			0.557	0.707
	1.0	0.1	0.619	0.938	0.871	0.452
		1.0			0.785	0.256
		10.0			0.688	0.001
	3.0	0.1	0.939	0.906	0.920	0.092
		1.0			0.876	0.017
		10.0			0.827	0.000
1.0	0.1	0.1	0.091	0.909	0.5	0.810
		1.0				0.741
		10.0				0.301
		100.0				0.000
	1.0	0.1	0.500	0.500	0.5	0.122
		1.0				0.050
		10.0				0.000
		100.0				0.000
	3.0	0.1	0.750	0.250	0.5	0.002
		1.0				0.000
		10.0				0.000
		100.0				0.000
10.0	0.1	0.1	0.0095	0.905	0.139	0.889
		1.0			0.175	0.876
		10.0			0.333	0.760
		100.0			0.455	0.182
	1.0	0.1	0.062	0.381	0.114	0.308
		1.0			0.138	0.267
		10.0			0.241	0.064
		100.0			0.320	0.000
	3.0	0.1	0.094	0.061	0.072	0.029
		1.0			0.084	0.019
		10.0			0.136	0.000
		100.0			0.177	0.000

grating Eq. (11.21). Since at steady-state conditions Q^+ approaches 1.0, the shaded area in Fig. 11.3 is equal to $K/(G^+/V^+)_{\max}$.

The transient response of a heat storage exchanger with a $C^+ = 1$ is presented in Fig. 11.4. The transient response of the hot fluid, as indicated by the value of ϵ_h, is shown to be strongly dependent on Bi_c for NTU = 0.1, but as the NTU is increased this dependency tends to disappear. In fact, for NTU = 3.0, the three curves for $\mathrm{Bi}_c = 0.1$, 1.0, and 10.0 coincide. It will also be noted that for specific values of C^+ and NTU, the value of the abscissa where the steady-state condition is reached is nearly independent of Bi_c. The response of the cold fluid, ϵ_c, seems to be a weak function of Bi_c for $C^+ = 1.0$. Figures 11.5 and 11.6 present similar curves for $C^+ = 0.1$ and $C^+ = 10.0$. For $C^+ = 10$, the trends are similar to those for $C^+ = 1.0$. The results are quite different for $C^+ = 0.1$, and indicate that as the NTU is changed the curves tend to keep their general shape. The influence of Bi_c is equally detectable for the range of NTUs examined.

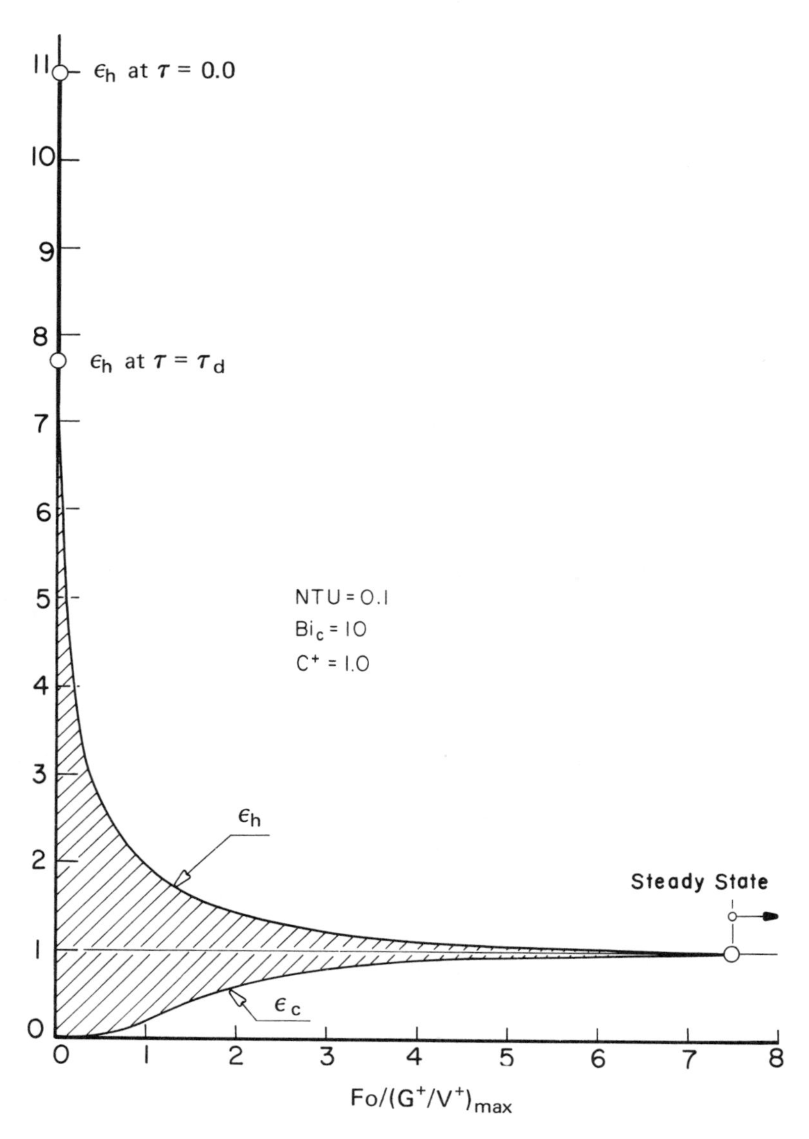

Figure 11.3 Typical representation of the results [5].

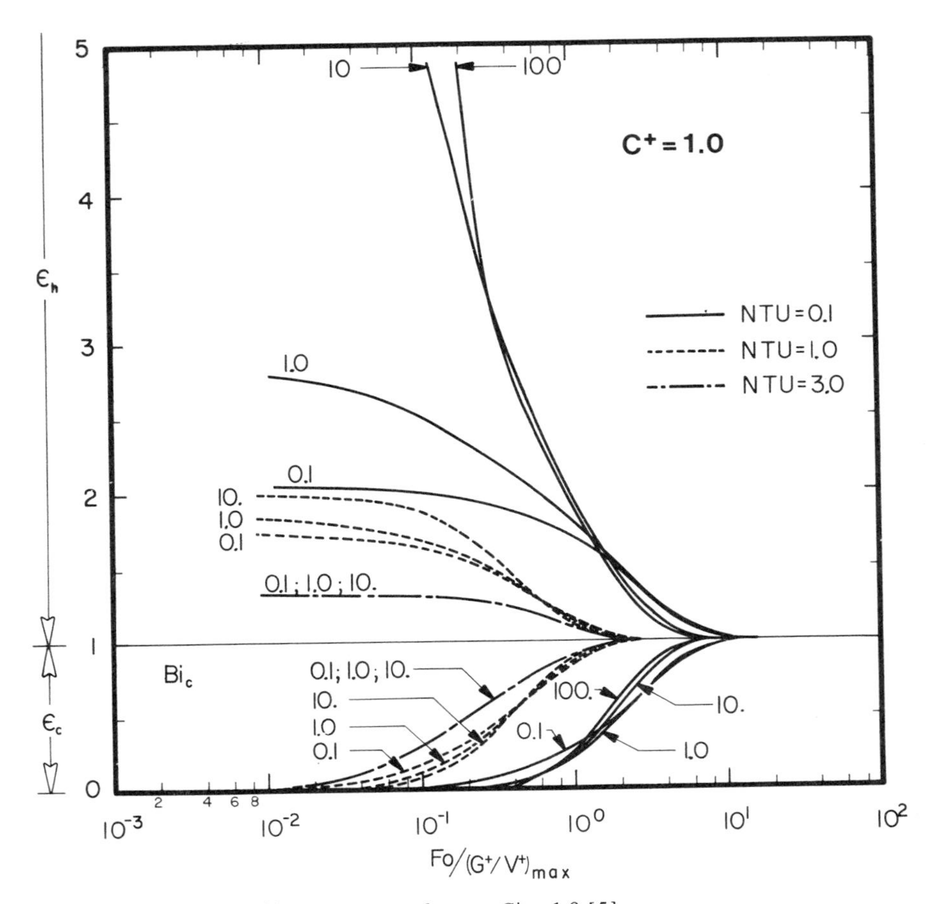

Figure 11.4 Response of heat storage exchanger, $C^+ = 1.0$ [5].

The results for a constant NTU, with varying Bi_c and C^+, are shown in Figs. 11.7 and 11.8. The trends are similar for the two NTUs considered. The lines representing ϵ_c for the various cases tend to overlap, but there is a clear distinction between the lines for ϵ_h, which indicates that large values of ϵ_h are obtained for the smaller values of C^+. The influence of the Bi_c over these results is similar to that discussed previously in conjunction with Figs. 11.4 through 11.6.

In Figs. 11.9 through 11.11, the fraction of the steady-state heat storage Q^+ is presented as a function of $\mathrm{Fo}/(G^+/V^+)_{\max}$ for various combinations of the other independent variables. For all cases, the constant NTU curves shift to the right as the NTU decreases. The influence of the Bi_c is similar to that discussed earlier, since ϵ_h, ϵ_c, and Q^+ are functionally related. It is important to realize that these results cover a wide range of possible physical situations, but the actual amount of heat stored and the transient fluid outlet temperatures can be evaluated only in conjunction with the steady-state quantities used to define ϵ_h, ϵ_c, and Q^+. Table 11.1 shows the steady-state parameters for the range of the variables covered in this study.

Example 11.1 A heat storage exchanger similar to that shown in Fig. 11.1 is 5.8 m (19.03 ft) long and 1.0 m (3.28 ft) wide. The channels are both 1.9 cm (0.748 in) high and are separated by a Feolite storage material 8 cm (3.15 in) thick. The hot gases flow through the unit at a mass flow rate of 0.5 kg/s per channel (1.10 lb_m/s per channel) and the mass rate of flow of the cold fluid is 0.05 kg/s per channel (0.11 lb_m/s per channel). The hot gases enter at 80°C (176°F), whereas the cold gas is at 10°C (50°F). The initial temperature of the unit is 10°C (50°F). Consider both gas streams to have the physical properties of air at 45°C (113°F) and determine the transient response of the heat storage unit.

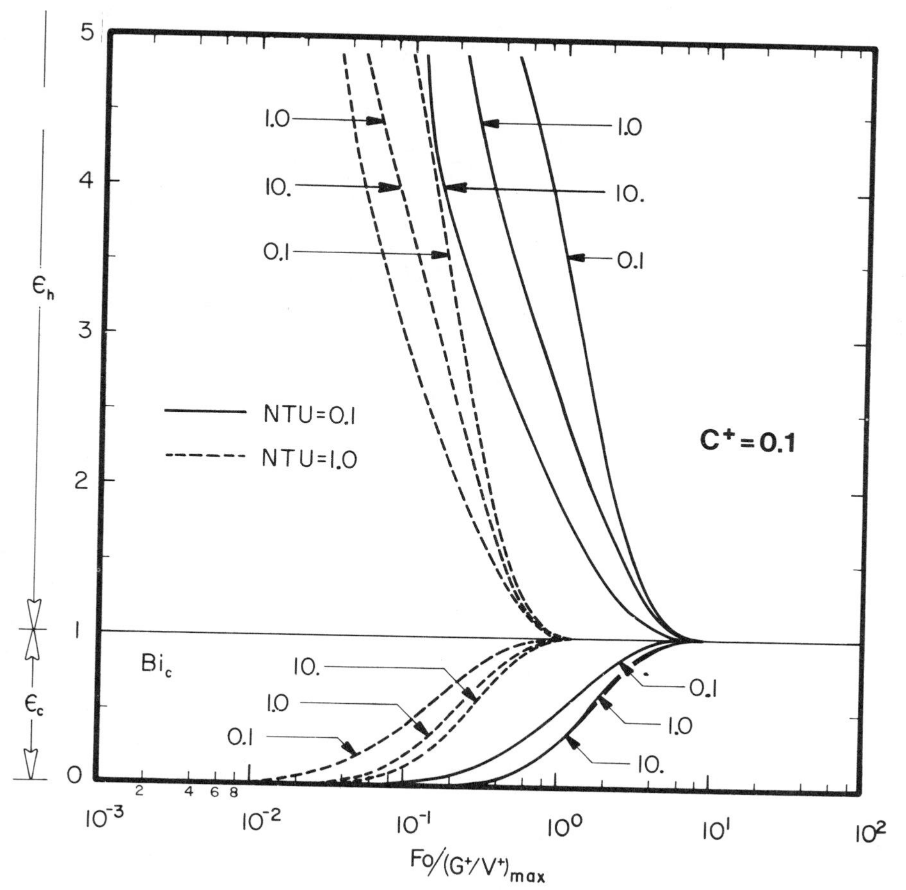

Figure 11.5 Response of heat storage exchanger, $C^+ = 0.1$ [5].

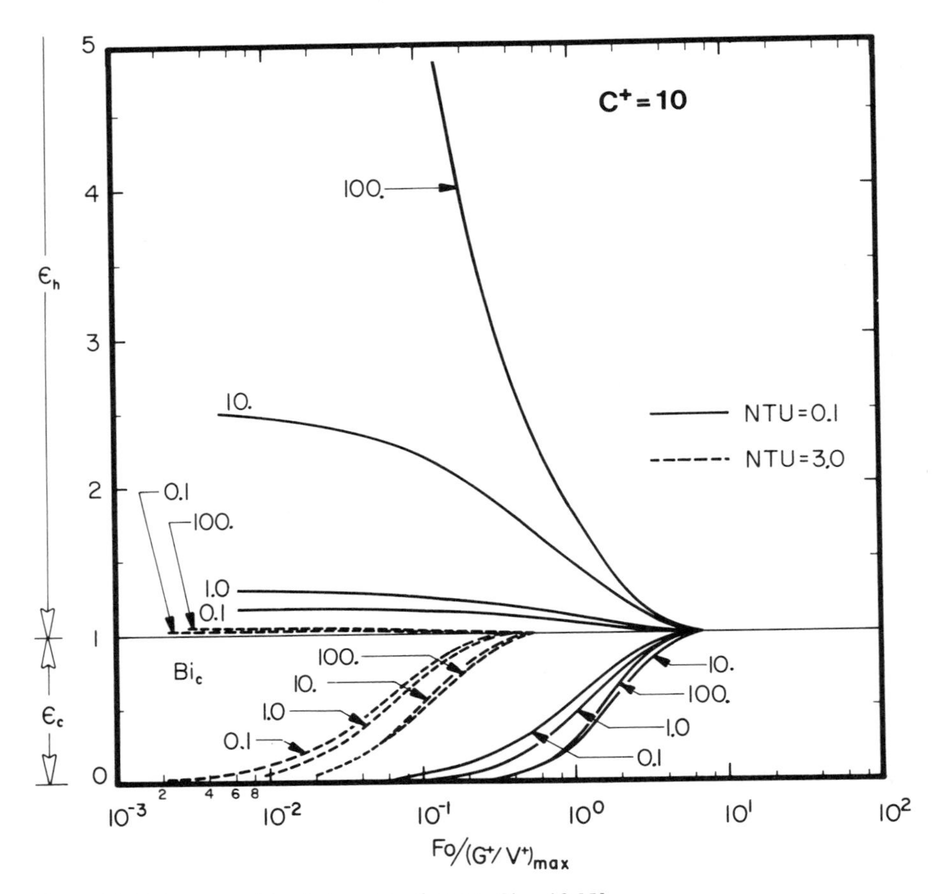

Figure 11.6 Response of heat storage exchanger, $C^+ = 10$ [5].

SOLUTION The physical properties of the gases and the Feolite are

Air @ 45°C:

$$c_f = 1.011 \text{ kJ/kg °C} \quad (0.24 \text{ Btu/lb}_m \text{ °F})$$

$$\rho_f = 1.106 \text{ kg/m}^3 \quad (0.069 \text{ lb}_m\text{/ft}^3)$$

$$\mu_f = 2.016 \times 10^{-5} \text{ kg/m s} \quad (4.877 \times 10^{-2} \text{ lb}_m\text{/ft h})$$

$$k_f = 0.0276 \text{ W/m °C} \quad (0.0159 \text{ Btu/ft h °F})$$

Feolite @ 45°C:

$$c_m = 0.92 \text{ kJ/kg °C} \quad (0.2197 \text{ Btu/lb}_m \text{ °F})$$

$$\rho_m = 3900 \text{ kg/m}^3 \quad (243.47 \text{ lb}_m\text{/ft}^3)$$

$$k_m = 2.1 \text{ W/m °C} \quad (1.213 \text{ Btu/ft h °F})$$

The Reynolds numbers for the flows are

Hot fluid: $$\mathrm{Re}_h = \frac{GD_h}{\mu} = \frac{(26.3)(0.038)}{(2.016 \times 10^{-5})} = 49{,}573$$

Cold fluid: $$\mathrm{Re}_c = 4957$$

Both flows will be considered to be turbulent, and the convective film coefficients are calculated using

$$Nu = 0.023(\mathrm{Re})^{0.8} Pr^{0.4}$$

The values obtained are

$$h_h = 84.4\ \mathrm{W/m^2\ °C} \quad \text{and} \quad h_c = 13.4\ \mathrm{W/m^2\ °C}$$

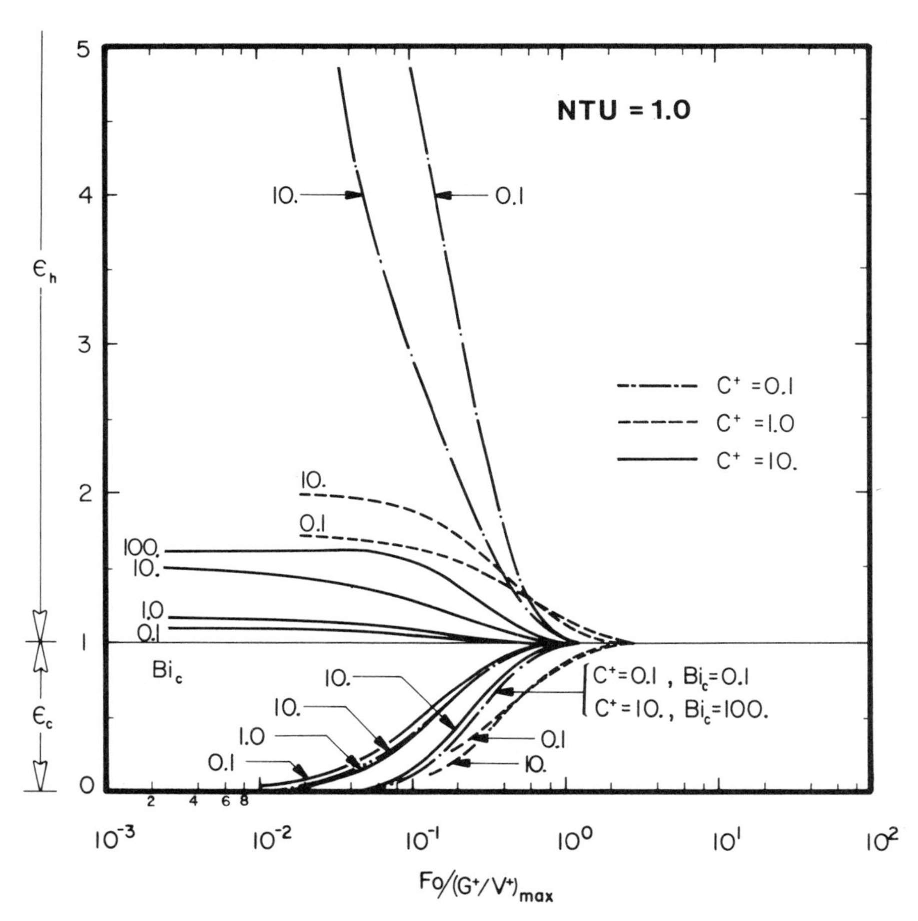

Figure 11.7 Response of heat storage exchanger, NTU = 1.0 [5].

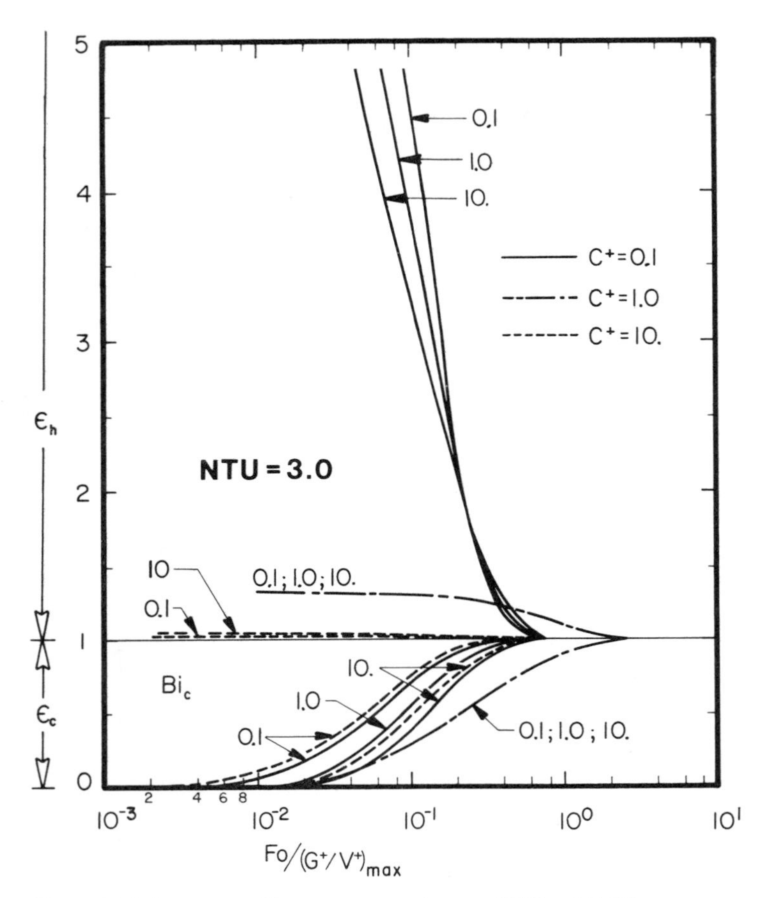

Figure 11.8 Response of heat storage exchanger, NTU = 3.0 [5].

The Biot numbers are

$$\mathrm{Bi}_h = \frac{h_h w}{k_m} = \frac{84.4(0.08)}{2.1} = 3.22$$

and $\mathrm{Bi}_c = 0.51$

The capacity ratio C^+ is

$$C^+ = \frac{E_c}{E_h} = 0.1$$

and $$\left(\frac{G^+}{V^+}\right)_{\max} = \frac{P_h k_m L}{\dot{m}_c c_c\, w} = \frac{(1.0)(2.1)(5.8)}{(0.025)(1011.0)(0.08)} = 6.024$$

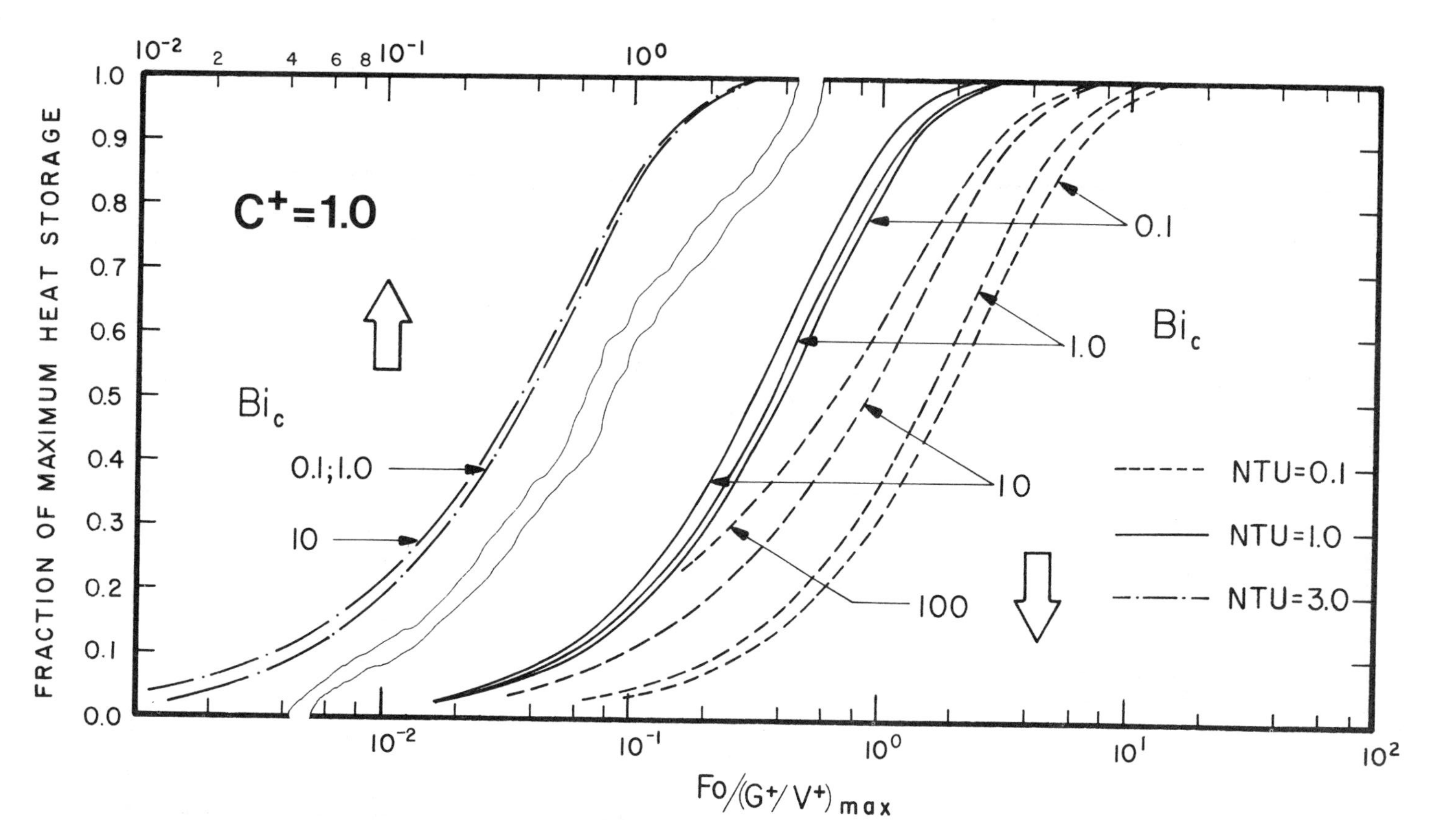

Figure 11.9 Fraction of steady-state heat storage, $C^+ = 1.0$ [5].

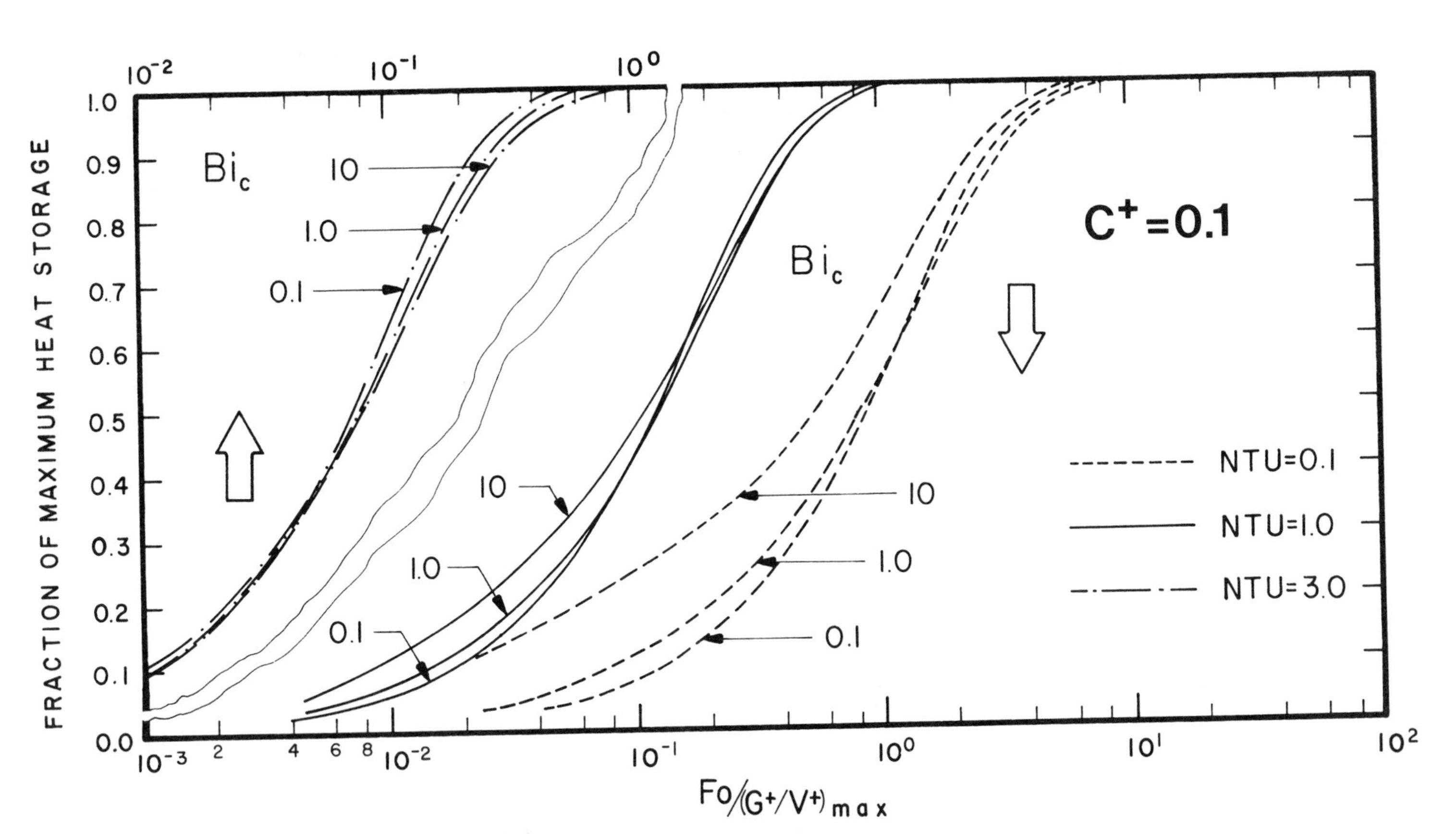

Figure 11.10 Fraction of steady-state heat storage, $C^{+} = 0.1$ [5].

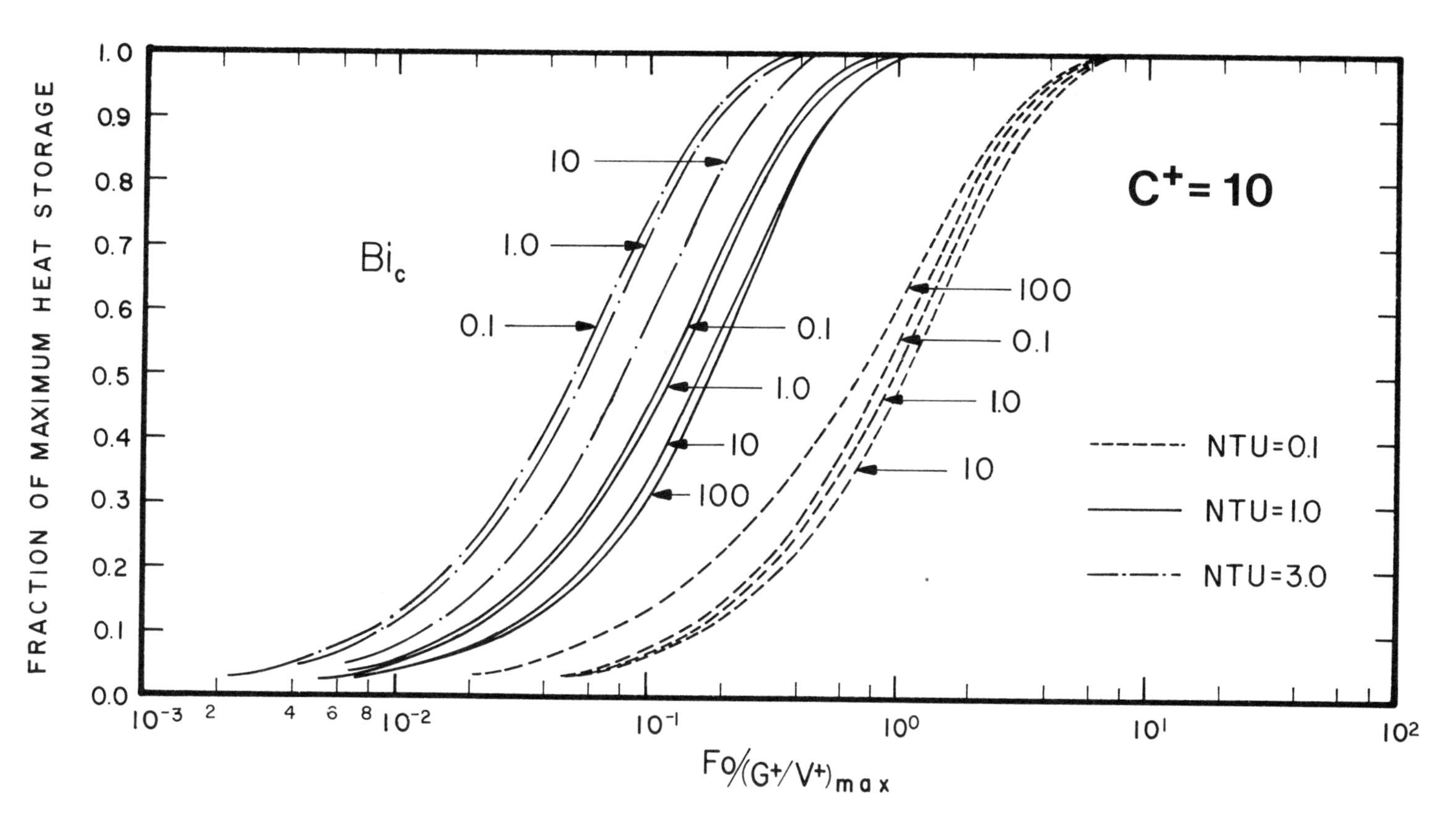

Figure 11.11 Fraction of steady-state heat storage, $C^+ = 10$ [5].

where $\dot{m}_c$ is the mass flow rate for one-half the fluid channel. The value of the NTU is

$$\text{NTU} = \left(\frac{G^+}{V^+}\right)_{\max} \left(\frac{1}{\text{Bi}_c} + \frac{1}{\text{Bi}_h} + 1\right)^{-1}$$

$$= 6.024 \left(\frac{1}{0.51} + \frac{1}{3.22} + 1\right)^{-1}$$

$$= \frac{6.024}{3.27} = 1.84$$

The value of $\text{Fo}/(G^+/V^+)_{\max}$ is

$$\frac{\text{Fo}}{(G^+/V^+)_{\max}} = \frac{\dot{m}_c c_c \tau}{(\rho c V)_m} = \frac{(0.025)(1011)\tau}{(3900.0)(920)(1)(0.08)(5.8)}$$

where τ is in seconds.

The results obtained from extrapolation of the data shown in Fig. 11.5 and interpolation of the data shown in Fig. 11.10 are shown in Figs. 11.12 and 11.13. The steady-state conditions of the heat storage exchanger are determined using Eqs. (11.18), (11.19), and (11.20). The results obtained are

Steady state:

Heat storage $Q_{ss} = 1.0 \times 10^5$ kJ (0.947 × 10^5 Btu)

Temperature of hot fluid out = 74.2°C (165.56°F)

Temperature of cold fluid out = 67.8°C (154.04°F)

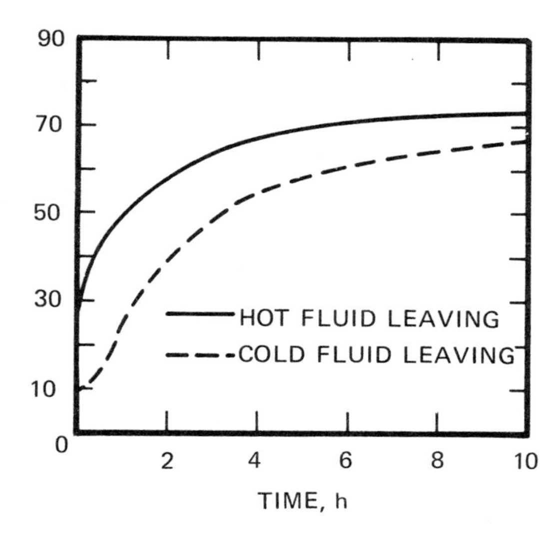

Figure 11.12 Response of fluid temperatures leaving.

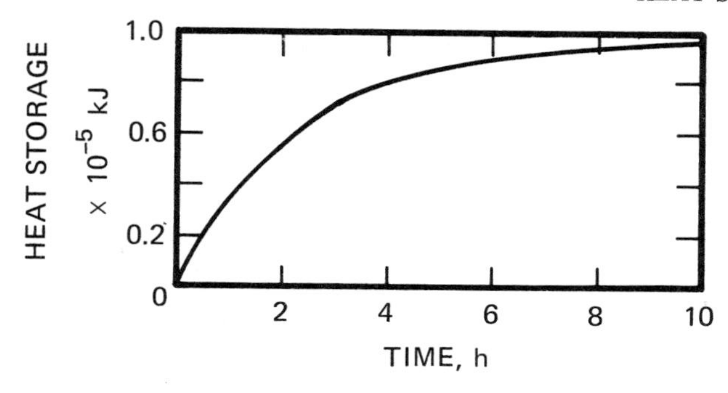

Figure 11.13 Heat storage.

11.2.4 Method of Superposition for Arbitrary Variations in Inlet Fluid Temperature for a Two-Fluid Heat Storage Exchanger

A very important extension of the results presented in Sec. 11.2.1 is possible because of the linear nature of the governing differential equations and the associated boundary conditions. The procedures to be discussed, usually referred to as Duhamel's method or the method of superposition, can be used to predict the transient response of a heat storage exchanger with an arbitrary varying fluid inlet temperature. The techniques to be used are quite similar to those described in Chap. 4. The time-varying fluid inlet temperature is approximated by a series of step changes in fluid inlet temperature. The total response of the unit is the sum of the responses of the unit to each of the step changes in the inlet fluid temperature. The procedure to be followed will be illustrated by allowing the inlet temperature of the hot fluid to experience a series of step changes while the cold fluid's inlet temperature remains constant.

As noted, the response of a heat storage exchanger to a step change in the inlet temperature of the hot fluid will be used to determine the response of a unit with an arbitrary timewise variation in the inlet temperature of the hot fluid. Since the geometry of the heat storage exchanger, the storage material, and the mass flow rates of the two fluid streams are fixed, the nondimensional steady-state outlet fluid temperatures, T_{ho}^{ss} and T_{co}^{ss}, and the mean steady-state temperature, $\bar{T}_m^{ss}$, will not change as a result of the change in the inlet fluid temperature. From the basic definitions previously introduced,

$$\epsilon_h = \frac{t_{hi} - t_{ho}}{t_{hi} - t_{ho}^{ss}} = \frac{T_{hi} - T_{ho}}{T_{hi} - T_{ho}^{ss}}$$

and since $T_{hi} = 1.0$, the expression for the nondimensional temperature of the hot fluid leaving can be expressed as

$$T_{ho} = 1.0 - \epsilon_h(1.0 - T_{ho}^{ss}) \tag{11.27}$$

Similarly,

$$\epsilon_c = \frac{t_{ci} - t_{co}}{t_{ci} - t_{co}^{ss}} = \frac{T_{ci} - T_{co}}{T_{ci} - T_{co}^{ss}}$$

The values of $T_{ci} = 0.0$, so that we obtain

$$T_{co} = \epsilon_c T_{co}^{ss} \tag{11.28}$$

The nondimensional temperature T is defined as

$$T \equiv \frac{t - t_{ci}}{t_{hi} - t_{ci}}$$

(*Note:* Since $t_{ci} = t_o$, the definition of T is consistent with that previously introduced.)

Thus
$$t_{ho} = t_{ci} + (t_{hi} - t_{ci})T_{ho}$$
$$= t_{ci} + (t_{hi} - t_{ci})[1.0 - \epsilon_h(1 - T_{ho}^{ss})] \tag{11.29}$$

and
$$t_{co} = t_{ci} + (t_{hi} - t_{ci})T_{co}$$
$$= t_{ci} + (t_{hi} - t_{ci})\epsilon_c T_{co}^{ss} \tag{11.30}$$

It is important to note that ϵ_h and ϵ_c are functions of the nondimensional time, Fo.

A typical timewise variation in the inlet temperature of the hot fluid is shown in Fig. 11.14. It is subdivided into step changes in the temperature, and the times at which the changes occur are identified by a superscript. Each step in inlet temperature of the hot fluid represents a subproblem of the basic type described in Sec. 11.2. The initial temperature distribution in the heat storage exchanger and the inlet temperature of the hot fluid is t_o for the first subproblem and 0 for all other subproblems. The

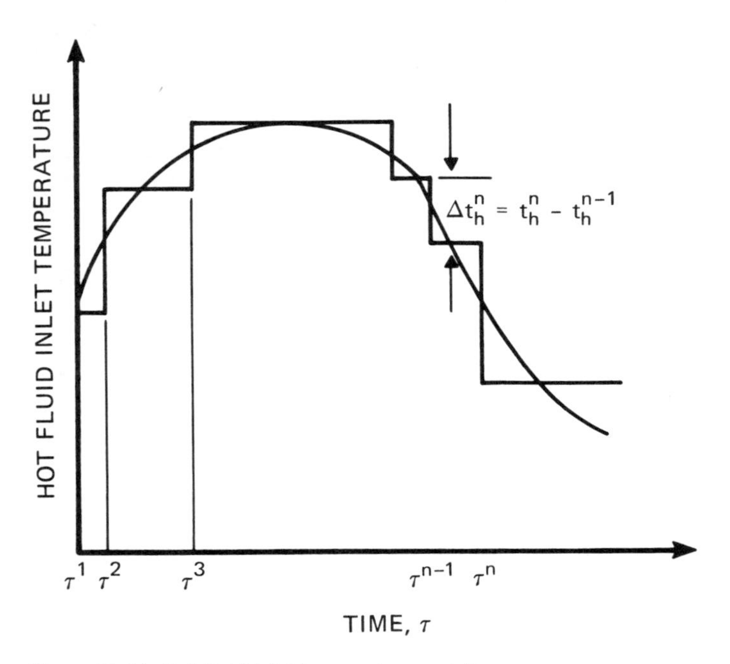

Figure 11.14 Hot fluid inlet temperature variation.

step change in the inlet temperature of the hot fluid at time τ^n is defined as

$$\Delta t_h^n = t_{hi}^n - t_{hi}^{n-1}$$

The contributions of each subprogram to the outlet temperature of the hot fluid stream are

$$\begin{aligned}
t_{ho}^1 &= t_o + \Delta t_h^1 [1.0 - \epsilon_h^1(1.0 - T_{ho}^{ss})] \\
t_{ho}^2 &= \Delta t_h^2 [1.0 - \epsilon_h^2(1.0 - T_{ho}^{ss})] \\
&\vdots \\
t_{ho}^n &= \Delta t_h^n [1.0 - \epsilon_h^n(1.0 - T_{ho}^{ss})]
\end{aligned} \tag{11.31}$$

where ϵ_h^n is evaluated at $\text{Fo}^n = \alpha(\tau - \tau^n)/w^2$. The outlet temperature of the hot fluid is

$$t_{ho} = t_{ho}^1 + t_{ho}^2 + \cdots + t_{ho}^n \tag{11.32}$$

The expression for the outlet temperature of the cold fluid is

$$t_{co} = t_{co}^1 + t_{co}^2 + t_{co}^3 + \cdots + t_{co}^n \tag{11.33}$$

where

$$\begin{aligned}
t_{co}^1 &= t_o + \Delta t_h^1 \epsilon_c^1 T_c^{ss} \\
t_{co}^2 &= \Delta t_h^2 \epsilon_c^2 T_c^{ss} \\
&\vdots \\
t_{co}^n &= \Delta t_h^n \epsilon_c^n T_c^{ss}
\end{aligned} \tag{11.34}$$

ϵ_c^n is again evaluated at $\text{Fo}^n = \alpha(\tau - \tau^n)/w^2$.

When the heat storage exchanger experiences a single step in the inlet temperature of the hot fluid, the heat storage per unit width is evaluated from the following relationship:

$$Q = Q^{ss}Q^+ = [wL\rho_m c_m(\bar{t}^{ss} - t_o)]Q^+ \tag{11.35}$$

From the initial condition $t_o = t_{ci}$, the basic expression for the nondimensional mean temperature of the storage material at steady-state conditions reduces to

$$\bar{T}_m^{ss} = \frac{\bar{t}^{ss} - t_{ci}}{t_{hi} - t_{ci}} = \frac{\bar{t}^{ss} - t_o}{t_{hi} - t_o} \tag{11.36}$$

The heat storage can be expressed as

$$Q = (wL\rho_m c_m \bar{T}_m^{ss})(t_{hi} - t_o)Q^+ \tag{11.37}$$

For a series of step changes in the inlet temperature of the hot fluid, the expression becomes

$$\begin{aligned}
Q &= Q^1 + Q^2 + Q^3 + \cdots + Q^n \\
&= (wL\rho_m c_m \bar{T}_m^{ss})[\Delta t_h^1 (Q^+)^1 + \Delta t_h^2 (Q^+)^2 + \cdots + \Delta t_h^n (Q^+)^n]
\end{aligned} \tag{11.38}$$

where $(Q^+)^n$ is evaluated at $\text{Fo}^n = \alpha(\tau - \tau^n)/w^2$.

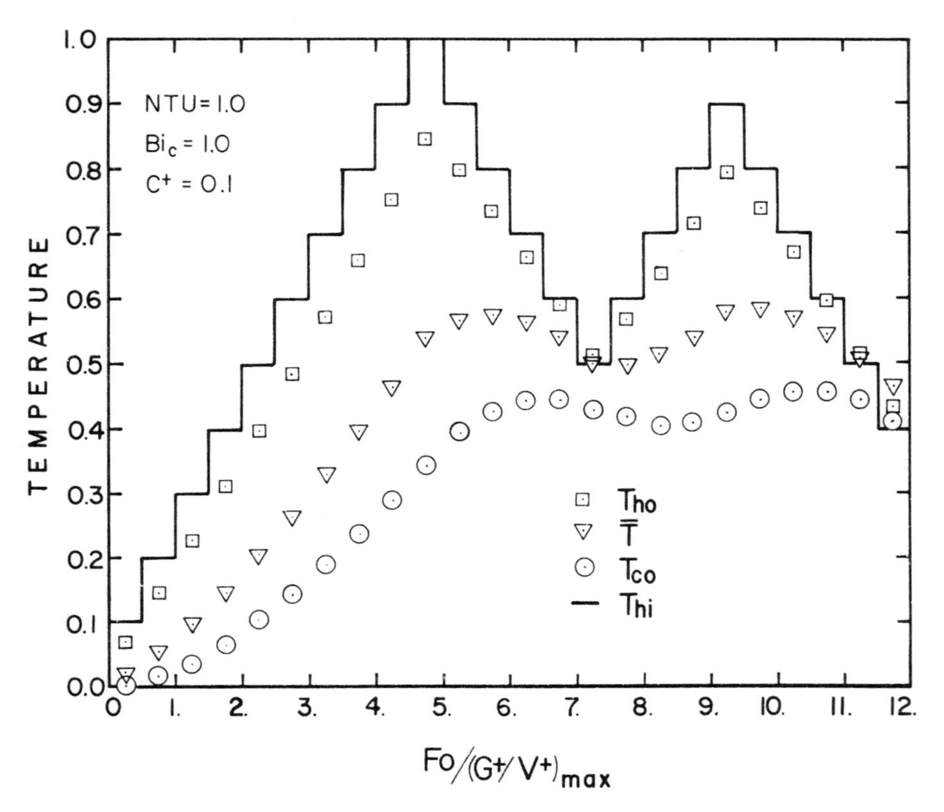

Figure 11.15 Response of the heat storage exchanger to a time variation in the hot fluid inlet temperature [5].

This procedure has been used to calculate the transient response of a heat storage exchanger when the hot fluid inlet temperature varies in the manner shown in Fig. 11.15. The responses of the unit, namely, the mean temperature of the storage material and the outlet temperatures for the hot and cold fluid streams, are also displayed in Fig. 11.15.

11.3 HEAT FLUX AND SINGLE-FLUID HEAT STORAGE EXCHANGER

11.3.1 Mathematical Model

The heat storage exchanger to be analyzed with one constant heat flux boundary condition and the other surface in contact with a fluid stream is shown in Fig. 11.16. The assumptions made in developing the mathematical model are identical to those employed in the two-fluid heat storage exchanger. They are listed in Sec. 11.2.1. The conservation of energy equations for the fluid stream and the storage material are

$$\text{Hot fluid:} \qquad \rho_f c_f S_f \left(\frac{\partial t_f}{\partial \tau} + v_f \frac{\partial t_f}{\partial x} \right) = hP(t_w - t_f) \tag{11.39}$$

$$\text{Storage material:} \qquad \frac{1}{\alpha} \frac{\partial t_m}{\partial \tau} = \frac{\partial^2 t_m}{\partial x^2} + \frac{\partial^2 t_m}{\partial y^2} \tag{11.40}$$

The initial and boundary conditions are

Initial conditions: $\tau = 0 \qquad t_f = t_m = t_o$

$$\tau = 0 \qquad t_m = t_f = t_o$$

$$\tau > 0 \qquad x = 0 \qquad t_f = t_{fi} \qquad \frac{\partial t_m}{\partial x} = 0 \text{ for } 0 < y < w$$

$$x = L \qquad \frac{\partial t_m}{\partial x} = 0 \text{ for } 0 < y < w$$

$$y = 0 \qquad -k_m \frac{\partial t_m}{\partial y} = q \text{ for } 0 \leqslant x \leqslant L$$

$$y = w \qquad -k_m \frac{\partial t_m}{\partial y} = h(t_w - t_f) \text{ for } 0 \leqslant x \leqslant L$$

(11.41)

where q is a constant heat flux. The transient term in Eq. (11.39) can be neglected.

The equations and boundary conditions are nondimensionalized by using the dimensionless groups introduced in Sec. 11.2.1. The dimensionless temperature,

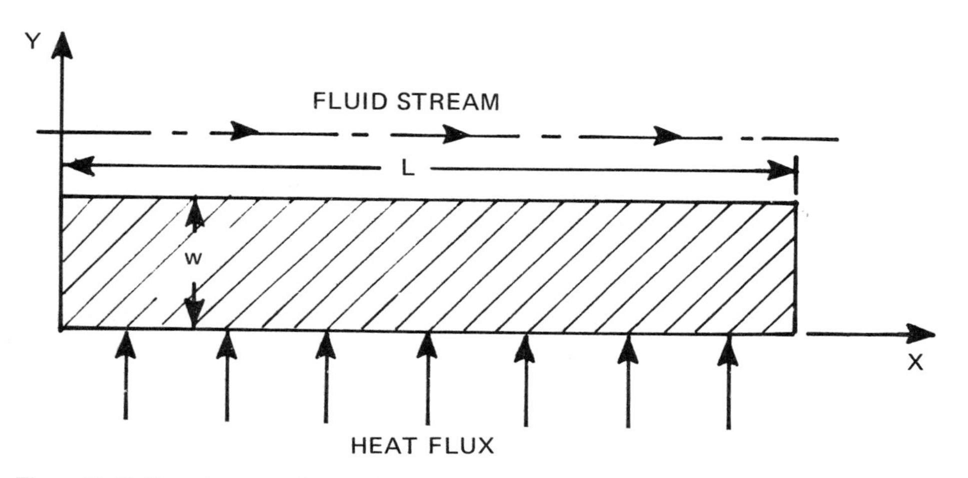

Figure 11.16 Heat storage exchanger, heat flux-fluid stream.

however, is defined as

$$T \equiv \frac{t - t_o}{t_{fo}^{ss} - t_o} \tag{11.42}$$

At steady-state conditions a heat balance of the unit indicates

$$qPL = \rho_f v_f S_f c_f (t_{fo}^{ss} - t_o) \tag{11.43}$$

The dimensionless temperature can be written as

$$T = \frac{t - t_o}{PLq/\dot{m}_f c_f} \tag{11.44}$$

The resulting expressions are

$$\text{Fluid:} \qquad \frac{\partial T_f}{\partial X} = \frac{\text{Bi}\, G^+}{V^+}(T_w - T_f) \tag{11.45}$$

$$\text{Material:} \qquad \frac{\partial T_m}{\partial \text{Fo}} = (V^+)^2 \left(\frac{\partial^2 T_m}{\partial X^2} + \frac{\partial^2 T_m}{\partial Y^2}\right) \tag{11.46}$$

The initial conditions are

$$\text{Fo} = 0 \qquad T_f = T_m = 0$$

and the boundary conditions are

$$\begin{aligned}
X &= 0 \qquad T_f = T_o = 0 \qquad & \frac{\partial T_m}{\partial X} &= 0 \text{ for } 0 < Y < 1 \\
X &= 1 & \frac{\partial T_m}{\partial X} &= 0 \text{ for } 0 < Y < 1 \\
Y &= 0 & \frac{\partial T_m}{\partial Y} &= -\frac{V^+}{G^+} \text{ for } 0 \leqslant X \leqslant 1 \\
Y &= 1 & \frac{\partial T_m}{\partial Y} &= -\text{Bi}(T_w - T_f) \text{ for } 0 \leqslant X \leqslant 1
\end{aligned} \tag{11.47}$$

The dimensionless steady-state temperature of the fluid leaving the heat storage exchanger is $T_{fo} = 1$. The steady-state temperature distribution in the storage material is

$$T_m^{ss} = \frac{V^+}{G^+}\left[\frac{1}{\text{Bi}} + (1 - Y)\right] + X \tag{11.48}$$

The mean temperature of the storage material is

$$\bar{T}_m^{ss} = \frac{V^+}{G^+}\left(\frac{1}{\text{Bi}} + \frac{1}{2}\right) + \frac{1}{2} \tag{11.49}$$

The nondimensional heat storage Q^+ is defined as the ratio of the actual heat stored to that stored when the unit is at steady-state conditions. This may be expressed as

$$Q^+ = \frac{Q}{\rho_m (PwL)c_m [(PLq/\dot{m}_f c_f)\bar{T}_m^{ss}]} \tag{11.50}$$

In many storage applications it is desirable to determine the fraction of the available energy that is stored. The nondimensional expression for this quantity is

$$\begin{aligned} A^+ &= \frac{\text{heat stored}}{\text{total heat into the system}} \\ &= \frac{Q}{qPL\tau} \\ &= \frac{Q^+ \bar{T}_m^{ss} G^+}{\text{Fo } V^+} \end{aligned} \tag{11.51}$$

The complete set of equations was solved simultaneously by finite difference techniques similar to those employed in obtaining the solution for the two-fluid heat storage exchanger. The computer program described by Schmidt and Szego [6] was used by Delpero [7] to obtain the results presented.

11.3.2 Results

If the heat capacity of the fluid is large, that is, G^+/V^+ approaches zero, the fluid temperature will remain unchanged as it passes through the heat storage exchanger. The nondimensional heat stored can be obtained from the rearrangement of the expressions given in Carslaw and Jaeger [8],

$$Q^+ = 1 - \sum_{n=1}^{\infty} \frac{4\text{Bi} \exp(-\alpha_n^2 \text{Fo})}{\cos(\alpha_n)(\text{Bi} + 2)\alpha_n^2 [\text{Bi}(\text{Bi} + 1.0) + \alpha_n^2]} \tag{11.52}$$

where α_n are the roots of $\alpha_n \tan \alpha_n = \text{Bi}$.

When the value of $(G^+/V^+)\text{Bi} > 0.1$, the unit must be represented by the finite conductivity model presented in Sec. 11.3.1. The results for a step change in heat flux have been obtained by Delpero [7]. The nondimensional fluid temperatures leaving the storage unit for the Biot number range $0.01 \leqslant \text{Bi} \leqslant 30.0$ are given in Figs. 11.17 and 11.18. The nondimensional heat storage for these conditions is presented in Figs. 11.19 and 11.20.

Example 11.2 The sun strikes a vertical concrete wall 3 m (9.84 ft) high, 5 m (16.40 ft) wide, and 0.15 m (0.492 ft) thick. The net energy transferred to the wall by the solar radiation during an 8-h period is 2.75×10^2 W/m^2 (87.18 Btu/ft^2 h). The inside of the wall is exposed to an environment of 21°C (69.8°F) with

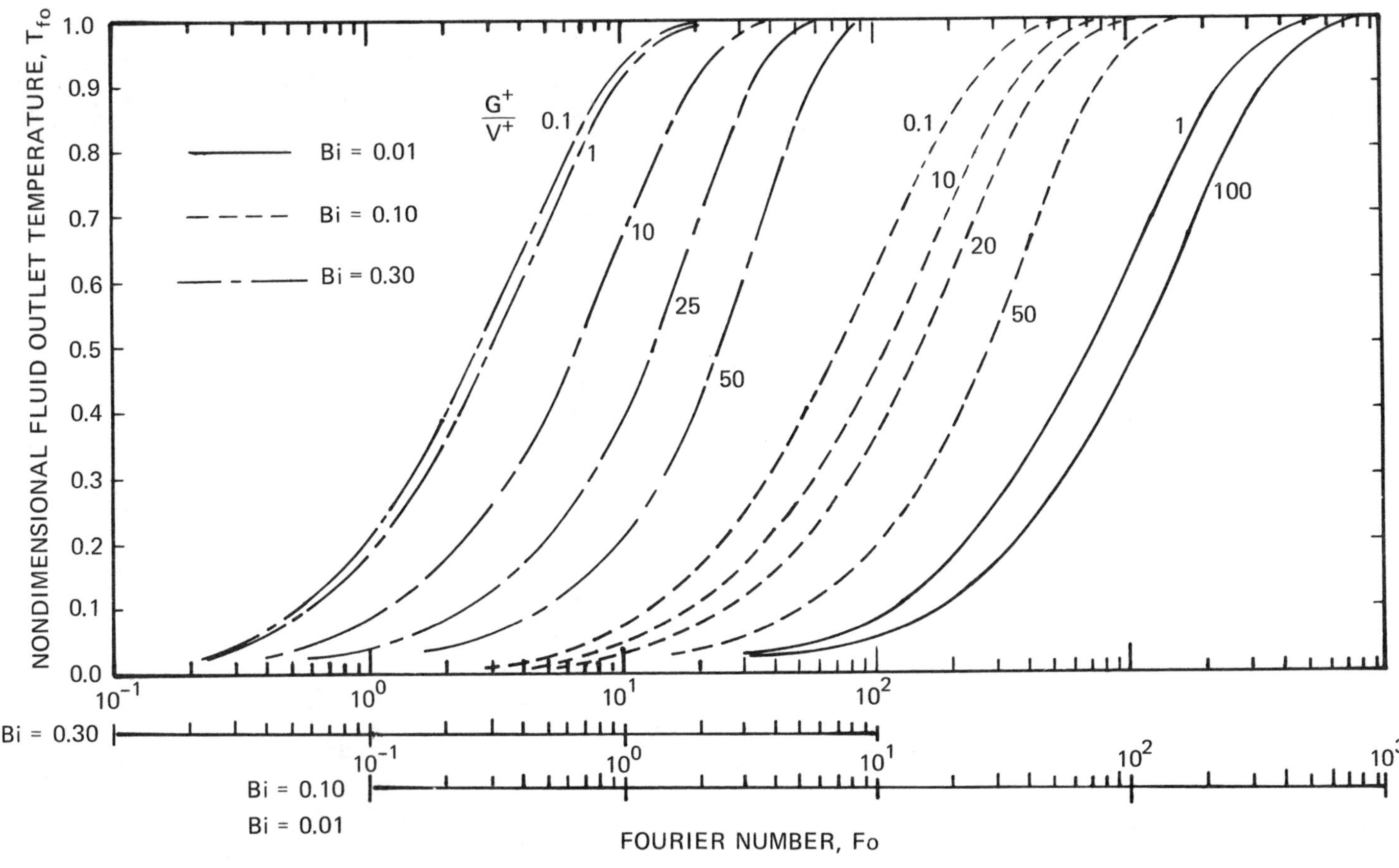

Figure 11.17 Nondimensional fluid outlet temperature.

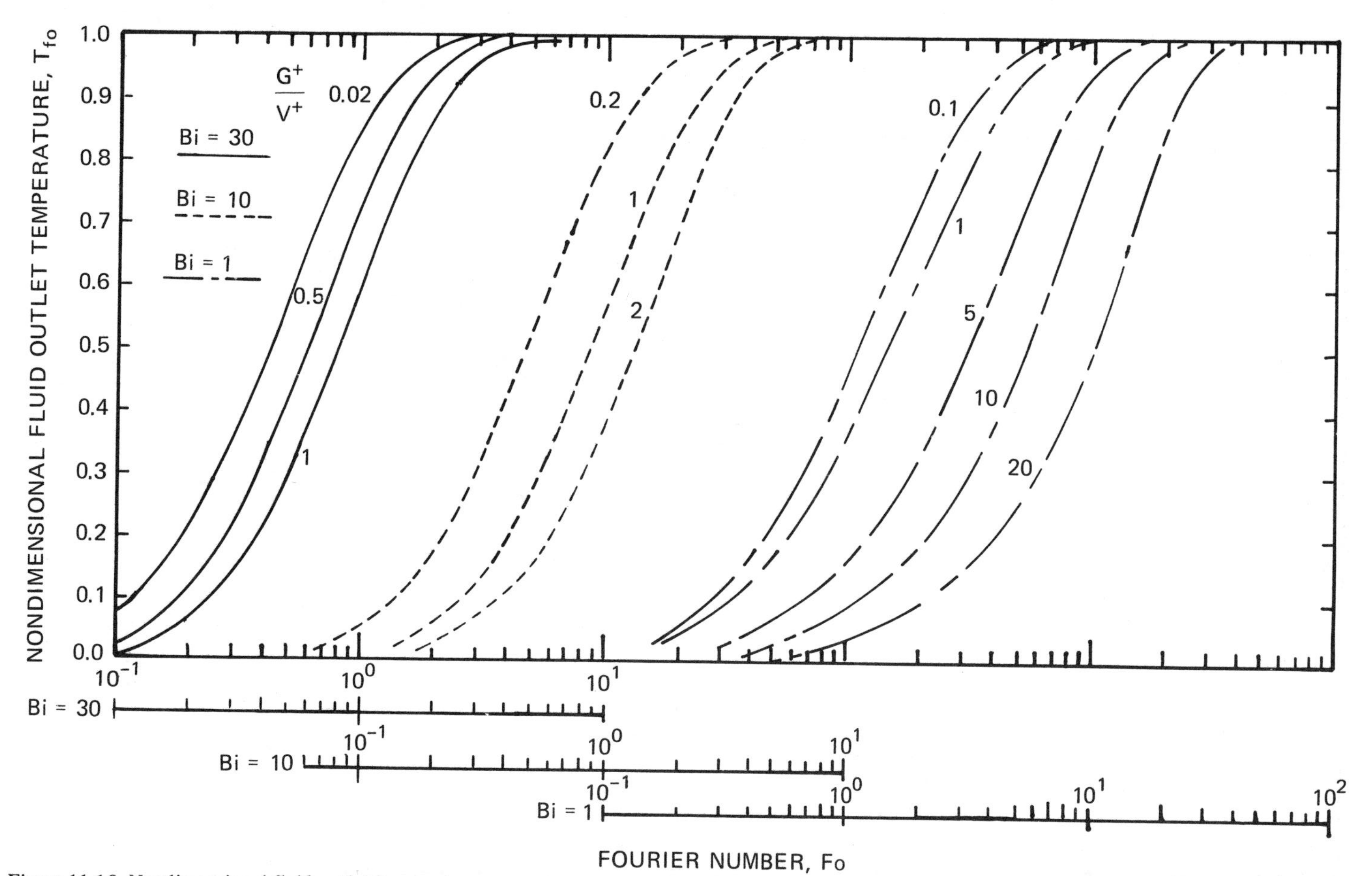

Figure 11.18 Nondimensional fluid outlet temperature.

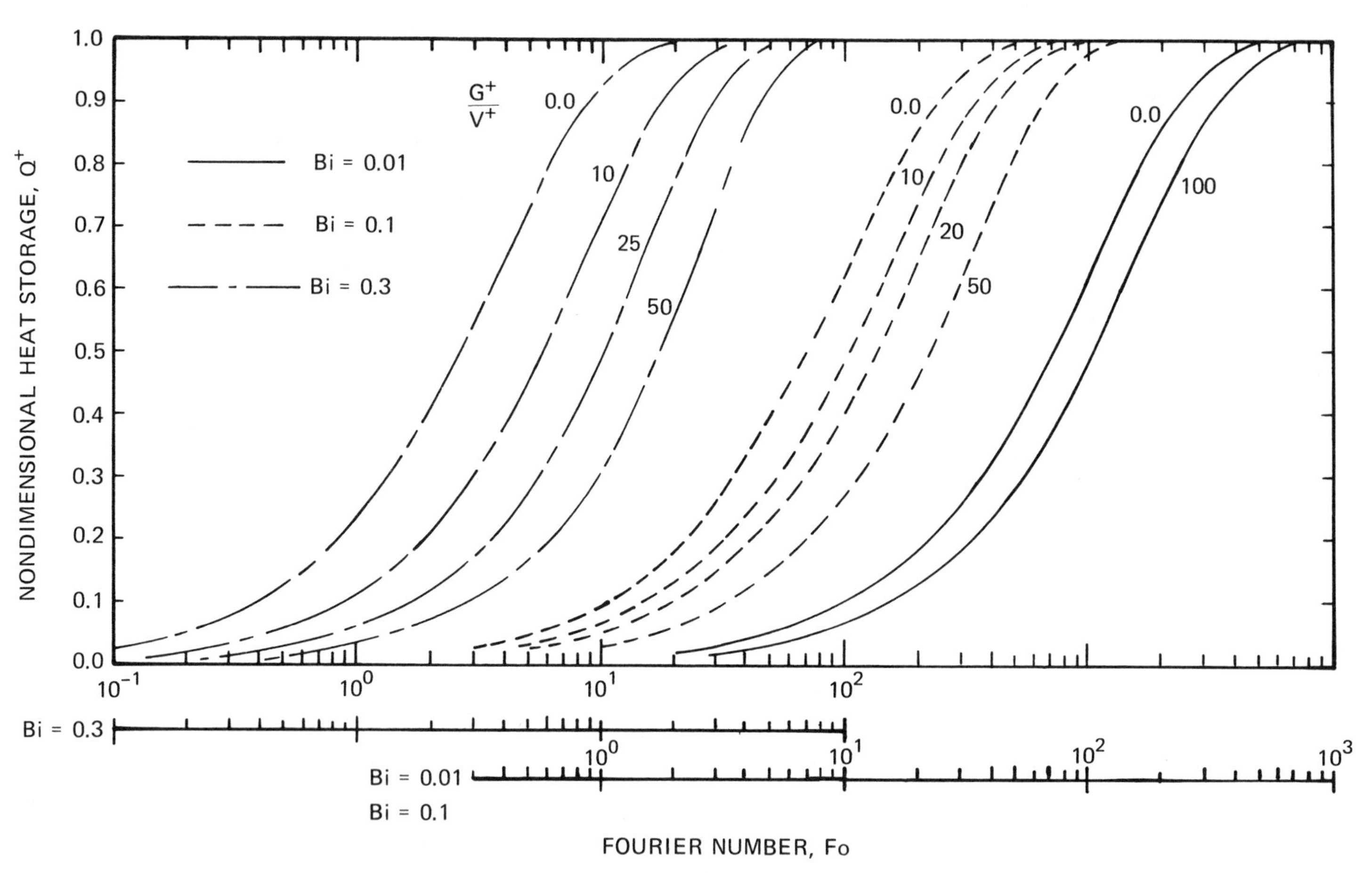

Figure 11.19 Nondimensional heat storage.

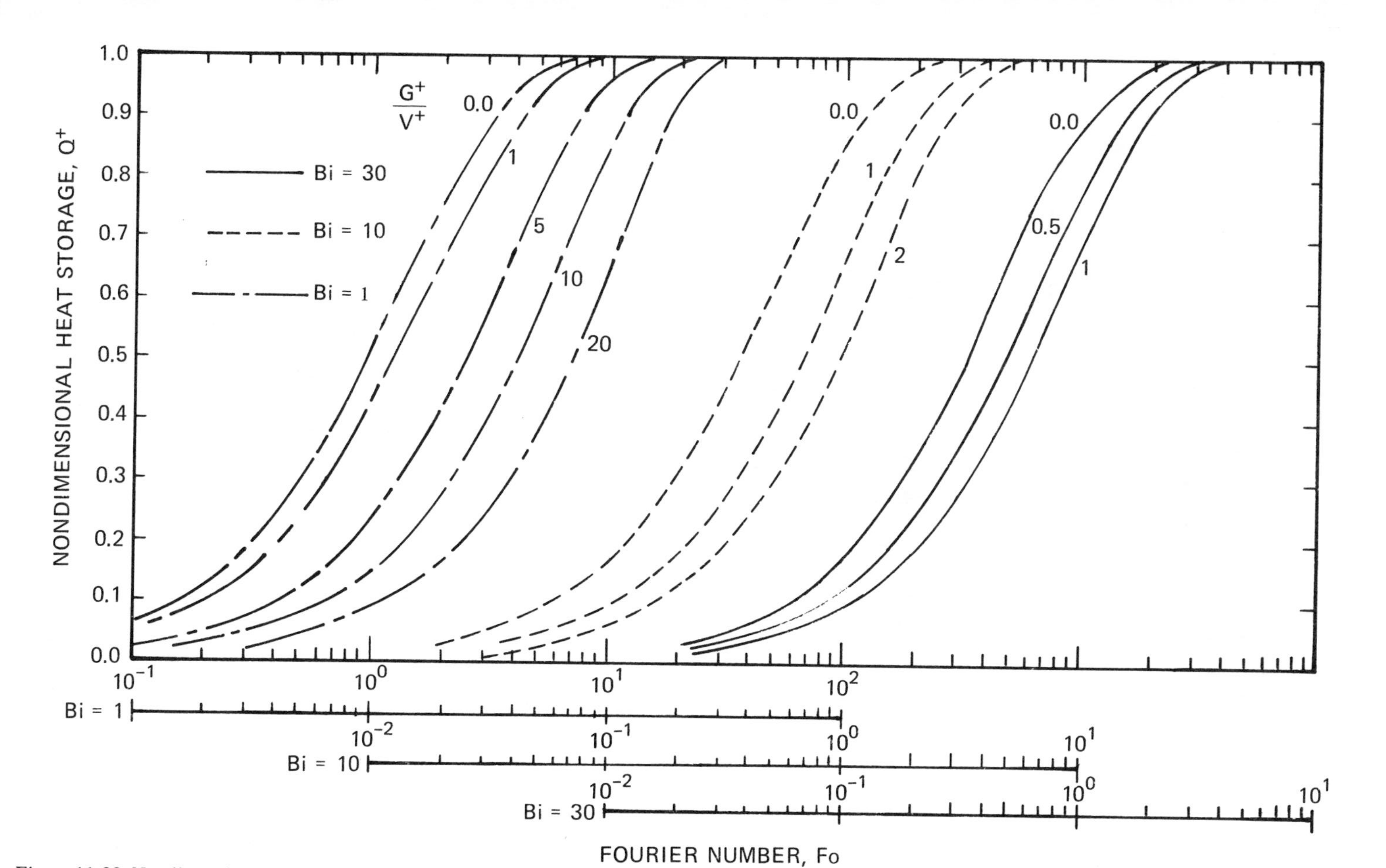

Figure 11.20 Nondimensional heat storage.

an average natural convective film coefficient of 5 W/m² °C (0.88 Btu/ft² h °F) assumed. The initial temperature of the wall is 21°C (69.8°F). It is desired to determine the total amount of heat stored in the wall during the 8-h period and the amount of energy that is transferred to the fluid.

SOLUTION The properties of the concrete are

$$\rho_m = 2.1 \times 10^3 \text{ kg/m}^3 \quad (131.1 \text{ lb}_\text{m}/\text{ft}^3)$$

$$c_m = 0.878 \text{ kJ/kg °C} \quad (0.210 \text{ Btu/lb}_\text{m} \text{ °F})$$

$$k_m = 1.1 \text{ W/m °C} \quad (0.636 \text{ Btu/ft h °F})$$

$$\alpha = 0.597 \times 10^{-6} \text{ m}^2/\text{s} \quad (2.31 \times 10^{-2} \text{ ft}^2/\text{h})$$

The room temperature is considered to remain constant so the infinite fluid heat capacity model can be used. The following dimensionless variables are needed:

$$\text{Bi} = \frac{hw}{k_m} = \frac{(5.0)(0.15)}{1.1} = 0.682$$

$$\text{Fo} = \frac{\tau\alpha}{w^2} = \frac{(8.0)(3600.0)(0.597 \times 10^{-6})}{(0.15)^2} = 0.764$$

The nondimensional heat storage is obtained from Eq. (11.52), $Q^+ = 0.341$. The amount of heat stored is

$$\begin{aligned} Q &= Q^+ \left[\rho_m (PwL) c_m \left(\frac{PLq}{\dot{m}_f c_f} \bar{T}_m^{ss}\right)\right] \\ &= Q^+ \left[(PwL) \frac{qw}{\alpha} \left(\frac{1}{\text{Bi}} + \frac{1}{2}\right)\right] \\ &= (0.341) \left[\frac{(5.0)(0.15)(3.0)(2.75 \times 10^2)(0.15)}{(0.597 \times 10^{-6})} \left(\frac{1}{0.682} + \frac{1}{2}\right)\right] \\ &= 1.04 \times 10^8 \text{ J} \quad (9.85 \times 10^4 \text{ Btu}) \end{aligned}$$

The amount of energy transferred to the fluid is

$$\begin{aligned} Q_f = Q_t - Q = PLq\tau - Q \\ &= (5.0)(3.0)(2.75 \times 10^2)(8.0)(3600) - 1.04 \times 10^8 \\ &= 1.188 \times 10^8 - 1.04 \times 10^8 \\ &= 1.48 \times 10^7 \text{ J} \quad (1.40 \times 10^4 \text{ Btu}) \end{aligned}$$

These calculations indicate that 87.54% of the energy striking the concrete wall is retained by the wall.

11.3.3 Methods of Superposition for Arbitrary Timewise Variations in Heat Flux

The mathematical model describing the heat storage exchanger is linear and allows superposition techniques to be used for the calculation of the transient response when the heat flux varies with time and the flow rate of the fluid remains constant. The nondimensional temperature of the fluid leaving the exchanger is

$$T_{fo} = \frac{t_{fo} - t_o}{qPL/\dot{m}_f c_f}$$

The temperature of the fluid leaving for a step change in heat flux is

$$t_{fo} = \left(\frac{qPL}{\dot{m}_f c_f}\right) T_{fo} + t_o$$

For the first step change t_o is the initial temperature of the storage unit, whereas for all additional steps the value of t_o is zero. The general expression is

$$t_{fo} = t_o + \left(\frac{PL}{\dot{m}_f c_f}\right) \sum_{i=1}^{n} \Delta q_i T_{fo}^i \tag{11.53}$$

The heat stored is

$$Q = [\rho_m c_m (PLw)] \left\{\frac{PL}{\dot{m}_f c_f} \frac{V^+}{G^+} \left[\left(\frac{1}{\text{Bi}} + \frac{1}{2}\right) + \frac{1}{2}\right]\right\} \sum_{i=1}^{n} \Delta q_i Q_i^+$$

$$= \frac{PLw^2}{\alpha} \left[\left(\frac{1}{\text{Bi}} + \frac{1}{2}\right) + \frac{G^+}{2V^+}\right] \sum_{i=1}^{n} \Delta q_i Q_i^+ \tag{11.54}$$

The Fourier number used in calculating T_{fo}^i and Q_i^+ is

$$\text{Fo}_i = \frac{\alpha(\tau - \tau^i)}{w^2}$$

Example 11.3 An energy storage unit constructed of Feolite utilizes off-peak electrical energy to heat an air stream. The heating elements are sandwiched between Feolite slabs that are 0.05 m (0.164 ft) thick, 0.5 m (1.64 ft) wide, and 4 m (13.123 ft) long. The Feolite is initially at a temperature of 20°C (68°F). The mass flow rate is held constant at 0.106 kg/s per channel (0.234 lb_m/s per channel). The convective film coefficient is 34.7 W/m² °C (6.11 Btu/ft² h °F). The temperature of the fluid entering the unit is 20°C (68°F), and the total heat generated by the heating elements is

$$0 < \tau \leqslant 8 \text{ h} \qquad q = 2.0 \times 10^4 \text{ W/m}^2 \quad (6340.0 \text{ Btu/ft}^2 \text{ h})$$

$$8 < \tau \leqslant 25 \text{ h} \qquad q = 0.0 \text{ W/m}^2$$

Determine the outlet temperature of the fluid and the amount of heat stored in the unit after 8 and 24 h of operation.

SOLUTION The properties of the Feolite are

$$k_m = 2.1 \text{ W/m °C} \quad (1.21 \text{ Btu/ft h °F})$$

$$\rho_m = 3.9 \times 10^3 \text{ kg/m}^3 \quad (243.48 \text{ lb}_\text{m}/\text{ft}^3)$$

$$c_m = 0.92 \text{ kJ/kg °C} \quad (0.2197 \text{ Btu/lb}_\text{m} \text{ °F})$$

$$\alpha = 5.85 \times 10^{-7} \text{ m}^2/\text{s} \quad (0.023 \text{ ft}^2/\text{h})$$

The specific heat of the air is

$$c_f = 1.01 \text{ kJ/kg °C} \quad (0.24 \text{ Btu/lb}_\text{m} \text{ °F})$$

The dimensionless groups are

$$V^+ = \frac{w}{L} = \frac{0.05}{4.0} = 0.0125$$

$$G^+ = \frac{Pk}{\dot{m}_f c_f} = \frac{(0.5)(2.1)}{(0.053)(1010)} = 1.96 \times 10^{-2}$$

$$\frac{G^+}{V^+} = \frac{1.96 \times 10^{-2}}{0.0125} = 1.568$$

$$\text{Bi} = \frac{hw}{k} = \frac{(34.7)(0.05)}{2.1} = 0.826$$

It should be noted that the analysis is restricted to a symmetrical section bounded by the center line of the flow channel and the plane of the heating element. The response of the unit can be obtained from Figs. 11.17 through 11.20.

After 8 h of operation,

$$\text{Fo} = \frac{\tau\alpha}{w^2} = \frac{(8.0)(3600.0)(5.85 \times 10^{-7})}{(0.05)^2} = 6.74$$

which yields $T_{fo} = 0.903$ and $Q^+ = 0.919$.

The temperature of the fluid leaving the unit is

$$t_{fo} = \frac{PL\,\Delta q}{\dot{m}_f c_f} T_{fo} + t_o$$

$$= \left[\frac{(0.5)(4.0)(1.0 \times 10^4)}{(0.053)(1010)}\right](0.903) + 20.0$$

$$= 357.4\text{°C} \quad (675.3\text{°F})$$

The amount of heat stored is

$$Q = \frac{PLw^2}{\alpha}\left[\left(\frac{1}{\text{Bi}} + \frac{1}{2}\right) + \frac{G^+}{2V^+}\right]\Delta q\, Q^+$$

$$= \frac{(0.5)(4.0)(0.05)^2}{5.85 \times 10^{-7}} \left[\left(\frac{1}{0.826} + \frac{1}{2} \right) + \frac{1.568}{2} \right] (1.0 \times 10^4)(0.919)$$

$$= 1.96 \times 10^5 \text{ kJ per semislab} \quad (1.86 \times 10^5 \text{ Btu per semislab})$$

At the completion of 24 h of operation,

$$\text{Fo}_1 = \frac{\tau\alpha}{w^2} = \frac{(24.0)(3600)(5.85 \times 10^{-7})}{(0.05)^2}$$

$$= 20.2$$

$$\text{Fo}_2 = \frac{(\tau - \tau_2)\alpha}{w^2} = \frac{(24.0 - 8.0)(3600)(5.85 \times 10^{-7})}{(0.05)^2}$$

$$= 13.48$$

From Figs. 11.17 and 11.19,

$$T_{fo}^1 = 0.996 \qquad Q_1^+ = 0.996$$

and

$$T_{fo}^2 = 0.985 \qquad Q_2^+ = 0.987$$

The temperature of the fluid leaving is obtained from Eq. (11.53).

$$t_{fo} = t_o + \left(\frac{PL}{\dot{m}_f c_f} \right) (\Delta q_1'' T_{fo}^1 + \Delta q_2'' T_{fo}^2)$$

$$= 20.0 + \left[\frac{(0.5)(4.0)}{(0.053)(1010)} \right] [1.0 \times 10^4)(0.996) + (-1.0 \times 10^4)(0.985)]$$

$$= 24.11°\text{C} \quad (75.4°\text{F})$$

The amount of stored heat in the unit after 24 h of operation is obtained from Eq. (11.54).

$$Q = \frac{PLw^2}{\alpha} \left[\left(\frac{1}{\text{Bi}} + \frac{1}{2} \right) + \frac{G^+}{2V^+} \right] (\Delta q_1 Q_1^+ + \Delta q_2 Q_2^+)$$

$$= \frac{(0.5)(4.0)(0.05)^2}{5.85 \times 10^{-7}} \left[\left(\frac{1}{0.826} + \frac{1}{2} \right) + \frac{1.568}{2} \right]$$

$$\times [(1.0 \times 10^4)(0.996) + (-1.0 \times 10^4)(0.987)]$$

$$= 1.92 \times 10^3 \text{ kJ} \quad (1.82 \times 10^3 \text{ Btu})$$

REFERENCES

1. F. W. Schmidt, "Numerical Simulation of the Thermal Behavior of Convective Heat Transfer Equipment," *Heat Exchanger: Design and Theory Sourcebook*, N. Afgan and E. U. Schlunder (eds.), Hemisphere, Washington, D.C., 1974, p. 491.

2. R. M. Cima and A. L. London, "The Transient Response of a Two Fluid Counterflow Heat Exchanger–The Gas Turbine Regenerator," *Trans. ASME*, vol. 80, 1958, p. 1169.
3. A. L. London, D. F. Lampsell and J. G. McGowan, "The Transient Response of Gas Turbine Plant Heat Exchangers–Additional Solutions for Regenerators of the Periodic and Direct Transfer Types," *J. Eng. Power, Trans. ASME*, ser. A, vol. 86, 1964, p. 127.
4. F. W. Schmidt and J. Szego, "Transient Behavior of Solid Sensible Heat Thermal Storage Units for Solar Energy Systems," *Future Energy Production–Heat and Mass Transfer Problems*, 1975 International Seminar, Yugoslavia, 1975.
5. J. Szego and F. W. Schmidt, "Transient Behavior of a Solid Sensible Heat Thermal Storage Exchanger," *J. Heat Transfer, Trans. ASME*, vol. 100, 1978, p. 148.
6. F. W. Schmidt and J. Szego, "Computer Program for the Prediction of the Transient Response of Solid Sensible Heat Storage Units," Mech. Eng. Report TESR-4, Pennsylvania State University, University Park, Pa., 1978.
7. P. Delpero, Private correspondence, 1978.
8. H. S. Carslaw and J. C. Jaeger, *Conduction of Heat in Solids*, Oxford University Press, London, 1959, p. 125.

CHAPTER

TWELVE

PACKED BEDS

12.1 INTRODUCTION

Many industrial processes involve an interaction between one or more fluids and one or more solids. In order to obtain a large ratio of surface area to volume, the fluids may be passed over a packed bed of the solid material. As the fluid flows over the solid, heat and mass transfer or a chemical reaction may occur. In many cases two or more of these phenomena may occur simultaneously. Examples of industrial application involving packed beds are catalytic reactors, pebble bed heaters, evaporators, absorbers, glass furnaces, and thermal energy storage units. Our discussion will be restricted to the use of packed beds in thermal energy storage and will utilize, whenever possible, theories and correlations developed for other packed bed applications.

The rate of heat transfer from or to the solid in the packed bed is a function of the physical properties of the fluid and solid, the local temperature of the fluid and surface of the solid, the mass rate of flow of the fluid, and the characteristics of the packed bed. The bed may be arranged in an orderly or a random fashion. Random packing is the most common arrangement and results when particles of the same approximate size and shape are deposited into a container. The characteristics of the packed bed are dependent on the shape and orientation of the packing material and the bed voidage. The void fraction or porosity is the fraction of the total volume occupied by the gas and is denoted by ϵ. The mechanism by which heat is transferred between the fluid and the solid is very complicated, partly because of the recirculating nature of the fluid flow and the inter- as well as intraparticle heat conduction effects. A major resistance to the transfer of heat is located at the innerface between the solid and the fluid and is inversely proportional to the convective heat transfer coefficient h. The temperature of the solid surface is dependent on the transient heat conduction from the surface to the interior of the solid. It is also dependent, to a lesser degree, on the interparticle conduction of heat when the adjacent solids come into direct physical contact with each other. The transfer of heat by or to the container walls will also influence the transient performance of the packed bed. Another factor influencing the rate of heat transfer is the mixing action within the fluid that results from the eddys created as the fluid flows through the complex set of flow passages and is named the dispersion effect.

In summary, it is pointed out that the use of packed beds as a thermal energy storage device has a certain attractiveness since the response of such a unit is relatively fast due to the large surface-to-volume ratio. A distinct disadvantage of packed beds is the very large pressure drop required to force the fluid through the bed. A regenerator may be 30 m high and still have a relatively low pressure drop. It would be totally impractical to have a packed bed of this length because of the extremely high pressure drop. Packed beds thus characteristically have a large frontal area and short length in order to hold the pressure drop to a reasonable value.

12.2 MATHEMATICAL MODELS: SINGLE BLOW

Three models for the determination of the transient response of packed bed heat storage units will be presented. The first is based on the assumption that the temperature of the fluid and the storage material are equal at any specific axial location. The second considers the storage material and fluid streams separately with intraparticle and dispersion effects neglected. The effect of these latter two items will be considered in the third model.

12.2.1 Negligible Thermal Resistance

If the thermal conductivity of the bed material and the convective heat transfer coefficient are very large, negligible resistance will be offered in the transfer of heat between the fluid and the bed material. The temperature of the fluid and the bed will then be equal. One energy balance that includes both the fluid and the storage material can be written for the packed bed. The following assumptions will be made when writing the energy balance:

1. No transverse heat transfer in the bed, that is, the container walls are considered to be perfectly insulated
2. Constant thermal and physical properties
3. Negligible rate of accumulation of energy by the fluid within the bed
4. A uniform initial temperature of the bed, t_o
5. Negligible thermal radiation effects

The resulting energy equation is

$$v_f \rho_f c_f \frac{\partial t}{\partial x} + \rho_m c_m (1 - \epsilon) \frac{\partial t}{\partial \tau} = k_m \frac{\partial^2 t}{\partial x^2} \tag{12.1}$$

The following nondimensional variables are introduced:

$$\text{Distance:} \qquad X = \frac{x}{k_m}(v_f \rho_f c_f)$$

$$\text{Temperature:} \qquad T = \frac{t - t_o}{t_{fi} - t_o} \tag{12.2}$$

$$\text{Time:} \qquad \Theta = \frac{\tau (v_f \rho_f c_f)^2}{k_m \rho_m c_m (1 - \epsilon)}$$

The nondimensional energy equation becomes

$$\frac{\partial T}{\partial X} + \frac{\partial T}{\partial \Theta} = \frac{\partial^2 T}{\partial X^2} \tag{12.3}$$

The boundary at the inlet to the bed is the Danckwert type

$$x = 0 \qquad -k_m \frac{\partial t}{\partial x} = v_f \rho_f c_f (t_{fi} - t)$$

In terms of nondimensional quantities, the inlet boundary conditions are

$$X = 0 \qquad \frac{\partial T}{\partial X} = T - T_{fi} \tag{12.4}$$

The solution to this problem has been presented by Raiz [1] and is

$$T[X, \Theta] = \frac{1}{2} \operatorname{erfc}\left(\frac{X - \Theta}{2\sqrt{\Theta}}\right) + \sqrt{\frac{\Theta}{\pi}} \exp\left[\frac{-(X - \Theta)^2}{4\Theta}\right] - \frac{1}{2}(1 + X + \Theta) \exp(X) \operatorname{erfc}\left(\frac{X + \Theta}{2\sqrt{\Theta}}\right) \tag{12.5}$$

The nondimensional temperature is plotted in Figs. 12.1 and 12.2. Raiz indicated that this model is much simpler and will give results almost identical to those obtained with the model to be presented in Sec. 12.2.2 when $\Theta < 10$.

12.2.2 Simplified Model

The basic relationship, which neglects intraparticle conduction and fluid dispersion, that governs the transient response of the packed bed shown in Fig. 12.3 is the same as that for the transfer of heat to a fluid flowing through porous media. Schumann [2] developed the basic differential equations and presented a solution for the case of a uniform initial solid temperature and a constant inlet fluid temperature. The following assumptions were made:

1. The bed material has infinite thermal conductivity in the transverse direction.
2. The bed material has zero thermal conductivity in the direction of flow.

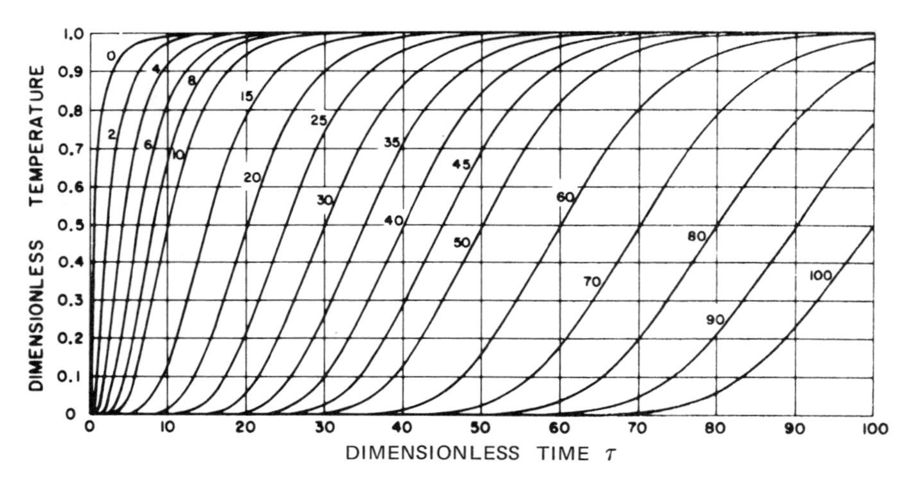

Figure 12.1 Nondimensional bed temperature [1].

3. There is no transverse heat transfer in the bed; that is, the container walls are considered to be perfectly insulated.
4. Constant thermal and physical properties are used.
5. There is a uniform convective heat transfer coefficient.
6. The fluid is in plug flow and the velocity does not vary as one moves in the flow direction.
7. The bed is initially at a uniform temperature.
8. Radiation effects are neglected.

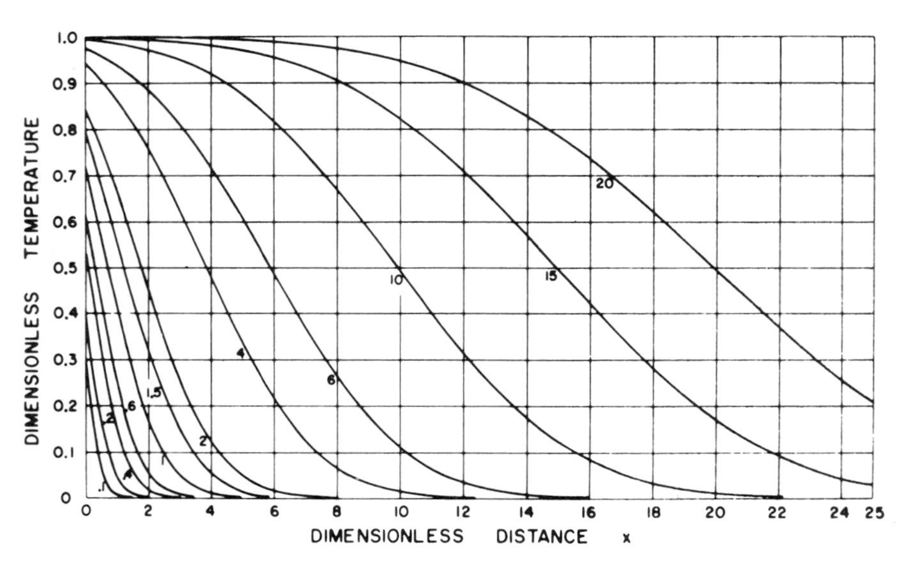

Figure 12.2 Nondimensional bed temperature [1].

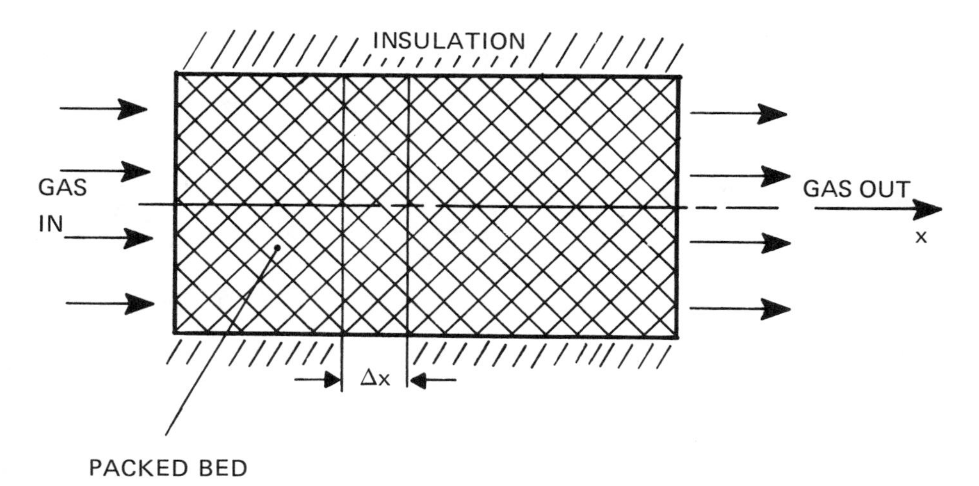

Figure 12.3 Packed bed.

The transient response of the bed is obtained by writing an energy balance for both the fluid and the packed bed material. A further simplification of the equation is obtained by neglecting the rate of accumulation of energy by the fluid contained within the volume. The final expression for the energy equation of the fluid is

$$\frac{\dot{m}_f c_f L}{hA} \frac{\partial t_f}{\partial x} = t_m - t_f \tag{12.6}$$

The rate at which energy is transferred from the fluid to the solid packed bed material must equal the rate of accumulation of energy within the packed bed in order to satisfy the conservation of energy:

$$\frac{S_{fr}(1-\epsilon)\rho_m c_m L}{hA} \frac{\partial t_m}{\partial \tau} = t_f - t_m \tag{12.7}$$

Before continuing with the development of the simplified mathematical model describing the packed bed, it is convenient to introduce a dimensionless length $\xi \equiv hAx/\dot{m}_f c_f L$ and a dimensionless time $\eta \equiv hA(\tau - x/v)/S_{fr}(1-\epsilon)\rho_m c_m L$. The term x/v represents the time required by the fluid particles to pass from the entrance of the bed to location x. This term is usually quite small when compared with the time scales of interest for the unit and can be neglected without introducing significant error. The dimensionless time is redefined as $\eta \equiv hA\tau/S_{fr}(1-\epsilon)\rho_m c_m L$. The non-dimensionalization process is completed by defining the dimensionless temperature as

$$T_f \equiv \frac{t_f - t_o}{t_{fi} - t_o} \quad \text{and} \quad T_m \equiv \frac{t_m - t_o}{t_{fi} - t_o}$$

The complete mathematical model of the packed bed in terms of the dimensionless parameters becomes

Fluid: $$\frac{\partial T_f}{\partial \xi} = T_m - T_f \tag{12.8}$$

Packed bed material: $$\frac{\partial T_m}{\partial \eta} = T_f - T_m \tag{12.9}$$

Boundary conditions: $\xi = 0 \qquad T_f = 0 \qquad T_m = 1 - \exp(-\eta)$

$$\eta = 0 \qquad T_m = 0.0 \tag{12.10}$$

This model is identical to that described in Sec. 2.3 for the rectangular slab heat storage unit.

The volumetric heat transfer coefficient h_v is used frequently in packed bed calculations. The interrelationship between the surface heat transfer coefficient h and h_v is

$$h_v S_{fr} = \frac{hA}{L} \tag{12.11}$$

The nondimensional length and time, expressed in terms of h_v, become

$$\xi = \frac{h_v S_{fr} x}{\dot{m} c_f} \qquad \eta = \frac{h_v \tau}{(1-\epsilon)\rho_m c_m} \tag{12.12}$$

As noted in Chap. 2, the solution of these equations was first presented by Anzelius [3], although Schumann [2] is usually the first reference cited. Others reporting on the solution to this problem were Furnas [4], Hausen [5], and more recently Larsen [6]. A complete evaluation of all the methods available up to 1954 was made by Klinkenberg [7].

The paper by Klinkenberg provided the source for the following relationships used to generate the results tabulated in Table 12.1:

$\eta < 2.0 \qquad \xi < 2.0$

$$T_f(\eta, \xi) = 1.0 - e^{-\eta-\xi} \sum_{N=1}^{N=\infty} \left(\frac{\xi^N}{N!} \sum_{k=0}^{k=N-1} \frac{\eta^k}{k!} \right) \tag{12.13}$$

$2.0 \leqslant \eta < 4.0 \qquad 2.0 \leqslant \xi < 4.0$

$$T_f(\eta, \xi) = 1.0 - \frac{1}{2}\left[1 + \operatorname{erf}\left(\sqrt{\xi} - \sqrt{\eta}\right)\right] - \frac{\xi^{1/4}}{\eta^{1/4} + \xi^{1/4}} e^{-\eta-\xi} I_o(2\sqrt{\eta\xi}) \tag{12.14}$$

$\eta \geqslant 4.0 \qquad \xi \geqslant 4.0$

$$T_f(\eta, \xi) = 1.0 - \frac{1}{2}\left[1 + \operatorname{erf}\left(\sqrt{\xi} - \sqrt{\eta} - \frac{1}{8\sqrt{\xi}} - \frac{1}{8\sqrt{\eta}}\right)\right] \tag{12.15}$$

Since the equations used in the description of the mathematical model of the packed bed are symmetric, the value of T_m may be obtained by using $T_m(A, B) = 1 - T_f(B, A)$.

Table 12.1 Nondimensional fluid temperature, T_f

Nondimensional Time, η	Nondimensional Distance, ξ									
	1	2	3	4	5	6	7	8	9	10
0	0.3679	0.1353	0.0569	0.0207	0.0075	0.0027	0.0010	0.0004	0.0001	0.0000
1	0.6543	0.3943	0.2248	0.1233	0.0656	0.0340	0.0173	0.0087	0.0043	0.0021
2	0.8174	0.6035	0.4146	0.2700	0.1685	0.1017	0.0596	0.0341	0.0191	0.0105
3	0.9063	0.7531	0.5833	0.4269	0.2982	0.2003	0.1302	0.0823	0.0507	0.0306
4	0.9529	0.8520	0.7170	0.5702	0.4339	0.3174	0.2242	0.1537	0.1026	0.0669
5	0.9767	0.9140	0.8150	0.6919	0.5628	0.4401	0.3323	0.2432	0.1730	0.1200
6	0.9887	0.9513	0.8828	0.7871	0.6747	0.5574	0.4449	0.3441	0.2587	0.1894
7	0.9945	0.9730	0.9278	0.8573	0.7659	0.6615	0.5532	0.4487	0.3537	0.2717
8	0.9974	0.9853	0.9565	0.9070	0.8363	0.7488	0.6509	0.5497	0.4518	0.3618
9	0.9988	0.9921	0.9744	0.9408	0.8885	0.8185	0.7346	0.6421	0.5469	0.4544
10	0.9994	0.9958	0.9852	0.9631	0.9257	0.8720	0.8032	0.7226	0.6347	0.5445
11	0.9997	0.9978	0.9915	0.9774	0.9516	0.9118	0.8574	0.7899	0.7123	0.6283
12	0.9999	0.9989	0.9952	0.9864	0.9690	0.9404	0.8989	0.8444	0.7782	0.7033
13	0.9999	0.9994	0.9974	0.9920	0.9805	0.9605	0.9297	0.8870	0.8326	0.7679
14	1.0000	0.9997	0.9986	0.9953	0.9879	0.9742	0.9519	0.9194	0.8760	0.8219
15	1.0000	0.9999	0.9992	0.9973	0.9926	0.9835	0.9677	0.9436	0.9097	0.8658
16	1.0000	0.9999	0.9996	0.9985	0.9956	0.9895	0.9786	0.9611	0.9354	0.9006
17	1.0000	1.0000	0.9998	0.9991	0.9974	0.9935	0.9861	0.9736	0.9545	0.9275
18	1.0000	1.0000	0.9999	0.9995	0.9985	0.9960	0.9910	0.9823	0.9684	0.9480
19	1.0000	1.0000	0.9999	0.9997	0.9991	0.9975	0.9943	0.9883	0.9784	0.9632
20	1.0000	1.0000	1.0000	0.9999	0.9995	0.9985	0.9964	0.9924	0.9854	0.9743
	11	12	13	14	15	16	17	18	19	20
0	0.0000	0.0000	0.0000	0.0000	0.0000	0.0000	0.0000	0.0000	0.0000	0.0000
1	0.0010	0.0005	0.0002	0.0001	0.0000	0.0000	0.0000	0.0000	0.0000	0.0000
2	0.0057	0.0031	0.0016	0.0008	0.0004	0.0002	0.0001	0.0001	0.0000	0.0000
3	0.0181	0.0105	0.0060	0.0034	0.0019	0.0011	0.0006	0.0003	0.0002	0.0001
4	0.0427	0.0267	0.0165	0.0100	0.0060	0.0035	0.0020	0.0012	0.0007	0.0004
5	0.0814	0.0541	0.0353	0.0226	0.0143	0.0089	0.0054	0.0033	0.0020	0.0012
6	0.1355	0.0948	0.0650	0.0438	0.0290	0.0189	0.0122	0.0077	0.0048	0.0030
7	0.2037	0.1493	0.1072	0.0755	0.0523	0.0356	0.0239	0.0158	0.0103	0.0066
8	0.2828	0.2161	0.1617	0.1187	0.0855	0.0606	0.0422	0.0290	0.0196	0.0131
9	0.3686	0.2924	0.2271	0.1729	0.1292	0.0949	0.0686	0.0488	0.0342	0.0237
10	0.4566	0.3745	0.3009	0.2369	0.1831	0.1390	0.1038	0.0763	0.0553	0.0395
11	0.5424	0.4585	0.3797	0.3083	0.2458	0.1924	0.1482	0.1123	0.0838	0.0616
12	0.6228	0.5406	0.4602	0.3843	0.3151	0.2538	0.2010	0.1566	0.1202	0.0909
13	0.6953	0.6179	0.5391	0.4617	0.3884	0.3211	0.2611	0.2089	0.1646	0.1278
14	0.7585	0.6882	0.6136	0.5376	0.4630	0.3921	0.3266	0.2678	0.2162	0.1720
15	0.8121	0.7501	0.6819	0.6097	0.5364	0.4642	0.3954	0.3316	0.2739	0.2230
16	0.8563	0.8032	0.7425	0.6761	0.6062	0.5352	0.4653	0.3985	0.3362	0.2796
17	0.8919	0.8475	0.7950	0.7355	0.6708	0.6030	0.5342	0.4663	0.4013	0.3405
18	0.9199	0.8837	0.8393	0.7874	0.7291	0.6660	0.6001	0.5332	0.4672	0.4039
19	0.9415	0.9126	0.8760	0.8317	0.7804	0.7232	0.6616	0.5974	0.5323	0.4681
20	0.9579	0.9353	0.9056	0.8686	0.8245	0.7738	0.7177	0.6575	0.5949	0.5315

This relationship and Table 12.1 may be used to determine the nondimensional fluid and storage material temperatures. Thus, in a packed bed at a nondimensional distance $\xi = 8$ and a nondimensional time $\eta = 10$, the nondimensional fluid temperature T_f is 0.7226 and the nondimensional storage material temperature T_m is 0.6382.

It is important for an engineer to be able to predict both the temperature of the fluid leaving a packed bed and the temperature distribution within the bed at a given time. If the assumptions listed at the beginning of this section are valid, these items can be accurately predicted by the results presented in Table 12.1. The computational procedure to be utilized is illustrated in Example 12.1.

Example 12.1 A packed bed composed of steel spheres 2.5 cm (0.984 in) in diameter has a frontal area of 0.08 m^2 (0.861 ft^2) and a length of 1 m (3.28 ft). The porosity of the bed is 0.37 and the total heat transfer surface area of the bed is 12.1 m^2 (130.24 ft^2). The bed is initially at a temperature of 10°C (50°F) and is to be heated by passing hot air at 80°C (176°F) and a mass flow rate of 0.065 kg/s (0.1433 lb$_m$/s) through it. The convective film coefficient is 70.3 W/m^2 °C (12.380 Btu/ft^2 h °F). It is required to estimate the temperature of the fluid leaving the unit and the temperature distribution within the bed 29 min after the start of heating.

SOLUTION The physical properties of the steel and air needed in these calculations are

Carbon steel @ 45.0°C (113°F):

$\rho_m = 7801.0$ kg/m^3 (487.01 lb$_m$/ft^3)

$c_m = 0.473$ kJ/kg °C (0.113 Btu/lb$_m$ °F)

Air @ 50.0°C (122°F):

$c_f = 1.007$ kJ/kg °C (0.24 Btu/lb$_m$ °F)

The nondimensional length at the exit of the packed bed is

$$\lambda = \frac{hA}{\dot{m}c_f} = \frac{(70.3)(12.1)}{(0.065)(1007.0)} = 13.0$$

and the nondimensional time is

$$\eta = \frac{hA\tau}{S_{fr}L(1-\epsilon)\rho_m c_m} = \frac{(70.3)(12.1)(29.0)(60.0)}{(0.08)(1.0)(1.0-0.37)(7801.0)(473.0)} = 7.959$$

The nondimensional temperature of the fluid leaving the packed bed is obtained from Table 12.1 and has the value 0.1592. The corresponding exit fluid temperature is 21.14°C (70.05°F). The temperature distribution in the bed is obtained by using $T_m(A, B) = 1 - T_f(B, A)$. The results are given in the following table:

x, m	ξ	T_m	t_m, °C	(°F)
0.0	0.0	0.9994	79.972	(175.95)
0.5	6.5	0.5996	51.872	(125.55)
1.0	13.0	0.1110	17.770	(64.99)

The temperature of the fluid leaving the packed bed could also have been estimated from the breakthrough curves given in Fig. 12.4. The value of ξ at the exit of the unit, $x = L$, is the nondimensional bed length, denoted as λ. In the operation of a packed bed for thermal energy storage, one of the items of interest is the fraction of the energy available in the hot stream that is stored. As the temperature of the fluid leaving the unit approaches the fluid inlet temperature, the amount of energy removed from the stream decreases. In order to ensure economically efficient use of the packed bed for energy storage, the designer may establish a maximum outlet fluid temperature; once the fluid temperature reaches that value, the hot gas stream is diverted to another packed bed storage unit. The breakthrough curves can be used to predict the length of time required for the outlet temperature of the fluid to reach its preselected value.

Example 12.2 The flow of the hot gases in the packed bed described in Example 12.1 is to be terminated when the temperature of the gas leaving the unit reaches 50°C (122°F). Determine the length of time required to reach this condition.

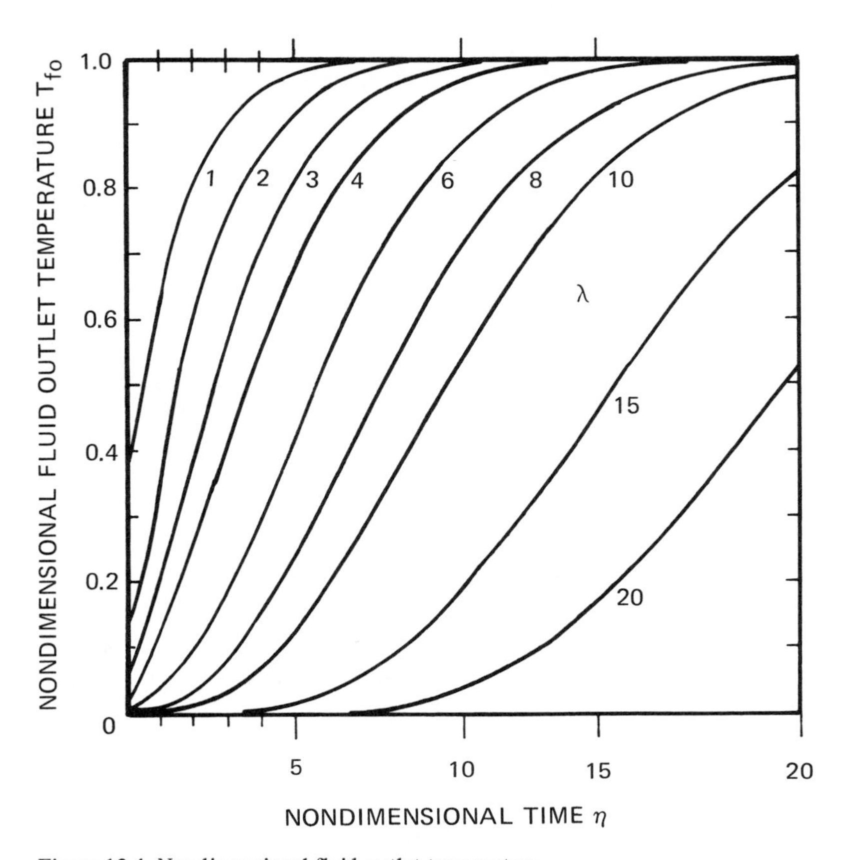

Figure 12.4 Nondimensional fluid outlet temperature.

SOLUTION From the data given in Example 12.1, the value of the nondimensional unit length $\lambda = 13.02$. The maximum nondimensional outlet fluid temperature is 0.571. The value of the nondimensional time obtained from Fig. 12.4 or Table 12.1 is $\eta = 13.43$, and the time required will be 48.94 min.

The amount of energy stored in the packed bed can be obtained by determining the average temperature of the bed, $Q = S_{fr}L(1-\epsilon)\rho_m c_m(t_{fi}-t_o)\bar{T}_m$, where

$$\bar{T}_m = \frac{1}{L}\int_0^L T_m\,dx$$

The energy stored may also be determined by using the temperature of the fluid leaving the unit. The appropriate expression is

$$Q = \dot{m}_f c_f \int_0^\tau (t_{fi}-t_{fo})\,d\tau = \dot{m}_f c_f(t_{fo}-t_o)\int_0^\tau (1-T_{fo})\,d\tau$$

The maximum possible heat storage will be obtained when the temperature of the packed bed is uniform and is equal to the temperature of the fluid entering the unit, $Q_{\max} = S_{fr}L(1-\epsilon)\rho_m c_m(t_{fi}-t_o)$.

The ratio of the actual heat storage to the maximum heat storage is denoted by Q^+, the nondimensional heat storage. Using the second expression for the actual heat storage, one obtains

$$Q^+ = \frac{\dot{m}_f c_f(t_{fi}-t_o)}{S_{fr}L(1-\epsilon)\rho_m c_m(t_{fi}-t_o)}\int_0^\tau (1-T_{fo})\,d\tau$$

$$= \frac{1}{\lambda}\int_0^\eta (1-T_{fo})\,d\eta \qquad (12.16)$$

The values of Q^+, a function of λ and η, are tabulated in Table 12.2 and plotted in Fig. 12.5. The use of these results to determine the heat storage in a packed bed is illustrated by Example 12.3.

Example 12.3 Determine the heat stored as a function of time in the packed bed described in Example 12.1.

SOLUTION The nondimensional length λ has been found to be 13.02. The nondimensional time $\eta = hA\tau/S_{fr}L(1-\epsilon)\rho_m c_m = 0.2744\tau$, where τ is in minutes. At $\tau = 29.0$, $\eta = 7.959$ and $Q^+ = 0.5889$. The maximum possible heat storage is

$$Q_{\max} = S_{fr}L(1-\epsilon)\rho_m c_m(t_{fi}-t_o)$$
$$= (0.08)(1.0)(1.0-0.37)(7801.0)(473.0)(80.0-10.0)$$
$$= 1.302 \times 10^4\ \text{kJ} \quad (1.234 \times 10^4\ \text{Btu})$$

Table 12.2 Nondimensional heat storage

Nondimensional Length, λ

Nondimensional Time, η	1	2	3	4	5	6	7	8	9	10
1	0.4762	0.3662	0.2887	0.2335	0.1936	0.1641	0.1418	0.1246	0.1109	0.0999
2	0.7324	0.6142	0.5153	0.4349	0.3708	0.3200	0.2795	0.2471	0.2209	0.1994
3	0.8658	0.7727	0.6814	0.5977	0.5244	0.4618	0.4091	0.3651	0.3283	0.2974
4	0.9336	0.8696	0.7970	0.7225	0.6511	0.5855	0.5269	0.4756	0.4311	0.3927
5	0.9675	0.9269	0.8741	0.8142	0.7512	0.6890	0.6301	0.5759	0.5270	0.4835
6	0.9841	0.9597	0.9237	0.8788	0.8271	0.7724	0.7175	0.6643	0.6143	0.5682
7	0.9921	0.9782	0.9547	0.9227	0.8827	0.8373	0.7889	0.7398	0.6914	0.6452
8	0.9960	0.9883	0.9736	0.9518	0.9221	0.8862	0.8456	0.8023	0.7578	0.7136
9	0.9978	0.9938	0.9849	0.9706	0.9494	0.9220	0.8894	0.8527	0.8134	0.7728
10	0.9987	0.9967	0.9915	0.9824	0.9677	0.9476	0.9222	0.8923	0.8588	0.8228
11	0.999	0.9982	0.9953	0.9897	0.9798	0.9654	0.9463	0.9226	0.8949	0.8641
12	0.999	0.9990	0.9974	0.9941	0.9877	0.9776	0.9635	0.9453	0.9231	0.8974
13	0.999	0.999	0.998	0.9967	0.9926	0.9858	0.9757	0.9620	0.9446	0.9238
14	1.000	0.999	0.999	0.9983	0.9957	0.9911	0.9840	0.9740	0.9607	0.9442
15	1.000	0.999	0.999	0.999	0.9976	0.9946	0.9897	0.9825	0.9726	0.9597
16	1.000	1.000	0.999	0.999	0.9988	0.9968	0.9935	0.9884	0.9811	0.9713
17	1.000	1.000	1.000	1.000	0.999	0.9982	0.9960	0.9924	0.9872	0.9799
18	1.000	1.000	1.000	1.000	0.999	0.9991	0.9976	0.9951	0.9914	0.9861
19	1.000	1.000	1.000	1.000	1.000	0.9996	0.9986	0.9969	0.9943	0.9905
20	1.000	1.000	1.000	1.000	1.000	0.9999	0.9993	0.9981	0.9963	0.9936
	11	12	13	14	15	16	17	18	19	20
1	0.0909	0.0833	0.0769	0.0714	0.0667	0.0625	0.0588	0.0556	0.0526	0.0500
2	0.1815	0.1665	0.1538	0.1428	0.1333	0.1250	0.1176	0.1111	0.1053	0.1000
3	0.2714	0.2493	0.2304	0.2141	0.1999	0.1875	0.1765	0.1667	0.1579	0.1500
4	0.3597	0.3312	0.3066	0.2851	0.2663	0.2498	0.2352	0.2222	0.2105	0.2000
5	0.4451	0.4113	0.3815	0.3554	0.3324	0.3120	0.2938	0.2776	0.2631	0.2500
6	0.5262	0.4885	0.4547	0.4245	0.3976	0.3736	0.3521	0.3329	0.3155	0.2999
7	0.6018	0.5617	0.5251	0.4918	0.4616	0.4345	0.4099	0.3878	0.3678	0.3496
8	0.6707	0.6299	0.5917	0.5563	0.5238	0.4940	0.4669	0.4421	0.4196	0.3991
9	0.7320	0.6921	0.6538	0.6174	0.5833	0.5517	0.5225	0.4956	0.4709	0.4482
10	0.7854	0.7477	0.7104	0.6742	0.6397	0.6069	0.5763	0.5477	0.5212	0.4967
11	0.8309	0.7963	0.7612	0.7262	0.6921	0.6591	0.6277	0.5980	0.5702	0.5442
12	0.8688	0.8380	0.8058	0.7729	0.7401	0.7077	0.6763	0.6462	0.6175	0.5904
13	0.8997	0.8730	0.8443	0.8142	0.7833	0.7523	0.7216	0.6916	0.6627	0.6350
14	0.9245	0.9019	0.8768	0.8499	0.8216	0.7925	0.7631	0.7339	0.7053	0.6775
15	0.9439	0.9252	0.9039	0.8803	0.8549	0.8282	0.8007	0.7729	0.7451	0.7177
16	0.9589	0.9438	0.9260	0.9058	0.8835	0.8595	0.8342	0.8082	0.7817	0.7551
17	0.9703	0.9583	0.9437	0.9267	0.9075	0.8864	0.8637	0.8397	0.8149	0.7896
18	0.9788	0.9694	0.9577	0.9437	0.9275	0.9092	0.8891	0.8675	0.8447	0.8210
19	0.9851	0.9778	0.9686	0.9573	0.9438	0.9283	0.9108	0.8916	0.8710	0.8492
20	0.9896	0.9841	0.9770	0.9680	0.9570	0.9440	0.9290	0.9123	0.8939	0.8742

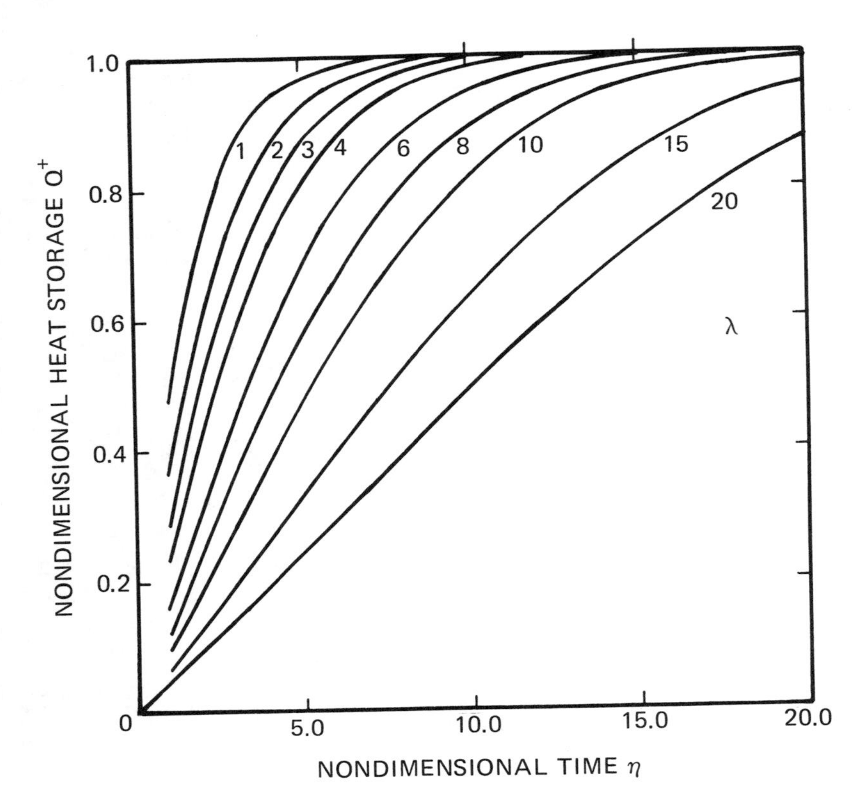

Figure 12.5 Nondimensional heat storage, packed bed.

The actual heat stored at the end of 29 min of operation is

$$Q = Q_{\text{max}} Q^+ = 1.302 \times 10^4(0.5889)$$
$$= 7.667 \times 10^3 \text{ kJ} \quad (7.267 \times 10^3 \text{ Btu})$$

The complete transient response of the packed bed is shown in Fig. 12.6.

12.2.3 Intraparticle Conduction and Dispersion Model

The model presented in Sec. 12.2.2 did not take into consideration the intraparticle conduction or the dispersion effects. For example, when intraparticle conduction is present, the storage particles are no longer assumed to be at a uniform temperature. The temperature of the surface in contact with the fluid will depend on the convective film coefficient and the intraparticle conduction. As the fluid flows through the packed bed, small eddys will be generated in the fluid, creating extensive mixing of the fluid particles. This mixing and the effect of axial molecular conduction of heat within the fluid are classified as the dispersion effects.

The conservation equations for a packed bed of spherical particles accounting for these effects are

$$\text{Fluids:} \qquad \frac{hA}{L}(t_w - t_f) - \dot{m}_f c_f \frac{\partial t_f}{\partial x} + k_f S_{fr} \epsilon \frac{\partial^2 t_f}{\partial x^2} = 0 \tag{12.17}$$

$$\text{Spherical particles:} \qquad \rho_m c_m \frac{\partial t_m}{\partial \tau} = k_m \left(\frac{\partial^2 t_m}{\partial r^2} + \frac{2}{r} \frac{\partial t_m}{\partial r} \right) \tag{12.18}$$

with $\quad r = r_o \qquad h(t_w - t_f) = -k_m \dfrac{\partial t_m}{\partial r}$

$$r = 0 \qquad \frac{\partial t_m}{\partial r} = 0 \tag{12.19}$$

The last term on the left-hand side of Eq. (12.17) represents the axial dispersion effects. The rate of accumulation of energy within the fluid was neglected when Eq. (12.17) was formulated.

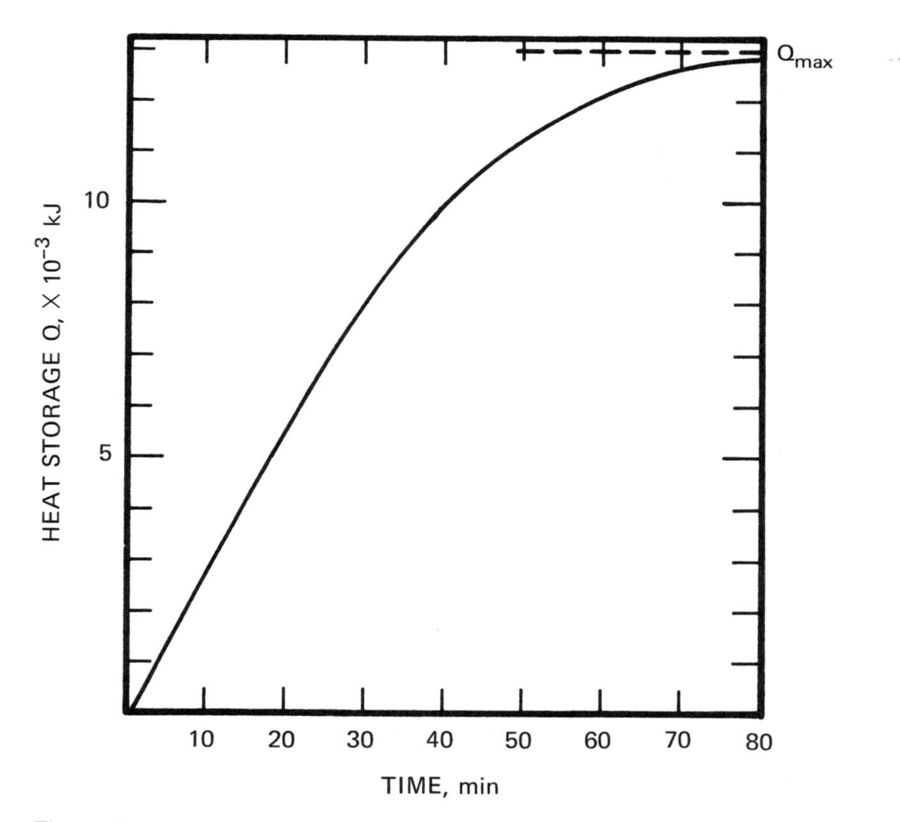

Figure 12.6 Heat storage in packed bed.

The complete set of equations is very difficult to solve in their present form. If only axial dispersion effects are considered, the fluid energy equation becomes

$$\frac{hA}{L}(t_w - t) - \dot{m}_f c_f \frac{\partial t_f}{\partial x} + k_f S_{fr} \epsilon \frac{\partial^2 t_f}{\partial x^2} = 0 \tag{12.20}$$

whereas the energy equation for the solids remains the same as that in the simplified model, Eq. (12.9),

$$S_{fr}(1-\epsilon)\rho_m c_{pm} L \frac{\partial t_m}{\partial \tau} = hA(t_f - t_w) \tag{12.9}$$

The solution to this model has been presented by Chao and Hoelscher [8] and Edwards and Richardson [9].

If only intraparticle conduction effects are considered, the energy equation for the fluid reduces to

$$\frac{hA}{L}(t_w - t_f) - \dot{m}_f c_f \frac{\partial t_f}{\partial x} = 0 \tag{12.21}$$

and together with Eq. (12.18), the boundary conditions Eq. (12.19) and

$$\tau = 0 \qquad t_f = t_w = t_o$$

$$\tau > 0 \qquad x = 0 \qquad t_f = t_{fi}$$

represent the model. An analytical solution for this problem has been presented by Rosen [10] and by Babcock, Green, and Perry [11], and finite difference solutions have been presented by Handley and Heggs [12] and by Leung and Quon [13].

A careful evaluation of the various methods available for the determination of the transient response of packed beds has been presented by Jeffreson [14]. The particles used were spherical in shape, and the intraparticle conduction and axial dispersion were considered. It was determined that there were three nondimensional parameters that were significant in determining the operating conditions under which the simplified model could be modified satisfactorily to account for intraparticle conduction and axial dispersion. These parameters are

$$\text{Biot number:} \qquad \text{Bi} = \frac{hr_o}{k_m}$$

$$\text{Heat capacity ratio:} \qquad V_H = \frac{\rho_m c_m (1-\epsilon)}{\rho_f c_f \epsilon}$$

$$\text{Thermal capacity ratio:} \qquad \beta = \frac{V_H}{V_H + 1} \tag{12.22}$$

$$\text{Peclet number:} \qquad \text{Pe} = \frac{(\rho_f c_f v_a)^2 S_{fr} L \epsilon}{k_f A}$$

If the thermal conductivity of the storage material is very large, in the limiting case approaching infinity, the Biot number is small and approaches zero as a limit. In the limiting case, the storage material will have a uniform temperature and the solution of the intraparticle conduction model will approach that of the simplified model. The heat capacity ratio V_H appears in the nondimensional energy equation for the storage material and is the ratio of the heat capacity of the solid to that of the fluid contained within the voids of the bed. The thermal capacitance ratio β is the ratio of the heat capacity of the storage material to the heat capacity of the complete bed, solid and fluid.

At large values of V_H, the transient response of the packed bed can be adequately predicted by the simplified model if an effective film coefficient h_e is used in the evaluation of the nondimensional length and time. The effective film coefficient takes into consideration the axial dispersion and the intraparticle effects and has the following form:

$$\frac{1}{h_e} = \frac{1}{h}\left(1 + \frac{\mathrm{Bi}}{5}\right)\beta^2 + \frac{k_f A}{(\rho_f c_f v_a)^2 S_{fr} L \epsilon} \qquad (12.23)$$

When the heat capacity ratio is of the order of one, the above correction cannot be used.

Example 12.4 Determine the temperature of the fluid leaving the packed bed described in Example 12.1 and take into account intraparticle conduction and axial dispersion.

SOLUTION The physical properties needed in these calculations are

Carbon steel @ 45°C:

$$\rho_m = 7801.0 \text{ kg/m}^3 \quad (487.01 \text{ lb}_\text{m}/\text{ft}^3)$$

$$c_m = 0.473 \text{ kJ/kg °C} \quad (0.113 \text{ Btu/lb}_\text{m} \text{ °F})$$

$$k_m = 43.0 \text{ W/m °C} \quad (24.845 \text{ Btu/ft h °F})$$

Air @ 50°C (122°F):

$$\rho_f = 1.106 \text{ kg/m}^3 \quad (0.069 \text{ lb}_\text{m}/\text{ft}^3)$$

$$c_f = 1.007 \text{ kJ/kg °C} \quad (0.24 \text{ Btu/lb}_\text{m} \text{ °F})$$

$$k_f = 0.0278 \text{ W/m °C} \quad (0.016 \text{ Btu/ft h °F})$$

The nondimensional groups are

$$\mathrm{Bi} = \frac{h r_o}{k_m} = \frac{70.3(0.0125)}{43.0} = 0.0204$$

$$V_H = \frac{\rho_m c_m (1 - \epsilon)}{\rho_f c_f \epsilon} = \frac{(7801)(0.473)(1 - 0.37)}{(1.106)(1.007)(0.37)} = 5641.1$$

$$\beta = \frac{V_H}{V_H + 1} = \frac{5641.1}{5641.1 + 1.0} = 0.9998$$

Since the value of the heat capacity ratio is large, the simplified model can be employed by using the effective film coefficient. The mean interstitial fluid velocity is

$$v_a = \frac{\dot{m}}{S_{fr}\epsilon\rho_f} = \frac{0.065}{(0.08)(0.37)(1.106)} = 1.985 \text{ m/s}$$

The effective film coefficient is calculated by

$$\frac{1}{h_e} = \frac{1}{h}\left(1 + \frac{\text{Bi}}{5}\right)\beta^2 + \frac{k_f A}{(\rho_f c_f v_a)^2 S_{fr} L\epsilon}$$

$$= \frac{1}{70.3}\left(1 + \frac{0.0204}{5}\right)(0.9998)^2$$

$$+ \frac{(0.0278)(12.1)}{[1.106(1007.1)(1.985)]^2(0.08)(1.0)(0.37)}$$

$$= 1.428 \times 10^{-2}$$

Thus, $h_e = 70.03$ W/m^2 °C. The value of λ equals 12.946 and η equals 7.928. The nondimensional temperature of the fluid leaving the packed bed is obtained from Table 12.1 and has the value of 0.1600. The corresponding temperature of the fluid leaving the bed equals 21.2°C (70.16°F).

12.3 ARBITRARY INLET FLUID AND INITIAL BED CONDITIONS

In the formulation of the mathematical model for a packed bed that was described in Sec. 12.2, it was assumed that the bed was initially at a uniform temperature t_o and that the inlet fluid temperature was held constant at t_{fi} during the period of energy storage. In many applications, the inlet temperature of the fluid is time varying and the bed is not initially at a uniform temperature. Fortunately, the differential equations describing the system are linear, and the method of superposition described in Chap. 4 can be used in conjunction with the results obtained in Sec. 12.2.2 to predict the performance of the unit under these more realistic operating conditions.

12.3.1 Arbitrary Time Variation in Fluid Inlet Temperature

Since the energy conservation equations (12.7) and (12.9) for the simplified model of the packed bed and fluid are linear, the actual fluid inlet temperature can be broken down into discrete steps and a method of superposition can be used to obtain the transient response of the bed. An arbitrarily varying inlet fluid temperature that has been subdivided is shown in Fig. 12.7. The bed is considered to be initially at a uniform temperature t_o.

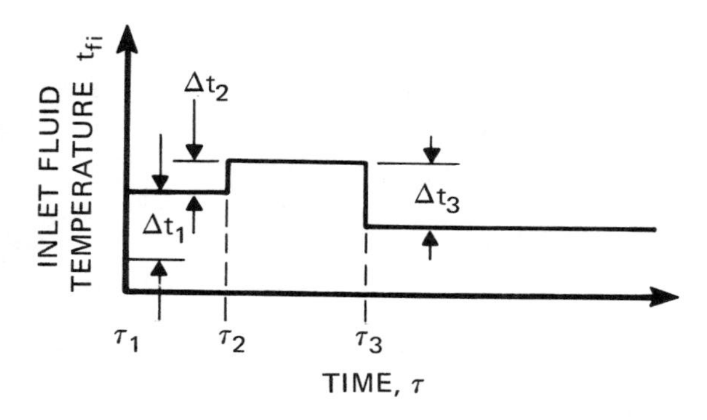

Figure 12.7 Inlet fluid temperature variation.

In Sec. 12.2.2, a solution was presented for the step change in inlet fluid temperature of a packed bed initially at a uniform temperature, Fig. 12.8. The nondimensional temperature

$$T_f(\eta, \xi) = \frac{t_f - t_o}{t_{fi} - t_o}$$

is a function of η and ξ, and the temperature of the fluid at the nondimensional length ξ is

$$t_f = t_o + (t_{fi} - t_o)T_f(\eta, \xi)$$

The quantity $t_{fi} - t_o$ will be identified as Δt.

A general expression for arbitrary variations in inlet fluid temperature containing both continuous and step changes was developed in Sec. 4.2. The general expressions are

Outlet fluid temperature:

$$t_{fo} = t_o + \sum_{j=1}^{J} (\Delta t_j)(T_f(\eta - \eta_j, \lambda)) + \int_0^{\eta} \frac{d(\Delta t)}{d\beta} (T_f(\eta - \beta, \lambda))\, d\beta \tag{12.24}$$

Heat storage:

$$Q = Mc_m\left[\sum_{j=1}^{J} \Delta t_j(Q^+(\eta - \eta_j, \lambda)) + \int_0^{\eta} \frac{d(\Delta t)}{d\gamma} Q^+(\eta - \gamma, \lambda)\, d\gamma\right] \tag{12.25}$$

where $M = S_{fr}L(1 - \epsilon)\rho_m$.

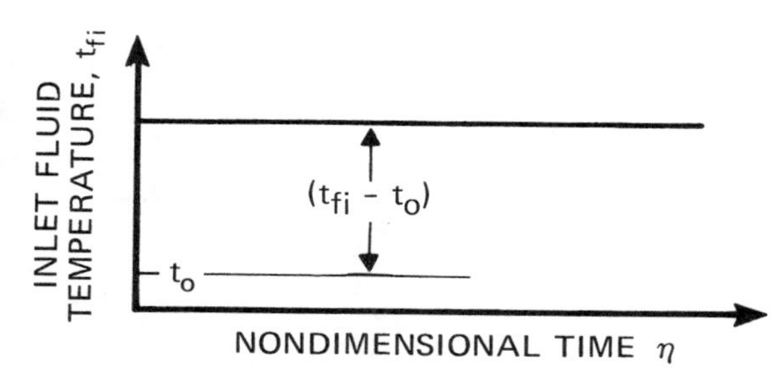

Figure 12.8 Step change in inlet temperature.

Example 12.5 The packed bed described in Example 12.1 is initially at a temperature of 10°C (50°F). For the first 10 min of operation the temperature of the gas entering the unit is at 80°C (176°F), after which the gas temperature drops to 70°C (158°F) for the next 10 min. When the unit has been operating 20 min, a further decrease in the gas temperature to 65°C (149°F) occurs. The inlet gas temperature variation is shown in Fig. 12.9*a*. Determine the temperature of the gas leaving the bed 29 min after the start of the storage process.

SOLUTION The fluid inlet temperature variation is subdivided into three steps as shown in Fig. 12.9*b*. The temperature steps are $\Delta t_1 = 70$°C, $\Delta t_2 = -10$°C, and $\Delta t_3 = -5$°C. The value of λ is 13.0 and the values of $(\eta - \eta_j)$ at 29 min are

Step	Δt, °C	$\eta - \eta_j$	T_{fj}	$(\Delta t_j)T_{fj}$
1	70	7.959	0.1592	11.144
2	−10	5.215	0.0405	−0.405
3	−5	2.47	0.0032	−0.016

The temperature of the fluid leaving the storage unit is

$$\begin{aligned} t_f &= t_o + (\Delta t_1)T_{f_1} + (\Delta t_2)T_{f_2} + (\Delta t_3)T_{f_3} \\ &= 10.0 + 11.144 - 0.405 - 0.016 \\ &= 20.723\text{°C} \quad (69.30\text{°F}) \end{aligned}$$

12.3.2 Arbitrary Variation in the Initial Temperature Distribution of the Packed Bed

In many instances, the packed bed has not reached an equilibrium state and thus the bed is not at a uniform temperature. The technique to be discussed in this section for the prediction of transient response of such units will be based on the principle of superposition. The inlet fluid temperature will be assumed to remain constant during the process, and the packed bed will be subdivided into sections that will be approximately at uniform temperatures.

A sketch of the subdivided bed temperature distribution is shown in Fig. 12.10*a*. Three subproblems, Fig. 12.10*b*, are needed to obtain the temperature of the fluid leaving the bed. In the first subproblem, the inlet temperature of the fluid is t_{fi}, and in all the other subproblems the inlet temperature of the fluid is zero.

The general expressions for the outlet fluid temperature and the heat storage were developed in Sec. 4.3. The outlet fluid temperature is

$$t_{fo} = t_{o1} + (t_{fi} - t_{o1})T_f\{\eta, \lambda\} + \sum_{j=1}^{J} (t_{oj} - t_{oj-1})(1.0 - T_f\{\eta, \lambda - \xi_j\})$$

$$+ \int_0^{\lambda} \frac{d(\Delta t_o)}{d\alpha} (1.0 - T_f\{\eta, \lambda - \alpha\})\, d\alpha \tag{12.26}$$

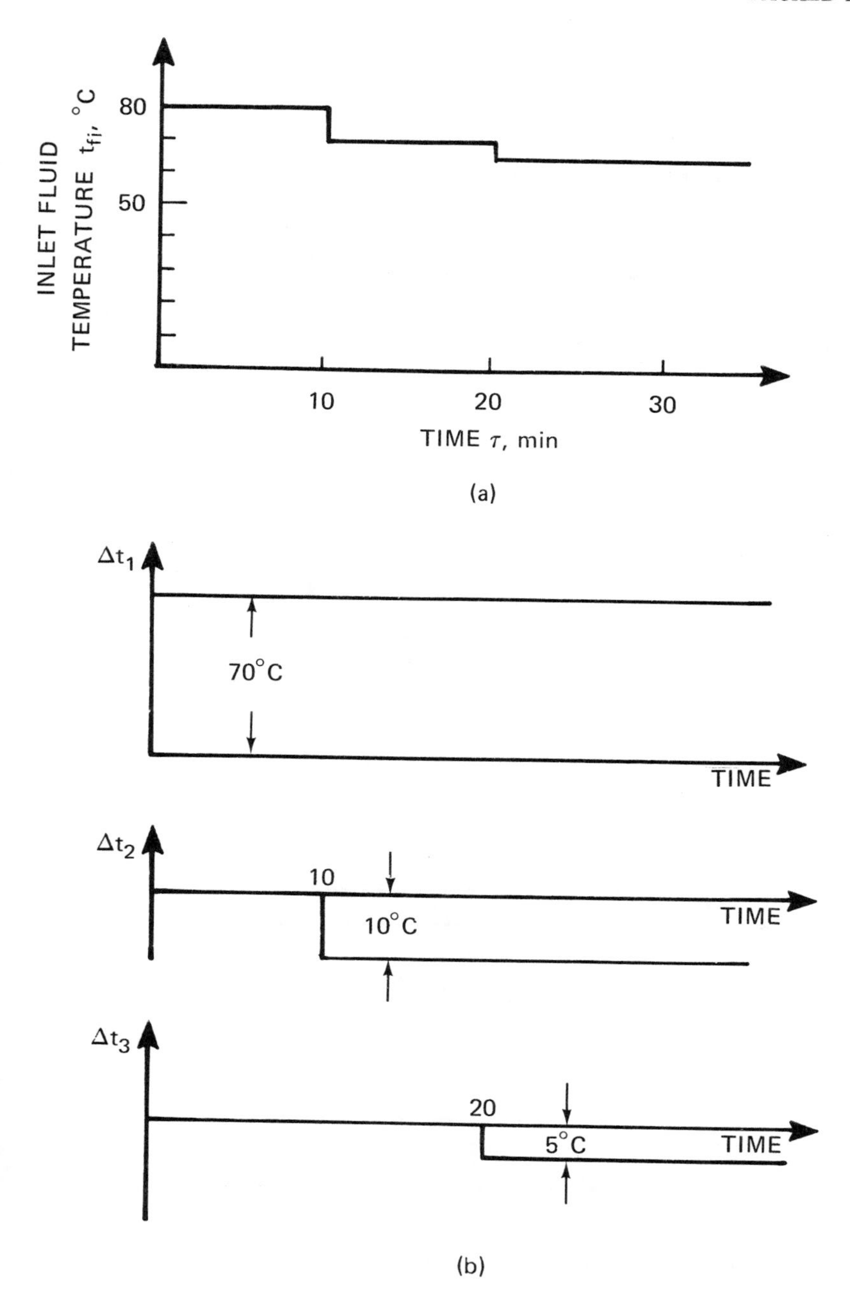

Figure 12.9 (*a*) Fluid inlet temperature variation. (*b*) Fluid temperature subproblems.

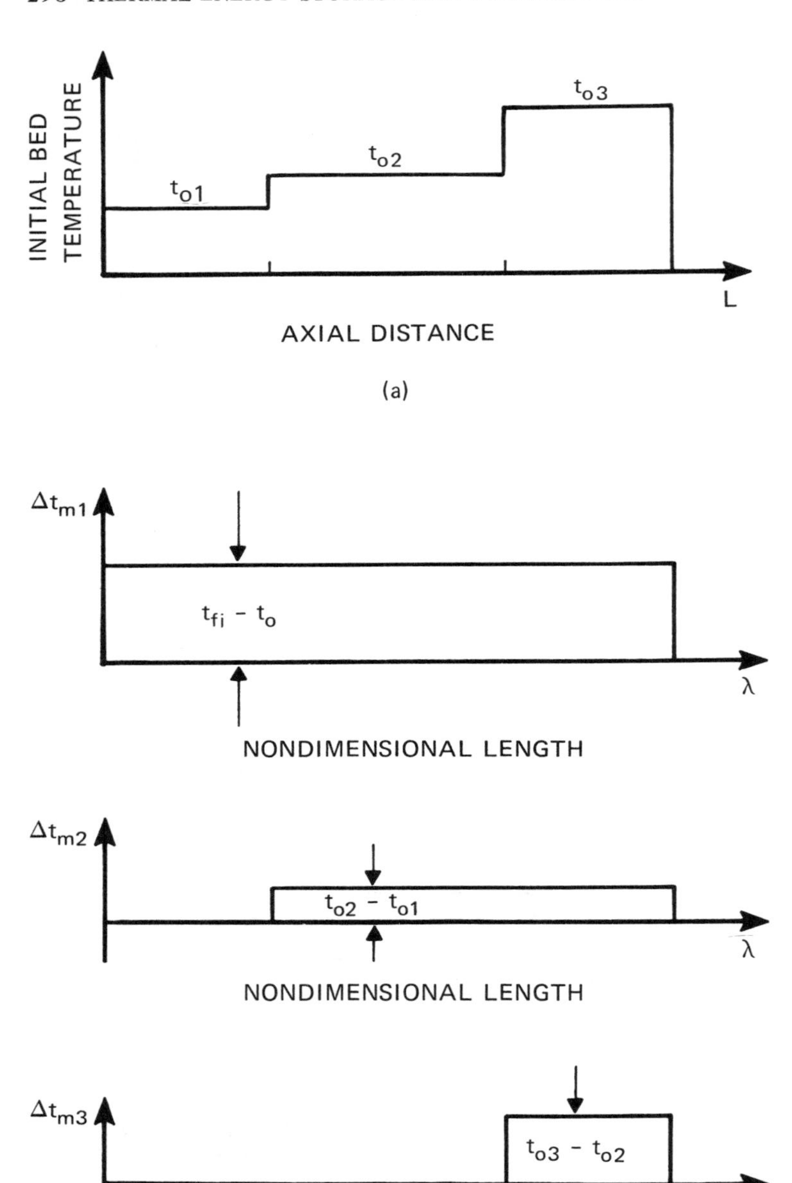

Figure 12.10 (*a*) Initial bed temperature distribution. (*b*) Nonuniform initial temperature superposition subproblems.

The heat storage is

$$Q = S_{fr}(1-\epsilon)\rho_m c_m \left[Q^+(\eta, \lambda)(t_{fi} - t_{o1})L - \sum_{j=2}^{J} Q^+(\eta, \lambda - \xi_j)(t_{oj} - t_{oj-1}) \right.$$

$$\left. \times (L - x_j) - \frac{\dot{m}_f c_f L}{hA} \int_0^\lambda Q^+(\eta, \lambda - \kappa) \frac{d(\Delta t_o)}{d\kappa} (\lambda - \kappa)\, d\kappa \right] \qquad (12.27)$$

Example 12.6 The packed bed described in Example 12.1 has the initial temperature distribution shown in Fig. 12.11. The inlet temperature of the fluid remains constant at 80.0°C (176°F). Determine the temperature of the fluid leaving 29 min after the start of the storage process.

SOLUTION Three subproblems are needed. The value of the nondimensional length of the storage unit is $\lambda = 13$. The first step in bed temperature occurs at $\xi = 3.25$ and the second at $\xi = 9.75$. The nondimensional time is $\eta = 7.959$. The fluid inlet temperature is 80°C (176°F) for the first subproblem and 0°C for all other subproblems. The temperature of the fluid leaving the packed bed at 29 min is

$$\begin{aligned} t_{fo} &= t_o + (t_{fi} - t_{o1})T_f(\eta, \lambda) + \Delta t_{o2}(1 - T_f(\eta, \lambda - \xi_2)) \\ &\quad + \Delta t_{o3}(1 - T_f(\eta, \lambda - \xi_3)) \\ &= 10.0 + (80.0 - 10.0)T_f(7.959, 13.0) \\ &\quad + (10.0)(1 - T_f(7.959, 9.75)) + (10.0)(1 - T_f(7.959, 3.251)) \\ &= 10.0 + (70.0)(0.1592) + (10.0)(0.6205) + (10.0)(0.055) \\ &= 27.90°\text{C} \quad (82.2°\text{F}) \end{aligned}$$

The values of $T_f(\eta, \lambda)$ were obtained from Table 12.1.

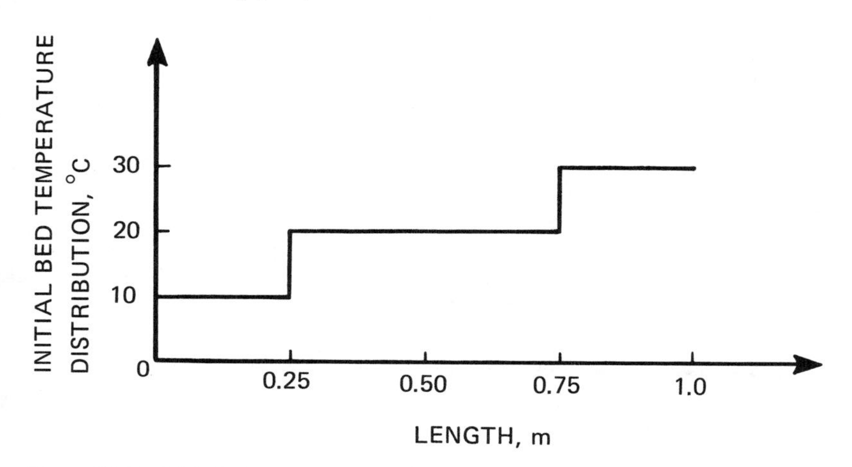

Figure 12.11 Initial bed temperature distribution.

12.3.3 Arbitrary Time Variation in Flow Rate

The determination of the transient response of a packed bed when the fluid flow rate varies with time can be obtained by the superposition technique described in Chap. 4. Only those situations in which the flow rate undergoes step changes will be considered here. The basic approach used will be to determine the temperature distribution in the bed at the instant the flow rate is changed. The technique described in Sec. 12.3.2, for arbitrary variations in the initial bed temperature distribution will then be used to determine the transient response until the next change in flow rate occurs. Note that after every change in velocity it is necessary to reevaluate the convective film coefficient and recalculate the pressure drop. The values of η and ξ must then be recalculated.

The procedure followed will be illustrated by using the temperature and flow variations shown in Fig. 12.12. The steps are as follows:

1. The convective film coefficient is determined at the mass rate of flow $\dot{m}_1$.
2. The temperature of the fluid leaving the packed bed during the time interval $0 \leqslant \tau \leqslant \tau_2$ is given by $t_{fo} = t_o - (t_{fi} - t_o)T_f(\eta_1, \lambda_1)$.
3. The temperature distribution in the packed bed at $\tau = \tau_2$ is calculated by $t_m = t_o - (t_{fi} - t_o)T_f(\xi, \eta_1)$.
4. The packed bed is subdivided into sections considered to be at a uniform temperature.
5. The convective film coefficient is determined at the new mass flow rate $\dot{m}_2$.
6. The temperature of the fluid leaving the packed bed for $\tau > \tau_2$ is given by Eq. (12.26), where η, λ, and ξ have been evaluated by using h_2.

12.3.4 Periodic Operation of Packed Bed Units

There are many industrial applications in which streams of hot and cold fluids are alternately passed through the packed bed. If the inlet fluids have a regular timewise variation, the packed bed, after a number of cycles, will reach a periodic operating condition.

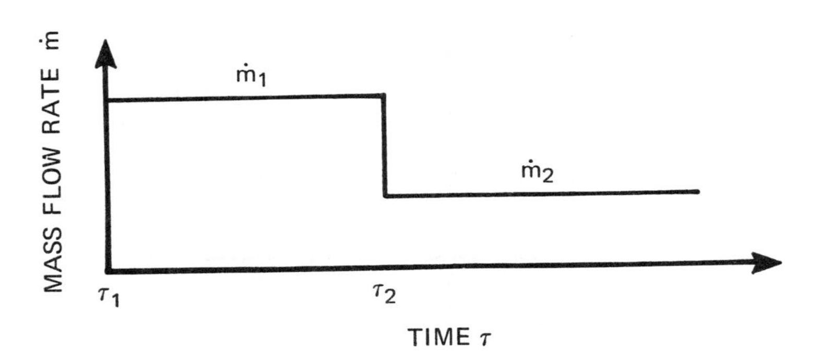

Figure 12.12 Variation in mass flow rates.

The prediction of the periodic operating conditions of packed beds can be obtained from the results presented in Chaps. 5 through 10, on the periodic operation of regenerators. The nondimensional variables for the packed bed will be

$$\lambda = \frac{hA}{\dot{m}_f c_f} \quad \text{and} \quad \pi = \frac{hA\tau_o}{S_{fr}(1-\epsilon)\rho_m c_m L}$$

where τ_o is the length of the heating or cooling period.

REFERENCES

1. M. Raiz, "Analytical Solutions for Single and Two-Phase Models of Packed Bed Thermal Energy Systems," *J. Heat Transfer, Trans. ASME*, vol. 99, 1977, p. 489.
2. T. E. W. Schumann, "Heat Transfer: A Liquid Flowing Through a Porous Prism," *J. Franklin Inst.*, vol. 208, 1929, p. 405.
3. A. Anzelius, "Uber Erwarmung Vermittals Durchstromender Medien," *Z. Agnew. Math. Mech.*, vol. 6, 1926, p. 291.
4. C. C. Furnas, "Heat Transfer from a Gas Stream to a Bed of Broken Solids," *Ind. Eng. Chem.*, vol. 22, 1930, p. 721.
5. H. Hausen, *Warmeubertragung im Gegenstrom, Gleichstrom und Kreuzstrom*, Springer-Verlag, Berlin, 1950.
6. F. W. Larsen, "Rapid Calculations of Temperature in a Regenerative Heat Exchanger Having Arbitrary Initial Solid and Entering Fluid Temperatures," *Int. J. Heat Mass Transfer*, vol. 10, 1967, p. 149.
7. A. Klinkenberg, "Heat Transfer in Cross-Flow Heat Exchangers and Packed Beds," *Ind. Eng. Chem.*, vol. 46, 1954, p. 2285.
8. R. Chao and H. E. Hoelscher, "Simultaneous Axial Dispersion and Absorption in a Packed Bed," *AIChE J.*, vol. 12, 1966, p. 271.
9. J. F. Edwards and J. F. Richardson, "Gas Dispersion in Packed Beds," *Chem. Eng. Sci.*, vol. 23, 1968, p. 109.
10. J. B. Rosen, "Kinetics of Fixed Bed System for Solid Diffusion into Spherical Particles," *J. Chem. Phys.*, vol. 20, 1952, p. 387.
11. R. E. Babcock, D. W. Green, and R. H. Perry, "Longitudinal Dispersion Mechanisms in Packed Beds," *AIChE J.*, vol. 12, 1966, p. 922.
12. D. Handley and P. J. Heggs, "The Effect of Thermal Conductivity of the Packing Material on Transient Heat Transfer in a Fixed Bed," *Int. J. Heat Mass Transfer*, vol. 12, 1969, p. 549.
13. P. K. Leung and D. Quon, "A Computer Model for the Regenerative Bed," *Can. J. Chem. Eng.*, vol. 43, 1965, p. 45.
14. C. P. Jeffreson, "Prediction of Breakthrough Curves in Packed Beds," *AIChE J.*, vol. 18, 1972, p. 409.

CHAPTER

THIRTEEN

DESIGN OPTIMIZATION

13.1 INTRODUCTION

The designer of a heat storage unit has many options with regard to the geometry and dimensions of the storage unit, the storage material used, the mass flow rate of the fluid, and the time duration when heat is stored or retrieved. The transient response and the overall performance of the unit will depend on the final design, and the design engineer may well ask. "Have I selected the best design?" The answer to this question is not easily obtained. First, one must be more specific with regard to the word "best." Does one want the unit to have the largest possible amount of energy stored during a given time interval; to store as much of the available energy as possible; to store the maximum amount of energy per unit volume or cost of the storage unit; or to store the maximum amount of the available energy per unit volume or cost of the storage unit? Many other criteria could also be used to determine if the design is the "best" possible.

Since individual or particular storage applications may have different objectives for the optimization of the design, it is very difficult to make any general statements. The objective of this chapter is to acquaint the designer with some of the tools available for use in making a rational decision concerning the best design. The single-fluid heat storage unit, initially at a uniform temperature and subjected to a step change in fluid inlet temperature, single-blow operating mode, will be used in this presentation. The method for predicting the transient response of these units has been presented in Chaps. 2 and 3. The finite conductivity model of the heat storage unit will be used in the discussions.

The amount of heat stored will be dependent on the mass rate of flow and the inlet temperature of the hot gas, the storage material and fluid used, and the length and thickness of the storage material. Since the quantity of energy stored is time dependent, it is necessary to know the transient response characteristics of the units. A typical solid sensible heat storage unit is shown in Fig. 13.1. The transient response of this unit has been presented by Schmidt and Szego [1] and is described in Chap. 3. The governing conservation of energy equations for the storage material and the fluid were nondimensionalized and solved using numerical techniques. The following nondimensional parameters were introduced for the flat slab storage unit:

Fourier number: $$\text{Fo} \equiv \frac{\alpha \tau}{w^2}$$

Biot number: $$\text{Bi} \equiv \frac{hw}{k_m}$$

and G^+/V^+, where $G^+ = P_h k_m / \dot{m}_f c_f$ and $V^+ = w/L$.

The results for the transient nondimensional outlet temperature T_{fo} and the fraction of the maximum possible heat stored Q^+ were presented in a series of curves and tables. In order to simplify the use of these results in the design of storage units, they were retained in the computer and a program was written to perform the necessary interpolations. The program has been named HSSF. For a given set of fluid flow conditions and storage unit configuration, the values of Fo, Bi, and G^+/V^+ can be computed and the program used to determine the nondimensional outlet fluid temperature and the fraction of the maximum possible heat that is stored.

Although this material will be presented with reference to heat storage, the results are also applicable to systems in which heat is transferred from the storage unit to the fluid. The only limitation is that the unit is initially in temperature equilibrium, at a uniform temperature. All heat quantities will be negative, since heat is rejected by the storage material.

13.2 COMPLEX OPTIMIZATION METHOD

The selection of a method of optimizing the design of the thermal energy storage unit was limited by the mathematical model describing the system. Seven variables are involved in the process. Since these are limited by either explicit or implicit constraints, applicable optimization methods were restricted, by practical considerations, to those that involve direct search techniques developed for nonlinear constrained optimization. Of the several such methods available, the "complex optimization method" of Box [2] was chosen. This selection was made for the following reasons:

1. The likelihood of the optimization routine to reach a global, rather than a local, optimum
2. The ease with which the optimization algorithm could be converted into a computer program
3. The nature of the method, which permitted a solution to be reached without requiring that the independent variables be scaled

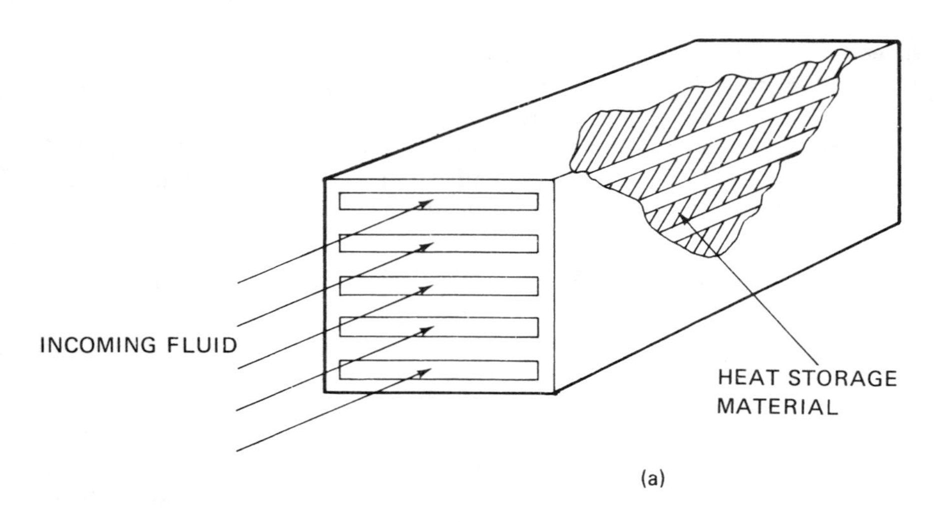

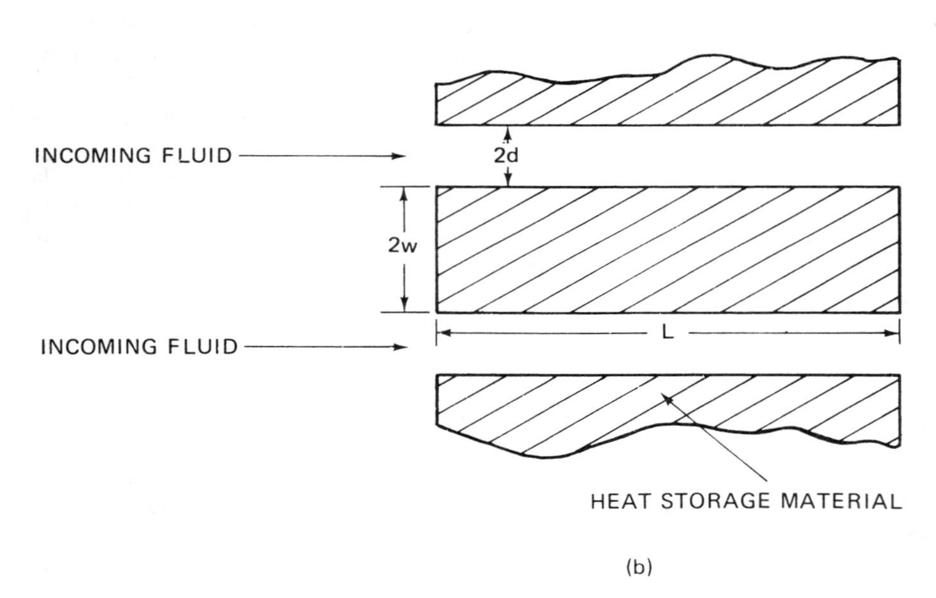

Figure 13.1 The solid sensible, single-fluid heat storage unit. (*a*) Three-dimensional view. (*b*) Cross-sectional view.

4. The ability of the technique to handle both explicit and implicit inequality constraints
5. The avoidance of time-consuming, and possibly error-inducing, gradient approximations
6. The computational efficiency of the method, especially when compared to Monte Carlo and random search techniques.

The complex method is particularly suited for convex search regions and has been used with considerable success in the optimization of the solutions to many practical problems. Examples of these have been presented by Adelman and Stevens [3].

The complex method of nonlinear constrained optimization is an adaptation of the "simplex method" originally proposed by Spendley, Hext, and Himsworth [4]. The objective is to maximize functions of the form $U(x_1, x_2, x_3, \ldots, x_n)$, where the x's are the independent variables. This maximization may be limited by explicit constraints of the form $a_i \leqslant x_i \leqslant b_i$, where $i = 1, 2, 3, \ldots, n$, and implicit constraints of the form

$$c_j \leqslant g_j(x_1, x_2, \ldots, x_n) \leqslant e_j$$

The Box complex method performs the optimization by forming a polygon of m vertices, $m \geqslant n + 1$, within the n-dimensional space consisting of the n independent variables and bounded by the explicit and implicit constraints placed on the problem. This space is to be searched to locate the combination of the n independent variables that will yield the optimum value of $U(x_1, x_2, x_3, \ldots, x_n)$. The polygon, termed a complex by Box, is made up of p vertices, $p = 1, 2, 3, \ldots, m$, each of which consists of a set of n independent variables, $x_{1p}, x_{2p}, \ldots, x_{np}$. The function to be optimized, U, is evaluated at each vertex. The vertex that yields the lowest value of U is replaced by a new vertex whose set of independent variables give a higher value of U. The replacement of lower-valued vertices with higher-valued vertices continues until the vertices of the complex converge to one value, which is the optimum for the system.

The following algorithm details the "complex optimization method" used by Somers [5] in the design optimization of the thermal energy storage unit. It is assumed throughout the discussion that the optimum being sought is a maximum value. The six steps of the algorithm have been labeled in the program listing.

1. The initial vertex of the complex is generated with the values of the n independent variables that make up the vertex and satisfies all system constaints, explicit and implicit. The generation procedure consists of a pattern search within the n-dimensional space defined by the independent variables and bounded by the constraints. Initially, every independent variable x_i is set at the value of its lower explicit constraint a_i. The optimization function $U(x_1, x_2, \ldots, x_n)$ is then evaluated for this set of values. If all implicit constraints are met, the set of independent variable values becomes the first vertex of the complex. If any implicit constraints are violated, however, the first independent variable, x_1, is increased in value by one-tenth of the difference between the lower and upper explicit constraints on x_1. Thus, the new value of x_1 is

$$x_1 = a_1 + \tfrac{1}{10}(b_1 - a_1) \tag{13.1}$$

The function U is reevaluated, and if implicit constraint violations still appear, the first independent variable is incremented upward again by one-tenth the difference between its lower and upper explicit constraints. This process continues as long as there are implicit constraint violations. If implicit constraint violations are still present when the upper explicit constraint, b_1, of the first independent variable is reached, x_1 is reset to the value found from Eq. (13.1) and x_2 is incremented upward by one-tenth the difference between its upper and lower constraints. A check for implicit constraint violations is made. If the function U still violates an implicit constraint, x_2 is again increased by one-tenth the difference between its upper and lower constraints. This procedure continues throughout all the n independent variables until either a set of independent variable values is found that satisfy all the implicit constraints, with this set of values becoming the initial vertex of the complex, or until the entire region of feasible values has been searched.

2. The remaining $m-1$ vertices of the initial complex are randomly generated by using an equation of the form

$$x_i = a_i + r(b_i - a_i) \tag{13.2}$$

where r is a randomly generated quantity between 0.0 and 1.0. The RAND subroutine, based on the RANDU subroutine supplied by IBM [6], was used in the computer program to generate uniformly distributed random numbers between 0.0 and 1.0 to be used for r. All vertices generated in this manner are checked for implicit constraint violations. If the pth randomly generated vertex, $2 \leqslant p \leqslant m$, violates an implicit constraint, a new pth vertex is developed. This is done by computing the geometric centroid, $(x_{1c}, x_{2c}, \ldots, x_{nc})$, of the $p-1$ randomly generated vertices that satisfied the implicit constraints and the previously known initial vertex of the complex:

$$x_{ic} = \frac{1}{p-1} \sum_{j=1}^{p-1} x_{ij} \qquad i = 1, 2, 3, \ldots, n \tag{13.3}$$

The subscript c denotes the centroid. The pth randomly generated vertex is then moved half the distance toward the centroid and once again checked against the implicit constraints for violations. If a violation still exists, the point is once again moved half the distance toward the centroid. This process is repeated until the point satisfies the implicit constraints of the problem. The point is then made a vertex of the initial complex. In this manner $m-1$ vertices, satisfying all explicit and implicit constraints, are found.

3. With all the vertices of the initial complex now generated, the optimization function $U_p(x_{1p}, x_{2p}, \ldots, x_{np})$ is evaluated at each pth vertex, $p = 1, 2, \ldots, m$. The vertex at which the function evaluation, U_p, is lowest is found, and the geometric centroid, $(x_{1c}, x_{2c}, \ldots, x_{nc})$, of all the vertices except for the lowest-valued vertex is calculated:

$$x_{ic} = \frac{1}{m-1} \sum_{j=1}^{m-1} x_{ij} \qquad i = 1, 2, 3, \ldots, n \tag{13.4}$$

4. The vertex that yielded the lowest optimization function value in step 3 is replaced. The replacement point is located on an extension of the line that passes through the vertex being replaced and the centroid. The location of the new point equals the distance between the centroid and the lowest-valued vertex times a reflection parameter β, where $\beta \geqslant 1$. If this new point violates an explicit constraint(s), the offending independent variable(s) is reset to a value that is within 10^{-6} of the con-

straint. Any implicit constraint violations are rectified by moving the new point halfway toward the centroid. If the violations are still present, the new point is again moved halfway toward the centroid, and the procedure is repeated until the implicit constraints are satisfied.

5. The value of $U(x_1, x_2, \ldots, x_n)$ is evaluated at the new point. If the function evaluation gives a quantity greater than that produced by the vertex to be replaced, the replacement is made and the new point becomes the mth vertex of the complex. If, however, the function value of the vertex that is to be replaced exceeds that of the newly generated point, the new point is moved half the distance toward the centroid and U is evaluated again. This process continues until the optimization function value U of the new point exceeds the value of the vertex that is being replaced. The new point then becomes the mth vertex of the complex. If, during this procedure, the new point should approach too closely to the centroid, within 10^{-5} times the original distance between the centroid and the newly generated point, the lowest-valued vertex is retained in the complex, the new point is discarded, and the next lowest-valued vertex is replaced, once again beginning with step 4.

6. The complex is considered to have converged to the maximum when five consecutive iterations of the algorithm produce complexes whose lowest-valued vertices yield optimization function values U that are all within 2×10^{-7} of each other.

13.3 OPTIMIZATION OF A SLAB HEAT STORAGE UNIT FOR SINGLE-BLOW OPERATING MODE

There are two logical parameters that one might wish to optimize in the design of the heat storage unit. One is the amount of heat being stored in the storage material. In terms of the nondimensional quantities, the value of Q^+ would thus be maximized. A second possibility would be to store as much of the available energy as possible. The nondimensional quantity A^+ will be introduced and defined as

$$A^+ = \frac{\text{heat stored}}{\text{available thermal energy in the fluid}} = \frac{Q}{\dot{m}_f c_f (t_{fi} - t_o)\tau} \tag{13.5}$$

The reference temperature used in determining the available energy was the initial temperature of the unit. From the basic definition for A^+ and Q^+, the following relationship can be found:

$$A^+ = \left(\frac{w c_m \rho_m L P_h}{\dot{m}_f c_f \tau}\right) Q^+ \tag{13.6}$$

or, in dimensionless form,

$$A^+ = \left(\frac{G^+}{V^+}\right) \frac{Q^+}{\text{Fo}} \tag{13.7}$$

Both A^+ and Q^+ are functions of Bi, Fo, and G^+/V^+ and can be determined by using the interpolation program HSSF. The modified Box complex computer program developed by Somers [5] allows the design to be optimized by maximizing either Q^+ or A^+. Because both Q^+ and A^+ are functionally related to the same dimensionless parameters, the remainder of this discussion will deal only with Q^+, although the conclusions and observations presented for Q^+ are equally applicable to A^+.

In dimensional form, the functional relationship for Q^+ can be written as $Q^+\{w, L, d, P_h, h, \dot{m}_f, \tau, t_o, t_{fi}, k_m, c_m, \rho_m, c_f, \rho_f\}$. This relationship for Q^+ has been simplified by noting that the cross-sectional area of the fluid flow channel S, the fluid velocity v, and the hydraulic diameter of the flow channel D_h can all be expressed as functions of other parameters; $S = dP_h$, $v = \dot{m}_f/\rho_f d$, and $D_h = 4d$. It is assumed that the heat storage unit's width is much greater than the semithickness of the fluid flow channel d, and that the unit is designed on a per-unit-width basis.

Although the modified Box complex method developed by Somers can be used to determine the value of any number of independent variables for the optimum design, the computational time becomes excessive as the number of variables increases. It was felt that the designer should make some decisions prior to the use of the program to optimize the design. The following items must be specified before the start of the optimization process:

1. The heat storage material
2. The fluid used to transport the thermal energy
3. The duration of the heat storage or retrieval process, τ
4. The temperature of the fluid entering the storage unit, t_{fi}
5. The initial temperature of the storage material, t_o
6. The semithickness of the thermal storage material, w
7. The mass flow rate of the fluid for one-half channel, $\dot{m}_f$

The first five items can usually be readily established, since they are closely related to the application for which the storage unit is being designed. The specification of the semithickness of the material, item 6, was considered to be realistic, since many common heat storage materials are supplied in the form of bricks or slabs that are of a standard thickness. The properties of these materials make it very difficult to subdivide them into thinner sections. As a result, for practical units composed of these materials, the unit's semithickness cannot vary arbitrarily, but must instead assume values that result when an integer number of bricks or slabs are used. The last item specified, $\dot{m}_f$, is selected as a matter of convenience since the total fluid flow rate is usually known and $\dot{m}_f$ can be determined once the number of channels are selected. It would appear, however, that in many cases one would like to have this item included in those selected by the optimization program.

In addition, certain constraints must be imposed on the heat storage unit design in the interests of practicality. They are as follows:

1. The maximum and minimum permissible lengths of the storage unit
2. The maximum and minimum semithicknesses of the fluid flow channel
3. The maximum allowable pressure drop, ΔP_{max}, in the fluid as it flows through the storage unit

4. The minimum amount of heat storage required of the unit, Q_{min}
5. The maximum outlet temperature of the fluid leaving the heat storage unit, t_{fo}

The maximization of Q^+ is achieved by selecting the proper geometry for the thermal energy storage unit. The specification of the previously noted items reduces the number of independent variables to three: the length of the heat storage unit L; the semithickness of the fluid flow channel d; and the heated perimeter of the flow channel P_h. The design optimization is performed on a per-unit-width, 1-cm basis, and thus the value of P_h is fixed at 1 cm. The nondimensional heat storage Q^+, which was initially a function of 12 variables, is now a function of only three: L, d, and h.

The convective film coefficient h is the only parameter that must be known before Q^+ can be determined. To calculate h, the expressions presented by Legkiy and Makarov [7] for convective heat transfer in ducts of rectangular cross section were utilized:

$$\begin{aligned} &\mathrm{Re}_L < 25 \times 10^3;\ \frac{L}{D_h} < 60: && h = 0.22 \frac{k_f}{L} \mathrm{Re}_L^{0.6} \left(\frac{L}{D_h}\right)^{0.08} \\ &\mathrm{Re}_L > 25 \times 10^3;\ \frac{L}{D_h} < 60: && h = 0.029 \frac{k_f}{L} \mathrm{Re}_L^{0.8} \left(\frac{L}{D_h}\right)^{0.08} \\ &\mathrm{Re}_D > 10^4;\ \frac{L}{D_h} > 60; && h = 0.018 \frac{k_f}{D_h} \mathrm{Re}_D^{0.8} \end{aligned} \tag{13.8}$$

This can be expressed as $h\{d, L, \dot{m}_f, \rho_f, k_f, \mu_f, c_f\}$.

To ensure that the thermal energy storage unit design will be practical in terms of size, performance, and economics, certain design constraints were placed on the two independent variables, L and d, as well as on functions involving these dependent variables. Explicit constraints consisted of minimum and maximum limitations on both the storage unit's length and the fluid channel's semithickness to ensure that the size of the heat storage unit would be reasonable.

Three implicit constraints were imposed on the dependent variables. The minimum and maximum values for the dimensionless fluid outlet temperature, T_{fo}, were stipulated. For situations in which heat is being stored, the maximum T_{fo} limitation is important because it prevents the design of a unit that would become inefficient toward the end of energy storage period. As the temperature of the storage material surface in contact with the fluid increases during the storage process, a point will be reached where the storage material's temperature is nearly the same as that of the incoming fluid. Thus, very little heat storage will take place, which in turn means that the outlet fluid temperature will be almost as high as the incoming fluid temperature. Therefore, by mandating a maximum outlet fluid temperature, the designer, in essence, is requiring a minimum heat storage rate at all times during the heat storage period. A minimum T_{fo} constraint performs primarily the same function during heat retrieval processes as the maximum T_{fo} requirement does during heat storage.

The second implicit constraint placed on the heat storage unit design was the requirement that the unit must store a minimum amount of heat, Q_{min}. This constraint is needed if a practical storage unit is to result from the optimization process. If no minimum amount of heat storage was required, many of the unit designs resulting from the optimization of Q^+ would approach zero length, since a unit of short length has a Q_{max} value almost equal to zero. The result would be that $Q^+ = Q/Q_{max}$ would almost instantaneously reach a value of 1.0 during the heat storage process.

The final implicit constraint imposes a maximum pressure drop limitation, ΔP_{max}, on the fluid as it flows through the heat storage unit. The equation used to compute the pressure drop is

$$\Delta P = \frac{\dot{m}_f^2 f}{2 g_c \rho_f} \left(\frac{L}{d^3}\right) \frac{1}{P_h^2} \tag{13.9}$$

As a functional relationship, this becomes $\Delta P\{L, d, \dot{m}_f, \rho_f, f\}$. The quantity f is the Fanning friction factor. The value of the friction factor may be calculated from expressions given by Kays [8] for smooth-walled ducts:

$$\begin{aligned} 5 \times 10^3 < \text{Re}_D < 3 \times 10^4: &\qquad f = 0.079\ \text{Re}_D^{-0.25} \\ 3 \times 10^4 < \text{Re}_D < 10^6: &\qquad f = 0.046\ \text{Re}_D^{-0.2} \end{aligned} \tag{13.10}$$

The friction factor is a function of the Reynolds number based on the hydraulic diameter, Re_D, which in turn is a function of $\dot{m}_f$ and μ_f. The expression for the friction factor clearly shows the effect that the maximum pressure drop constraint has on the two independent variables, L and d. Since there is an upper bound on the value of the pressure drop, the length of the heat storage unit L is limited. Also, the value of the fluid channel semithickness d is restrained from becoming too small.

The explicit constraints used in the forthcoming analysis unless otherwise stated were

Length: $0.2 \leqslant L \leqslant 10.0$ m

Semichannel thickness: $0.5 \leqslant d \leqslant 2$ cm

The implicit constraints were

Fluid outlet temperature: $0.2 \leqslant T_{fo} \leqslant 0.875$

Minimum heat storage: $Q_{min} \leqslant Q$

Pressure drop: $0 \leqslant \Delta P \leqslant 12.7$ cm H_2O (5 in H_2O)

13.4 RESULTS FROM OPTIMIZATION STUDY OF SLAB HEAT STORAGE UNITS

The computer program developed by Somers permits the optimum design of a thermal energy storage unit to be obtained for virtually an infinite combination of operating conditions and design constraints. The information derived from an optimization analysis must be presented in a concise manner, yet it also must be in as convenient a form as possible for use by a designer. A graphical or tabulated representation of the data satisfies both of these requirements.

The influence of variations in the mass flow rate, the maximum allowable pressure drop, and the outlet fluid temperature were evaluated by Schmidt et al. [9] for three different storage materials, Feolite, concrete, and cast iron, and are tabulated in Tables 13.1 through 13.3. The time, $\tau = 60$ min, and the semithickness, $w = 4$ cm, were held constant. The constraints placed on the other variables are given in the tables.

The effect of changes in the flow rates on the optimum designed heat storage unit with Feolite as the storage material is shown in Table 13.1. An optimum unit, with a mass flow rate of 4.0 g/s cm of unit width, a nondimensional outlet fluid temperature of $T_{fo} = 0.875$, and a pressure drop of 12.7 cm of water, was selected. The influence of variations in the flow rate can be obtained by comparing runs F-1, F-2, and F-3. Increasing the flow rate will cause a slight decrease per unit length in Q, Q^+, and A^+, and the total volume of material for the optimum design will be decreased. The effects are not dramatic, and one can conclude that the connection of three units in parallel, each with a flow rate of 2.0 g/s cm of width, will result in the storage of approximately 4% more heat than a single channel, 2.98 times longer, with a flow rate of 6.0 g/s cm of width.

The energy storage is a combined convection-conduction process. Increasing the fluid velocity in the channels by allowing a larger maximum pressure drop will affect only the convective part of the process; the amount of energy stored will not be in direct proportion to the pressure drop. As an example, a 10-fold increase in the pressure drop, from 1.27 to 12.7 cm of water, will increase the total heat storage by 26%, the Q^+ by 21.7%, and A^+ by 25.8%. These comparisons are obtained from runs F-1, F-4, and F-5.

A decrease in the maximum permissible fluid outlet temperature will increase the length of the heat storage unit, the total amount of heat stored, and the fraction of the total available energy stored. The value of Q^+ and thus the average temperature of the storage unit will, however, be decreased, as shown by runs F-1, F-6, and F-7.

The most unexpected results of this study are those associated with a comparison of the different storage materials. In the past all comparisons of storage materials were made with geometrically identical units. The results usually indicated that those units constructed of cast iron and Feolite have similar storage characteristics, whereas those with concrete have much less desirable storage properties. The relative merits of the different materials were found to be dependent, to a certain extent, on the length of time during which storage took place. Quite different trends were found when the optimum design was used. A comparison of the results for the 60-min storage duration presented in Tables 13.1 through 13.3 indicates that for the standard conditions (F-1, C-1, and CI-1), the concrete storage unit is about twice as long as the Feolite unit but is able to store more heat, and has better utilization of the storage material, since Q^+ is greater and stores more of the available energy. When one considers installation cost, it is obvious that the use of concrete as a storage material must be given serious consideration. It must, however, again be emphasized that the trends are dependent on the duration of heat storage or removal, and longer or shorter storage duration can be expected to alter the relative merits of the different storage materials. The effects of changes in mass flow rates, maximum pressure drop, and maximum outlet temperature

Table 13.1 Optimum heat storage units: Feolite [9]

Duration of active storage = 40 min
Inlet temperature = 100°C
Initial storage unit temperature = 20°C
Semithickness of storage material = 4 cm

Fluid: Air
Density: 1.06 kg/m³
Specific heat: 1.01 kJ/kg °C
Thermal conductivity: 2.88×10^{-2} W/m °C
Kinematic viscosity: 0.188×10^{-4} m²/s

Storage material: Feolite
Density: 3.9×10^{3} kg/m³
Specific heat: 0.92 kJ/kg °C
Thermal conductivity: 0.021×10^{2} W/m °C
Thermal diffusivity: 0.58×10^{-6} m²/s

Constraints: Length: $0.2 \leqslant L \leqslant 10$ m
Semithickness of channel: $0.5 \leqslant d \leqslant 2$ cm

	Run number						
	F-1	F-2	F-3	F-4	F-5	F-6	F-7
Mass flow rate, g/s cm width	4	2	6	4	4	4	4
Semithickness of channel, cm	1.01	0.53	1.47	1.72	0.807	1.26	1.63
Length, m	3.55	1.78	5.30	3.49	1.61	6.87	14.94
Outlet temperature, T^+	0.875	0.875	0.875	0.875	0.875	0.75	0.50
Heat storage, kJ/cm width	292	150	432	247	311	478	768
Q^+	0.718	0.732	0.709	0.616	0.750	0.606	0.448
A^+	0.252	0.258	0.248	0.213	0.268	0.412	0.662
ΔP, cm water	12.7	12.7	12.7	2.54	25.4	12.7	12.7
Air hp/cm width	0.0063	0.0031	0.0094	0.00126	0.0126	0.0063	0.0063

Table 13.2 Optimized heat storage unit: concrete [9]

Duration of active storage = 60 min
Inlet temperature = 100°C
Initial storage unit temperature = 20°C
Semithickness of storage material = 4 cm

Fluid: Air
Density: 1.06 kg/m^3
Specific heat: 1.01 kJ/kg °C
Thermal conductivity: 2.88×10^{-2} W/m °C
Kinematic viscosity: 0.188×10^{-4} m^2/s

Storage material: Concrete
Density: 2.1×10^3 kg/m^2
Specific heat: 0.878 kJ/kg °C
Thermal conductivity: 0.011×10^2 W/m °C
Thermal diffusivity: 0.6×10^{-6} m^2/s

Constraints: Length: $0.2 \leqslant L \leqslant 30$ m
Semithickness of channel: $0.5 \leqslant d \leqslant 2.5$ cm

	Run number						
	C-1	C-2	C-3	C-4	C-5	C-6	C-7
Mass flow rate, g/s cm width	4	2	6	4	4	4	4
Semithickness of channel, cm	1.279	0.671	1.864	2.154	1.021	1.566	1.982
Length, m	7.17	3.61	10.70	6.86	7.31	13.16	26.70
Outlet temperature, T^+	0.875	0.875	0.875	0.875	0.875	0.75	0.50
Heat storage, kJ/cm width	329	168	488	286	345	525	813
Q^+	0.778	0.788	0.773	0.706	0.800	0.675	0.515
A^+	0.284	0.289	0.280	0.246	0.297	0.452	0.700
ΔP, cm water	12.7	12.7	12.7	2.54	25.4	12.7	12.7
Air hp/cm width	0.0063	0.0031	0.0094	0.0013	0.0126	0.0063	0.0063

Table 13.3 Optimized heat storage unit: cast iron [9]

Duration of active storage = 60 min
Inlet temperature = 100°C
Initial storage unit temperature = 20°C
Semithickness of storage material = 4 cm

Fluid: Air
Density: 1.06 kg/m^3
Specific heat: 1.01 kJ/kg °C
Thermal conductivity: 2.88×10^{-2} W/m °C
Kinematic viscosity: 0.188×10^{-4} m^2/s

Storage material: Cast Iron
Density: 7.26×10^3 kg/m^3
Specific heat: 0.42 kJ/kg °C
Thermal conductivity: 0.518×10^2 W/m °C
Thermal diffusivity: 0.17×10^{-4} m^2/s

Constraints: Length: $0.2 \leqslant L \leqslant 20$ m
Semithickness of channel: $0.5 \leqslant d \leqslant 2$ cm

	Run number						
	CI-1	CI-2	CI-3	CI-4	CI-5	CI-6	CI-7
Mass flow rate, g/s cm width	4	2	6	4	4	4	4
Semithickness of channel, cm	1.108	0.586	1.609	1.816	0.900	1.326	1.652
Length, m	4.66	2.40	6.89	4.11	5.00	7.99	15.46
Outlet temperature, T^+	0.875	0.875	0.875	0.875	0.875	0.75	0.50
Heat storage, kJ/cm width	379	198	554	295	420	558	782
Q^+	0.836	0.848	0.828	0.739	0.865	0.718	0.520
A^+	0.326	0.340	0.318	0.254	0.362	0.480	0.674
ΔP, cm water	12.7	12.7	12.7	2.54	25.4	12.7	12.7
Air hp/cm width	0.0063	0.0031	0.0094	0.0013	0.0126	0.0063	0.0063

for concrete and cast iron are essentially the same as those previously noted for Feolite.

For each set of operating conditions and design constraints to be studied, two different graphs may be prepared. The first of these two graphs presents the ratio of heat stored to the total amount of heat available, A^+, versus the time duration of the heat storage process, τ. This graph is made for a given energy transporting fluid and heat storage material. The mass flow rate of the fluid and the maximum allowable pressure drop in the fluid as it passes through the flow channel, ΔP_{max}, are also specified before the graph is developed. Figure 3.2 is a plot of A^+ versus τ for a thermal energy storage unit composed of Feolite with air flowing through one-half the channel with a mass flow rate of 4 g/s cm width. The masimum allowable pressure drop in the channel is 12.7 cm of water.

For each storage material semithickness, w, two lines of particular significance to the designer can be plotted on the graph of A^+ versus τ; The upper line represents the maximum permissible length constraint imposed on the thermal energy storage unit design. The maximum dimensionless outlet fluid temperature limitation, T_{fo}, is indicated by the lower line. These two boundary lines may be observed on Fig. 13.2 for storage unit semithicknesses of 2, 4, and 8 cm. On this figure the upper line represents a maximum unit length limitation of 10.0 m, whereas the lower line represents a maximum dimensionless outlet fluid temperature of 0.875. Similar curves could be constructed for different length units and nondimensional outlet temperatures. As the length is decreased the upper curve moves downward and to the left, whereas the other curve moves upward and to the left as the value of T_{fo} is reduced. Also shown on the figure are lines of constant total heat storage in kilojoules per centimeter of width.

The region on the A^+ versus τ graph that contains all feasible heat storage unit designs for a particular set of operating conditions and design constraints is bounded by the maximum outlet fluid temperature line, the line signifying the maximum storage unit length, and the heat storage line with a value that corresponds to the minimum amount of heat storage required of the storage unit, Q_{min}.

The second of the two graphs used to present the optimum heat storage unit designs for a given set of operating conditions and construction specifications is a plot of the ratio of the amount of heat stored to the maximum amount of heat that the storage unit could contain, Q^+, against the time duration of the storage process, τ. Again the plot is made for a specified energy transporting fluid, heat storage material, fluid mass flow rate, and maximum allowable fluid pressure drop. The Q^+ versus τ graph is also made for a specific storage material semithickness. Figures 13.3 through 13.5 show plots of Q^+ versus τ for air flowing at a mass flow rate of 4 g/s cm width through a storage unit composed of Feolite, with the maximum air pressure drop restricted to 12.7 cm of water. The corresponding storage material semithicknesses are 2, 4, and 8 cm, respectively.

On the graphs where Q^+ is plotted against τ, the same three lines bound the area in which the optimum heat storage unit design will be found; the lines representing the maximum storage unit length, the maximum dimensionless outlet fluid temperature, and the minimum quantity of heat that must be stored by the unit. On the Q^+ versus τ plots the positions of the first two lines reverse themselves from those found on the

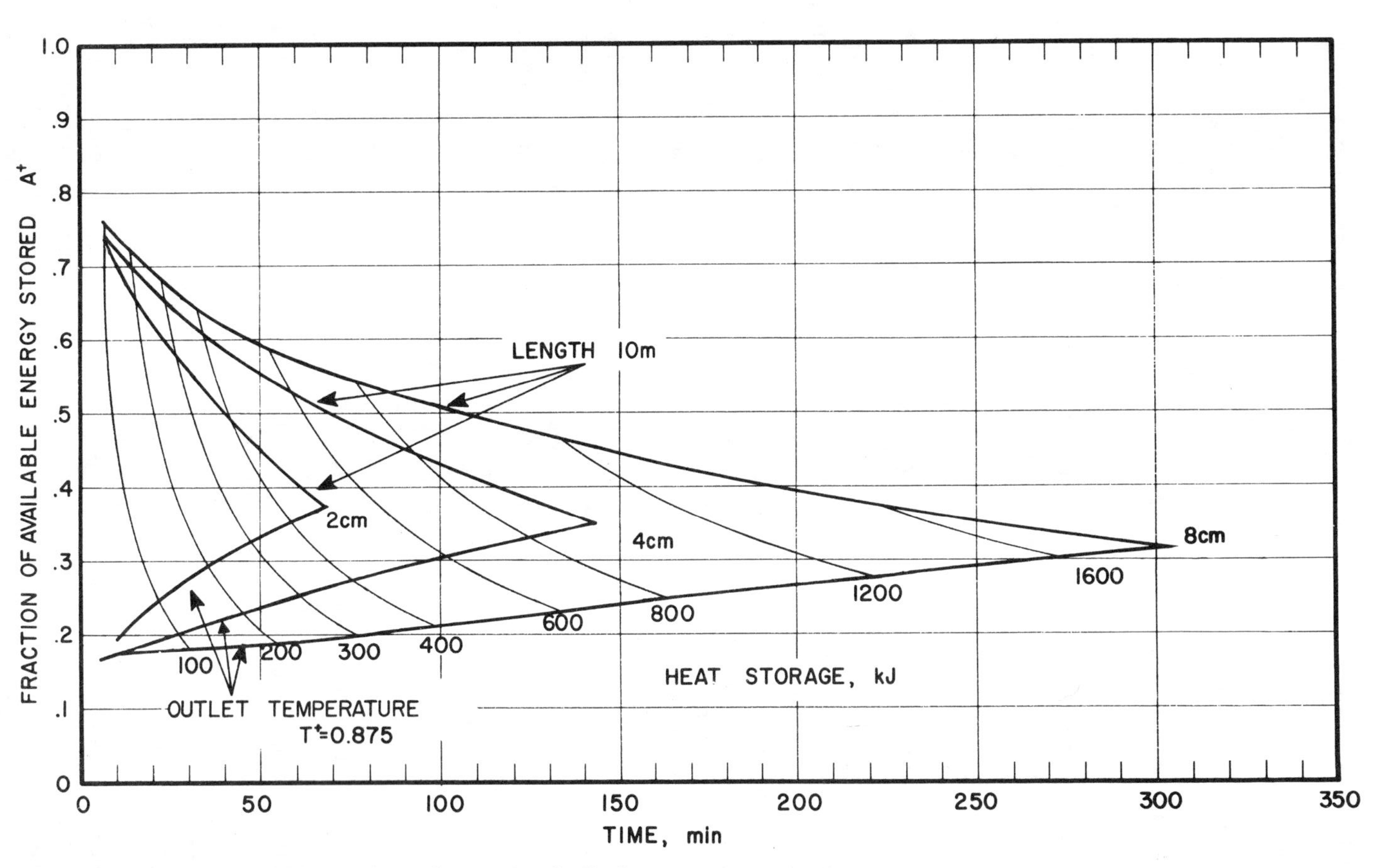

Figure 13.2 Fraction of available energy stored versus time for Feolite meat storage units [9].

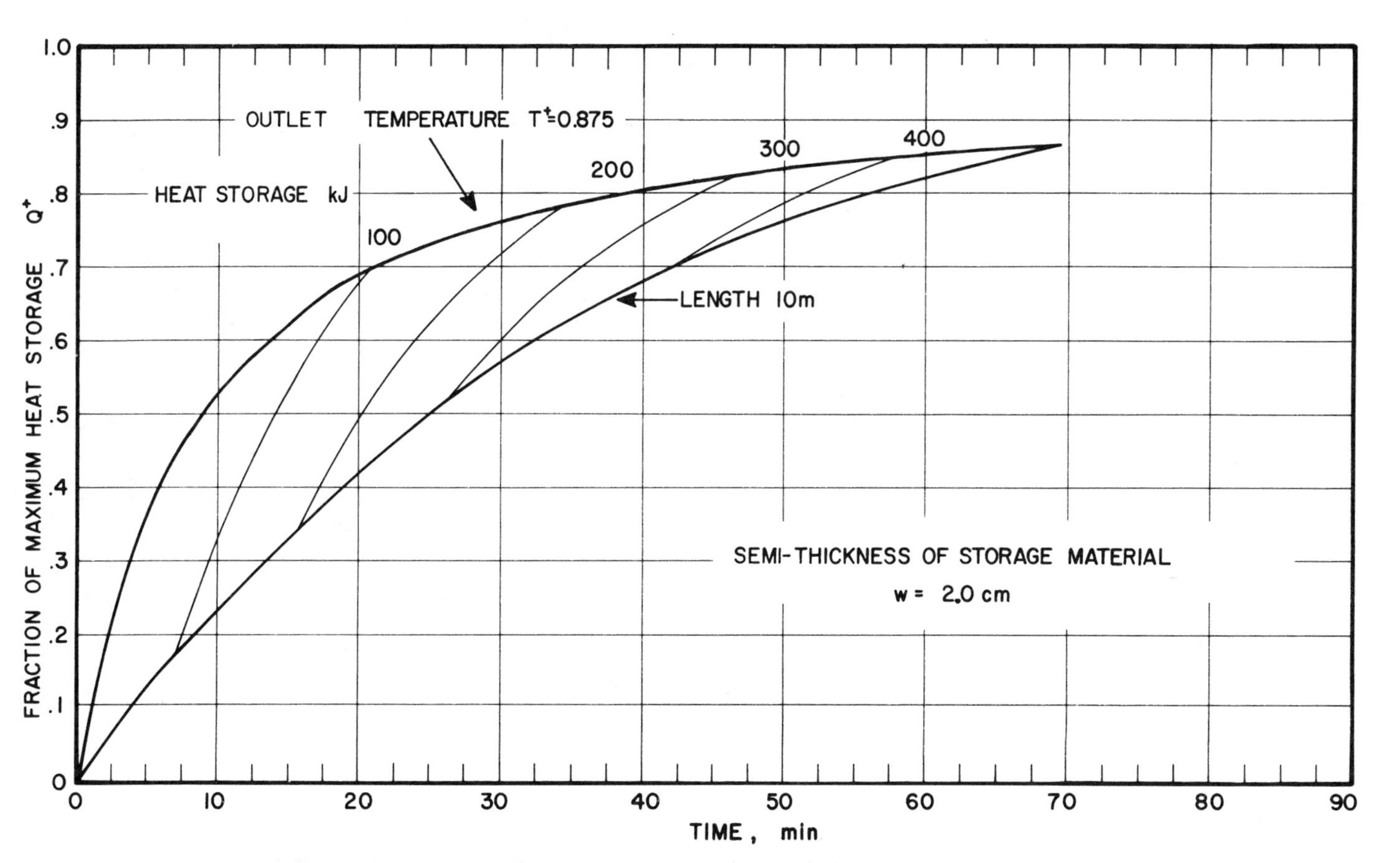

Figure 13.3 Fraction of maximum heat storage versus time for Feolite, 2 cm semithickness [9].

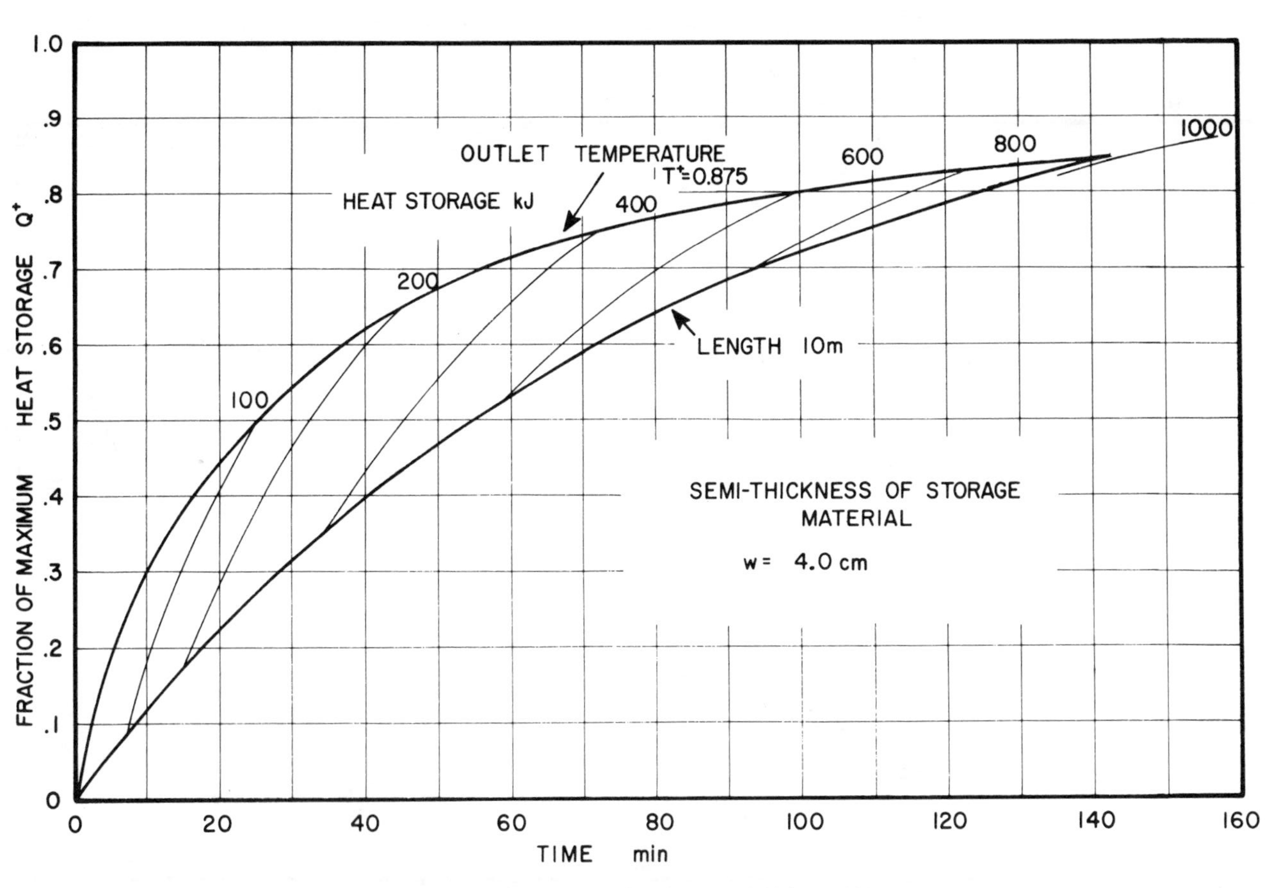

Figure 13.4 Fraction of maximum heat storage versus time for Feolite, 4 cm semithickness [9].

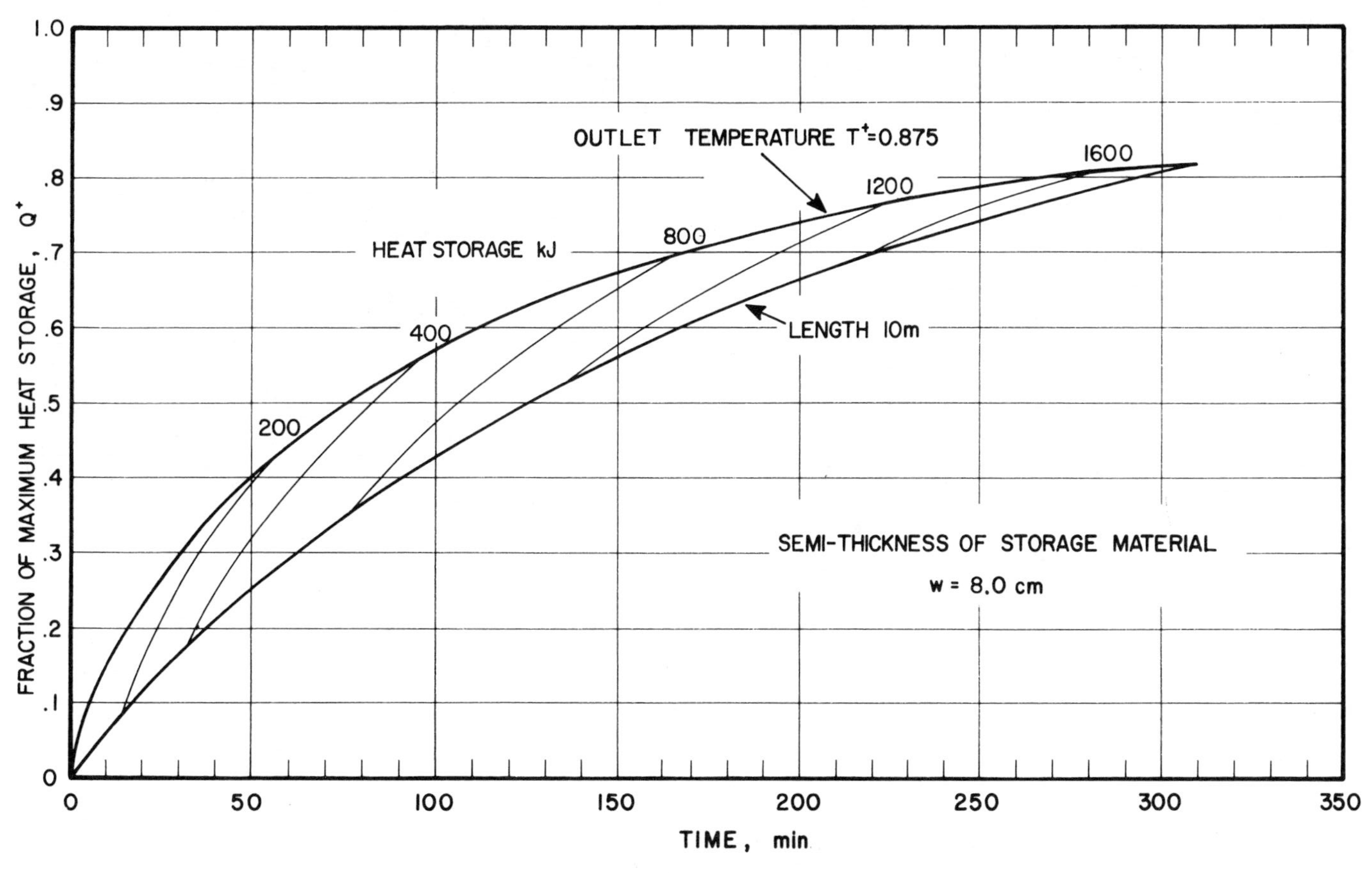

Figure 13.5 Fraction of maximum heat storage versus time for Feolite, 8 cm semithickness [9].

A^+ plot. The line due to the outlet fluid temperature constraint is above the line describing the maximum permissible unit length. This may be seen on Figs. 13.3 through 13.5. On all of these figures, the maximum dimensionless outlet fluid temperature is 0.875, and the maximum storage unit length has been set at 10.0 m.

The method used to generate these two types of curves has been described by Somers [5]. It is not, however, necessary to generate the complete set of curves to obtain the optimum design of a heat storage unit for a given storage duration. It is usually sufficient to use the optimization program to determine the limit of A^+ and Q^+ of the unit. These points fall on the maximum unit length line and the maximum outlet fluid temperature line as previously noted. The procedure used to find these points will now be described.

Several pieces of information must be supplied to the optimization program before the points on the curves can be determined. These data include the pertinent physical properties of the energy transporting fluid and the heat storage material, and the initial conditions of the heat storage system, including the mass flow rate of the fluid, the temperature of the fluid entering the storage unit, and the initial temperature of the heat storage material. In addition, the semithickness of the heat storage material, the maximal allowable pressure drop of the fluid as it passes through the fluid channel, and the maximum and minimum values for the fluid flow channel semithickness should be specified. Values for the maximum and minimum length of the storage unit, the maximum fluid outlet temperature, and the minimum amount of heat storage also should be decided upon at this point.

The operating condition that falls on the maximum fluid outlet temperature constraint is obtained by optimizing the thermal energy storage unit design on the quantity Q^+. The maximum fluid outlet temperature desired is specified for the optimization program. In order to ensure that the outlet fluid temperature constraint predominates in this series of optimization runs, the maximum storage unit length constraint should be set at a value greater than the actual maximum storage unit length desired. Also, the minimum heat storage requirement specified should be as low as possible. With the operating conditions and design constraints thus determined, an optimization run is made for the appropriate storage duration. The designed optimization run, with Q^+ as the optimized function, will always arrive at a final thermal energy storage unit design where outlet fluid temperature is equal to the maximum outlet fluid temperature constraint. The maximum value of Q^+ and the minimum value of A^+ are thus determined.

The operating conditions at the maximum heat storage unit length constraint are found by performing an optimization of the quantity A^+. The actual maximum storage unit length constraint is supplied to the optimization program. The maximum outlet fluid temperature constraint is set at the same value as the incoming fluid, $T_{fo} = 1.0$. The minimum heat storage is set as low as possible. With these input data, an optimization run is made and the values of A^+ and Q^+ are determined for a unit having the maximum allowable length. As a result, the maximum value of A^+ and the minimum value of Q^+ are found.

The two bounds of A^+ and Q^+ have been determined for a given storage duration. If the amount of heat stored is less than that desired, a longer storage duration is necessary and the calculations must be repeated. If the heat stored is within acceptable values, the designer can proceed by applying economic considerations to establish the final design. These considerations must take into account the cost of the storage material or the capital investment and the value of the energy removed from the fluid stream. The largest return on the initial investment is secured when Q^+ is a maximum. The most efficient use of the energy available in the fluid stream is obtained at the maximum A^+. As will be shown in the example problem, once the operating condition is selected, the design of the unit is fixed.

The use of these results to establish the design of a heat storage unit will be illustrated by several examples. Feolite will be used as the storage material.

From Fig. 13.2, it can be seen that as the thickness of the storage material increases the bounded region expands. If the designer selects a given storage duration τ, he or she can immediately determine, through the use of Fig. 13.2, the operating ranges of A^+ and Q that are possible while still satisfying the constraints established. As an example, for a storage duration of 80 min it is impossible to design a unit with 2 cm semithickness and still satisfy the constraints. A unit of 4 cm semithickness will have the following limits on A^+ and Q:

$$0.275 \leqslant A^+ \leqslant 0.475$$

and $$430 \leqslant Q \leqslant 635 \text{ kJ/cm of width}$$

If the semithickness is 8 cm, the limits are

$$0.21 \leqslant A^+ \leqslant 0.54$$

and $$330 \leqslant Q \leqslant 84 \text{ kJ/cm of width}$$

The fraction of the maximum possible heat storage may be determined from Figs. 13.3 through 13.5. Once again, lines of constant outlet temperature, length, and heat storage are plotted. It should be noted that the lines reverse their position, which indicates that the maximum value of Q^+ is determined by the outlet temperature constraint. The range over which Q^+ may vary can be determined by viewing the appropriate figure; for a 4 cm semithickness,

$$0.64 < Q^+ < 0.77$$

and for an 8 cm semithickness,

$$0.365 < Q^+ < 0.515$$

Example 13.1 A Feolite storage unit with a semithickness of 8 cm has been selected for an application in which the storage duration is 90 min. The mass flow rate is 4 g/s per half channel per cm of width, and each half channel of the unit must have the capability of storing a minimum of 800 kJ/cm of width. The constraints placed on the system are the same as those previously listed. Determine the dimensions of the unit for an optimum design.

SOLUTION From Figs. 13.2 and 13.4 it is found that the unit has a $Q^+ = 0.426$ and $A^+ = 0.459$. The length can be determined by noting that

$$\frac{Q^+}{A^+} = \frac{\dot{m}_f c_f \tau}{\rho_m c_m P_h L w}$$

which yields

$$L = \frac{\dot{m}_f c_f \tau}{\rho_m c_m P_h} \frac{A^+}{w Q^+}$$

For the stated conditions, the length of the optimum unit is 8.19 m. The semithickness of the channel can be obtained from the pressure drop

$$\Delta P = 4f \frac{L}{D} \frac{v^2}{2} \frac{\rho_f}{g_c}$$

$$= 48.0f \frac{\dot{m}_f^2 L}{d^3} \text{ cm } H_2O$$

$$d = \left(\frac{48.0 f \dot{m}_f^2 L}{\Delta P} \right)^{1/3}$$

where L is in meters and all other quantities have units of length in centimeters. For the stated flow conditions and a smooth channel assumption, the expression for the friction factor is obtained by using the previous stated relationships. The Reynolds number for the flow conditions, $Re_D = 4\dot{m}_f/\mu_f$, has the value of 8×10^4, and f is

$$f = 0.046(\text{Re})^{-0.2} = 0.046 \left(\frac{4\dot{m}_f}{\mu_f} \right)^{-0.2}$$

which for the specific conditions of this illustration gives a value of 0.0048. The value of the semithickness of the channel obtained is 1.34 cm. The sizing of the unit dimensions for the optimum design is

Length 8.19 m
Thickness 8 cm
Semithickness of channel 1.34 cm

13.5 OPTIMIZATION OF A PACKED BED HEAT STORAGE UNIT FOR SINGLE-BLOW OPERATING MODE

Solid sensible heat storage units can be constructed in many different geometrical configurations. To obtain the most efficient utilization of space, a storage unit should possess a large convective heat transfer surface area-to-volume ratio. A configuration possessing this characteristic is a packed bed.

In the packed bed, the material in particulate form is enclosed in a container, and the fluid is forced through the voids of the bed. Large rates of heat transfer may be achieved by the large ratio of the heat transfer surface area to the volume. This characteristic allows fast and efficient heat storage and retrieval without extensive investments in bed materials and complicated flow distribution systems.

The rate of heat transfer from or to the solid in the packed bed is a function of the physical and thermal properties of the fluid and solid, the temperature difference between the fluid and the surface of the solid, the mass rate of flow of the fluid, and the geometric characteristics of the packed bed material. The last is dependent on the shape and orientation of the packing material and on the bed voidage. The packing is usually random, where particles of roughly the same size and shape are deposited in an arbitrary manner into the container.

There are several variables that determine the performance of a packed bed thermal energy storage unit. These variables can be divided into three groups: those connected with the bed construction, those describing the characteristics of the flowing fluid as it passes through the bed, and those associated with the transient response of the bed material.

The packed bed geometry characteristics are described by the size, shape, and packing of the particles, the bed length, and the geometric configuration of the container. The second group includes the fluid properties and the mass velocity. This information together with the description of the bed geometry is sufficient to determine all fluid dynamic relationships that occur in the unit. Of major interest are the pressure drop experienced by the fluid as it passes through the bed and the velocity distribution and mixing in the fluid, which directly influence the convective heat transfer coefficient.

The last group of variables describes the initial thermal state of the bed, the inlet temperature of the fluid, the physical and thermal properties of the bed material, and the convective heat transfer coefficient. The total heat storage Q can be expressed in the following functional form:

$$Q = Q\{L, G, D_p, h, \tau, t_o, t_{fi}, \epsilon, \rho_m, c_m, c_f, \rho_f\}$$

Two quantities will be optimized in this study. One is A^+, the ratio of the amount of heat stored in the bed to the amount of thermal energy available in the entering fluid. This may be expressed as

$$A^+ = \frac{Q}{G\tau c_f(t_{fi} - t_o)} \qquad (13.11)$$

The second item is the ratio of the heat stored in the bed to the maximum amount of heat that would be stored if the bed reached a uniform temperature equal to that of the fluid entering the unit. This quantity is again expressed as Q^+ and has the following form for a packed bed:

$$Q^+ = \frac{Q}{\rho_m c_m L(1 - \epsilon)(t_{fi} - t_o)} \qquad (13.12)$$

These items are essentially the same as those used previously in the optimization of the flat slab storage unit configuration. It should be noted that the heat storage used in these definitions is for a packed bed with a unit frontal area, S_{fr}.

As noted previously, the complex optimization procedure arrives at the final design of the bed by manipulating some of the values of the parameters that describe the operating characteristics of the unit. Thus, before the optimization process can be initiated, certain information must be supplied about the operating conditions and the geometry of the storage unit. The two parameters that can be used in the optimization process, A^+ and Q^+, are related as follows:

$$\frac{A^+}{Q^+} = \frac{\rho_m c_m L(1-\epsilon)}{c_f G \tau} \tag{13.13}$$

These simple relationships allow the discussion to be restricted to the optimization of Q^+.

A number of variables will be specified before the optimization process is begun. The most common items defined are

1. Bed material
2. Type and size of particles
3. Porosity
4. Energy transporting fluid
5. Duration of the storage process
6. Temperature of the fluid entering the storage bed
7. Initial bed temperature

The number of unspecified dimensional parameters necessary to predict the transient response of the heat storage unit is reduced to five: $S_{fr}, L, \dot{m}_f, h$, and τ. This number can be reduced further by relating S_{fr} and $\dot{m}_f$ and by using the mass velocity G.

$$G = \frac{\dot{m}_f}{S_{fr}} \tag{13.14}$$

The convective heat transfer coefficient h is related to the other parameters. The exact expressions to be used in the evaluation of h will be presented in Chap. 14.

The remaining independent variables are the length of the packed bed, L, and the mass velocity of the energy transporting fluid, G. These two will be varied within the limits specified by the explicit and implicit constraints to obtain the maximum possible value of Q^+.

In order to ensure that the packed bed heat storage unit will be practical in terms of size, performance, and economics, certain constraints must be imposed on the design:

1. The maximum and minimum permissible length of the bed
2. The maximum and minimum permissible mass velocity of the energy transporting fluid
3. The maximum allowable pressure drop occurring as the fluid flows through the bed

4. The minimum amount of heat storage in the bed
5. The maximum fluid outlet temperature

These constraints are quite similar to those imposed in the optimization of the slab heat storage unit in Sec. 13.3.

13.6 RESULTS FROM OPTIMIZATION STUDY OF PACKED BED STORAGE UNITS

The mathematical model of the packed bed used to determine the transient response for the optimization study described in this section was that presented in Sec. 12.2.2. Beds composed of two different storage materials, cast iron and granite, will be discussed.

The pressure drop across the bed is calculated by using the relationship proposed by Ergun [10],

$$\frac{g_c\,\Delta P}{L} = 1.75\,\frac{1-\epsilon}{\epsilon^3}\,G\,\frac{v}{D_p} + 150.0\,\frac{(1-\epsilon)^2}{\epsilon^3}\,\mu\,\frac{v}{D_p^2} \tag{13.15}$$

where G, the mass velocity, is $\dot{m}_f/S_{fr}$.

This expression is valid for Reynolds numbers from 1 to 1.3×10^4. A more detailed discussion of pressure drop calculations in packed beds is presented in Chap. 14.

The accuracy of the optimized packed bed design is very dependent on the ability to estimate the convective heat transfer coefficients. In Chap. 14 the different correlations that may be used for packed beds are presented. For the carbon steel bed, the correlation proposed by Handley and Heggs [11] was selected. The convective film coefficient is

$$h = \frac{jGc_f}{(Pr)^{2/3}} \tag{13.16}$$

where $j = 0.26(\mathrm{Re}_m)$. $\qquad$ (13.17)

The Reynolds number Re_m is based on the superficial mean fluid velocity, $v = \dot{m}_f/S_{fr}\rho_f$, and the hydraulic diameter. The void volume of the bed divided by the surface area of the packing material is the hydraulic diameter,

$$D_h = \frac{2D_p\epsilon}{3(1-\epsilon)} \tag{13.18}$$

where D_p is the diameter of an equivalent sphere with a volume equal to that of the particle in the bed. The value of the convective film coefficient is corrected to take into account axial dispersion and intraparticle effects by using the method described in Sec. 12.2.3.

When the thermal conductivity of the particle is very low, the intraparticle conduction effects become dominant. The relationships discussed in Sec. 12.2.3 do not account adequately for this effect. As a result, the correlations obtained experimentally by Löf and Hawley [12] were used for the calculation of the convective film coefficients in the granite beds:

$$h = 108.7 D_p^{0.3} G^{0.7} \qquad \text{W/m}^2\,{}^\circ\text{C} \tag{13.19}$$

where the equivalent diameter is defined as

$$D_p = \frac{6\,(\text{net volume of particles})^{1/3}}{\pi\,(\text{number of particles})}$$

Parszewski [13] obtained results for the optimized design of carbon steel and granite packed bed storage units. Typical results for A^+ and Q^+ are shown in Figs. 13.6 through 13.9. The pressure drop across the bed was 35 cm of water and $\epsilon = 0.37$.

The operating conditions for the unit with the maximum allowable fluid outlet temperature is obtained by optiminizating the design for Q^+. To ensure that the design does have the desired outlet fluid temperature, the heat storage constraint is set very low while the maximum length constraint is set much greater than that for a practical unit. The program is then constrained by the outlet fluid temperature.

The design boundary imposed by the maximum packed bed length constraint can be obtained directly from the following pressure drop equation:

$$\Delta P = 1.02 \times 10^{-2} L \left[\frac{1.75 G^2 (1-\epsilon)}{\rho_f D_p \epsilon^3} + \frac{150 \mu G (1-\epsilon)^2}{\rho_f D_p^2 \epsilon^3} \right] \qquad \text{cm } H_2O \tag{13.20}$$

The mass velocity is the only unknown in the equation and is therefore easily calculated. The next step is to calculate the convective heat transfer coefficient from one of the correlations already presented. The values of the nondimensional time η and nondimensional length λ are determined and the value of Q^+ is calculated from Eq. (12.16) or the tabulated data. Finally, the value of A^+ is calculated from Eq. (13.13).

The effect of changes in particle size can be seen from the data in Table 13.4. The pressure drop and the storage duration were held constant. When the optimization is performed to obtain the maximum A^+, the design is limited by the length constraint. An increase in particle size causes an increase in the mass velocity of the fluid. This results in a decrease in the value A^+, whereas both Q^+ and Q increase. The decrease in A^+ is more noticeable in the granite bed, but the changes in Q^+ and Q are more pronounced in the carbon steel bed.

The influence of changes in the allowable pressure drop can be determined by holding the particle diameter and time duration for the storage process constant. The same trends observed in the preceding paragraph are noted when the pressure drop is increased and can be seen from Table 13.5. The mass velocity Q^+ and Q increase, whereas the A^+ decreases. It is important to note that a fourfold increase in the allowable pressure drop results in an increase in Q of less than 75% for the carbon steel bed and approximately 50% for the granite bed.

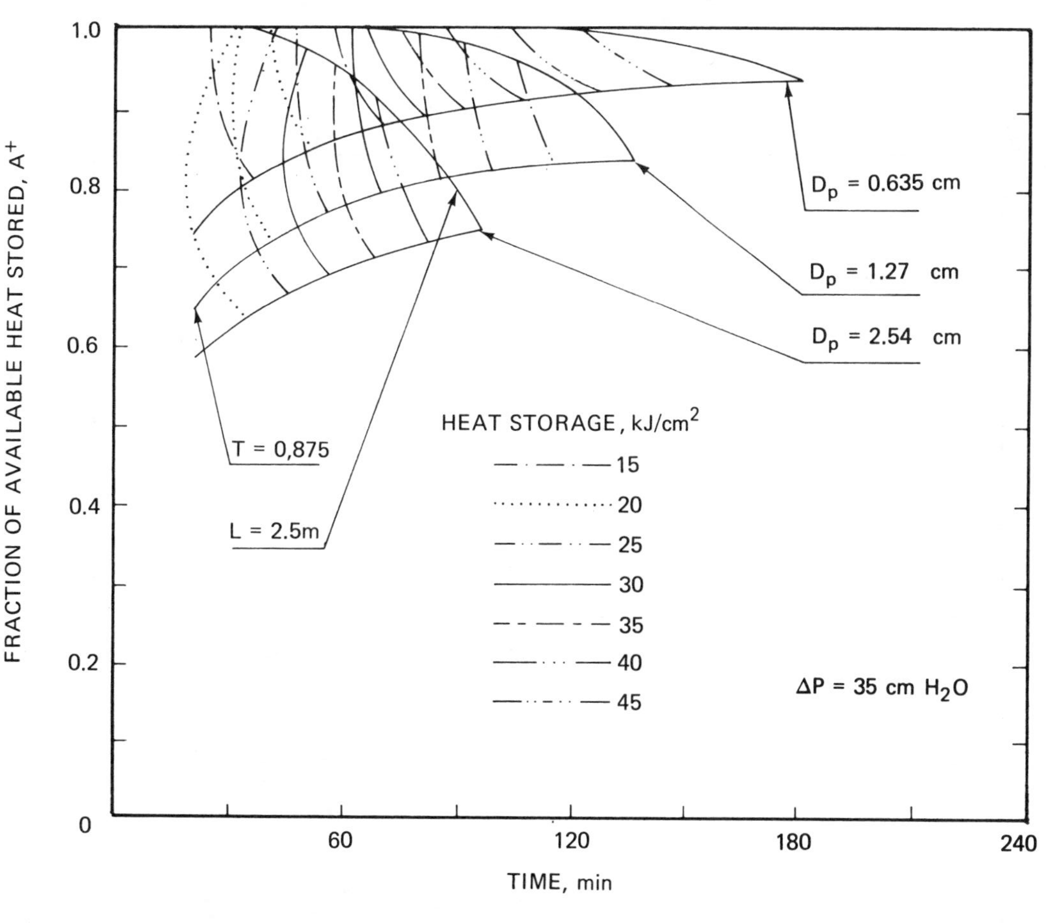

Figure 13.6 A^+ versus time (carbon steel) [13].

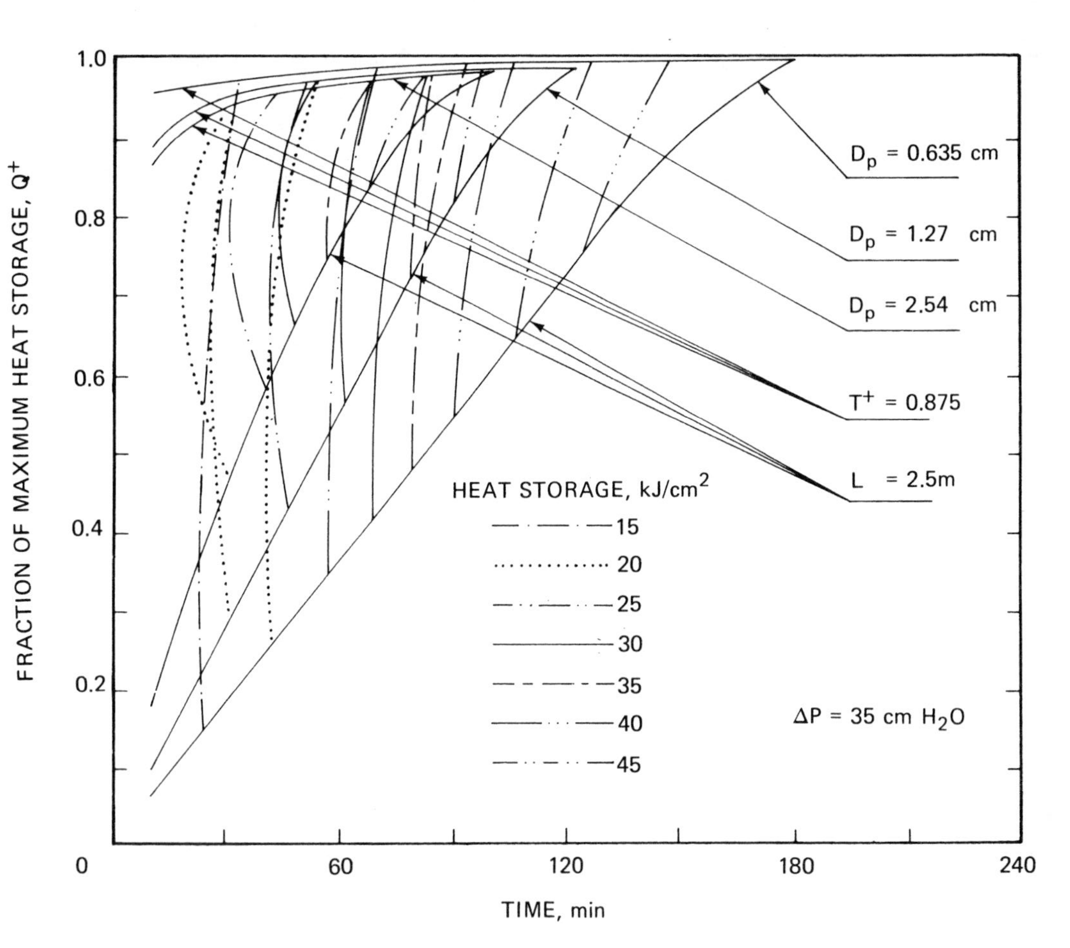

Figure 13.7 Q^+ versus time (carbon steel) [13].

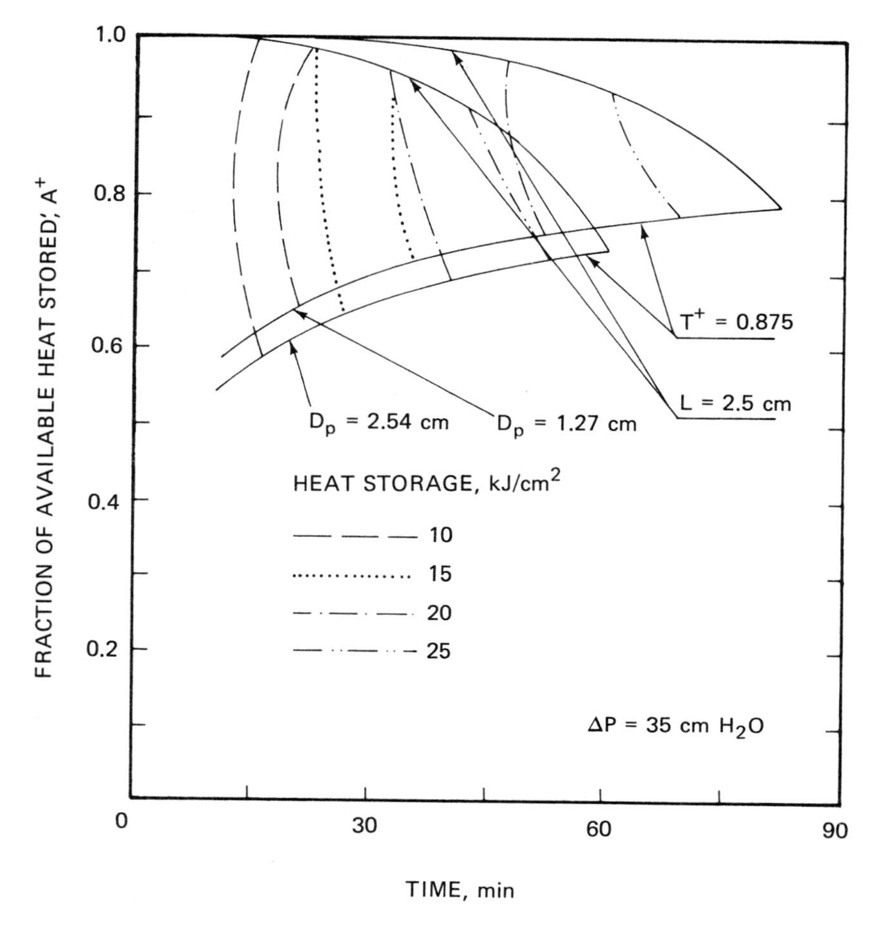

Figure 13.8 A^+ versus time (granite) [13].

An increase in the maximum allowable fluid outlet temperature when the particle size, pressure drop, and time duration are held constant yields the results shown in Table 13.6. From these results, it can be seen that a decrease in the fluid outlet temperature constraint causes a significant increase in the length of the optimum designed packed bed. This would be expected, since a lower outlet temperature constraint means that more heat must be removed from the fluid as it passes through the bed. In order to maintain the same pressure drop, the mass velocity must therefore be decreased. It is also important to note that changes in A^+ and Q^+ are once again in opposite directions. Thus, one becomes aware that a lower fluid outlet temperature constraint will yield a better utilization of the available energy but a poor utilization of the storage capacity of the material.

As the duration of the storage process increases, the packed bed lengthens as

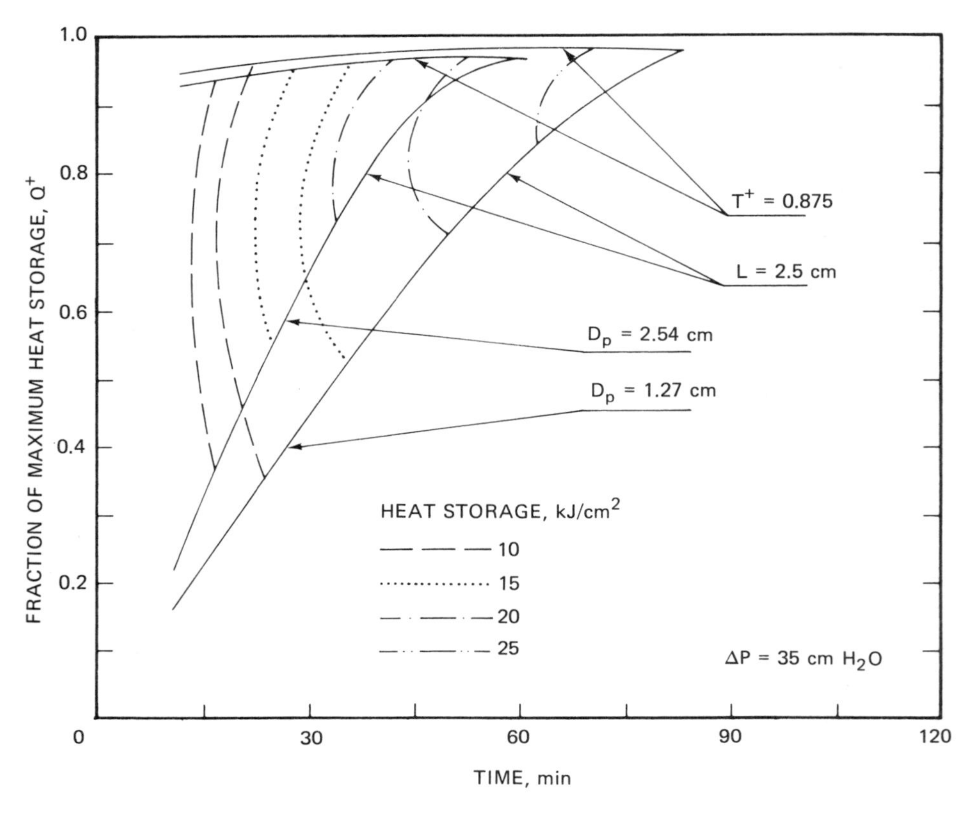

Figure 13.9 Q^+ versus time (granite) [13].

indicated in Table 13.7. The values of A^+ and Q^+ increase by an amount considerably smaller than the increase experienced by Q.

A comparison of the effect of bed material on the optimum design of a heat storage unit is presented in Table 13.8. The carbon steel bed has much better storage characteristics than the granite bed. When material costs are considered, however, the attractiveness of the granite bed increases dramatically.

Table 13.4 Effect of particle diameter variations,
$\tau = 60$ min, $\Delta P = 35$ cm water

Material	D_p, cm	T_{fo}	G, kg/m² s	Q, MJ/m²	A^+	Q^+
Carbon steel	1.270	0.0	0.895	260.0	1.0	0.558
Carbon steel	2.540	0.268	1.31	363.0	0.955	0.786
Granite	1.270	0.491	0.896	215.0	0.827	0.766
Granite	2.540	0.722	1.31	222.0	0.583	0.791

Table 13.5 Effect of pressure drop constraint variations,
$\tau = 60$ min, $D_p = 1.27$ cm, $A^+ = A^+_{max}$

Material	ΔP, cm water	T_{fo}	G, kg/m² s	Q, MJ/m²	A^+	Q^+
Carbon steel	35.0	0	0.895	260.0	1.0	0.558
Carbon steel	70.0	0.130	1.29	370.0	0.988	0.802
Carbon steel	140	0.813	1.86	451.0	0.846	0.979
Granite	15.0	0.179	0.571	166.0	0.999	0.590
Granite	35.0	0.491	0.900	215.0	0.827	0.766
Granite	70	0.784	1.26	251.0	0.688	0.893

Table 13.6 Effect of T constraint variation,
$\tau = 60$ min, $\Delta P = 35$ cm water, $D_p = 1.27$ cm, $Q^+ = Q_{max}$

Material	T_{fo}	L, m	G, kg/m² s	Q, MJ/m²	A^+	Q^+
Carbon steel	0.875	1.47	1.18	269	0.786	0.984
Carbon steel	0.625	1.60	1.10	289	0.859	0.923
Carbon steel	0.500	1.72	1.09	292	0.925	0.916
Granite	0.875	2.01	1.00	222	0.762	0.982
Granite	0.625	2.29	0.939	240	0.882	0.935
Granite	0.500	2.40	0.918	244	0.917	0.906

Table 13.7 Effect of the time duration on the storage process,
$D_p = 1.27$ cm, $\Delta P = 35$ cm water, $T_{fo} = 0.875$

Material	τ, min	L, m	G, kg/m² s	Q, MJ/m²	A^+	Q^+
Carbon steel	15.0	0.531	1.99	95.6	0.663	0.968
Carbon steel	60.0	1.47	1.18	269.0	0.786	0.984
Carbon steel	120.0	2.56	0.89	478.8	0.935	1.0
Granite	15.0	0.72	1.70	77.8	0.630	0.962
Granite	60.0	2.01	1.00	222.0	0.762	0.982
Granite	120.0	3.30	0.78	368.0	0.845	0.987

Table 13.8 Comparison of optimum designs of carbon steel and granite packed beds,
$D_p = 1.27$ cm, $\Delta P = 35$ cm water, $\tau = 60$ min, $T = 0.875$

Material	L, m	G, kg/m² s	Q, MJ/m²	A^+	Q^+	Q/L, MJ/m³	A^+/L, 1/m	Q^+/L, 1/m
Carbon steel	1.47	1.18	269.0	0.786	0.984	183.0	0.535	0.669
Granite	2.01	1.00	222.0	0.763	0.982	111.0	0.379	0.488

REFERENCES

1. F. W. Schmidt and J. Szego, "Transient Response of Solid Sensible Heat Thermal Storage Units–Single Fluid," *J. Heat Transfer, Trans. ASME*, vol. 98, 1976, pp. 471–477.
2. M. J. Box, "A New Method of Constrained Optimization and a Comparison with Other Methods," *Computer J.*, vol. 8, no. 1, 1965, pp. 42–52.
3. A. Adelman and W. F. Stevens, "Process Optimization by the Complex Method," *AIChE J.*, vol. 18, no. 1, 1972, pp. 20–24.
4. W. Spendley, G. R. Hext, and F. R. Himsworth, "Sequential Applications of Simplex Designs in Optimization and Evolutionary Optimization," *Technometrics*, vol. 4, 1962, pp. 441–461.
5. R. R. Somers, "The Design Optimization of a Single Fluid, Solid Sensible, Heat Storage Unit," M.S. thesis, Pennsylvania State University, University Park, Pa., 1976.
6. IBM System 360 Scientific Subroutine Package, Version III, Programmer's Manual, GH20-0205-4.
7. V. M. Legkiy and A. S. Makarov, "Heat Transfer in the Thermal Entry Region of Circular Pipes and Rectangular Ducts with Stabilized Turbulent Flow of Air," *Heat Transfer–Soviet Res.*, vol. 3, no. 6, 1971, pp. 4–5.
8. W. M. Kays, *Convective Heat and Mass Transfer*, McGraw-Hill, New York, 1966, p. 73.
9. F. W. Schmidt, R. R. Somers, II, J. Szego, and D. H. Laananen, "Design Optimization of a Single Fluid, Solid Sensible Heat Storage Unit," *J. Heat Transfer, Trans. ASME*, vol. 99, 1977, pp. 174–179.
10. S. Ergun, "Fluid Flows Through Packed Columns," *Chem. Eng. Proc.*, vol. 48, 1952, p. 89.
11. D. Handley and P. J. Heggs, "Momentum and Heat Transfer Mechanism in Regular Shaped Packing," *Trans. Inst. Chem. Eng.*, vol. 46, 1968, p. T251.
12. G. O. G. Löf and R. W. Hawley, "Unsteady State Heat Transfer Between Air and Loose Solids," *Ind. Eng. Chem.*, vol. 40, 1948, p. 1061.
13. M. Parszewski, "Design Optimization of a Packed Bed Heat Storage Unit," M.S. thesis, Pennsylvania State University, University Park, Pa., 1978.

CHAPTER

FOURTEEN

HEAT TRANSFER AND PRESSURE DROP CORRELATIONS

14.1 INTRODUCTION

The ability to predict accurately the transient response of a thermal energy storage unit depends not only on formulating a mathematical model that adequately describes the phenomenon but also on the use of accurate values for the physical properties of the fluid and storage material and for the convective heat transfer coefficient. The practicability of the storage unit design in many situations will be governed by the allowable pressure drop. Two general heat storage configurations have been discussed. One of them involves the flow of the energy transporting fluid in channels and in the second configuration the energy transporting fluid passes through a packed bed. Correlations for the prediction of the heat transfer coefficient and the pressure drop for each type of unit will be presented in this chapter. Those readers specifically concerned with regenerators are referred to the very complete coverage of this subject matter by Hausen [1].

14.2 HEAT TRANSFER AND PRESSURE DROP CORRELATIONS FOR CHANNEL FLOW

A very careful analysis of the thermal and hydraulic aspects of the flow is required before an accurate estimatation of the convective heat transfer coefficient and the pressure drop can be made. The classification of flow as either laminar or turbulent is usually the first item determined and is made on the basis of the value of the Reynolds number for the flow. In the calculation of the Reynolds number, the hydraulic diameter is used as the characteristic length. The pressure drop is influenced by the inlet and exit sections as well as by the length of the channel. If the channel is very short, the axial velocity profile will vary as one moves downstream from the entrance and the velocity profile is said to be developing. The convective film coefficient will be influenced by the thermal boundary conditions at the channel wall as well as any changes that occur in the velocity profile as the flow moves downstream. All of these factors that affect the pressure drop and the heat transfer coefficient are described in detail in Kays and London [2], the *Handbook of Heat Transfer* [3], and Shah and London [4].

14.2.1 Pressure Drop Calculation

The pressure drop in the flow configuration shown in Fig. 14.1 is

$$\Delta P = \frac{G^2}{2g_c\rho_{f1}}\Bigg[\underbrace{(K_c + 1 - \alpha^2)}_{\text{I}} + \underbrace{2\left(\frac{\rho_{f1}}{\rho_{f2}} - 1\right)}_{\text{II}} + \underbrace{\frac{4fL}{D_h}\frac{\rho_{f1}}{\rho_{fm}}}_{\text{III}} - \underbrace{(1 - \alpha^2 - K_e)\frac{\rho_{f1}}{\rho_{f2}}}_{\text{IV}}\Bigg] \tag{14.1}$$

where K_c and K_e are pressure loss coefficients and α is the ratio of the total cross-section flow area in the storage unit to the flow cross-sectional area at location 1 or 2 outside the heat storage unit. If each flow channel has its own inlet and exit sections, α is equal to 1. The pressure loss through the entrance section is indicated by the first term on the right-hand side, labeled I. The pressure drop associated with the acceleration of the fluid as it flows through the unit because of decreases in fluid density is indicated by II, and III indicates the pressure lost through the heat storage channel. The pressure drop associated with the exit effects is indicated by IV. The values of K_c and K_e can be obtained from Fig. 14.2*a* for multicircular heat storage units and from Fig. 14.2*b* for a multirectangular cross-sectional unit.

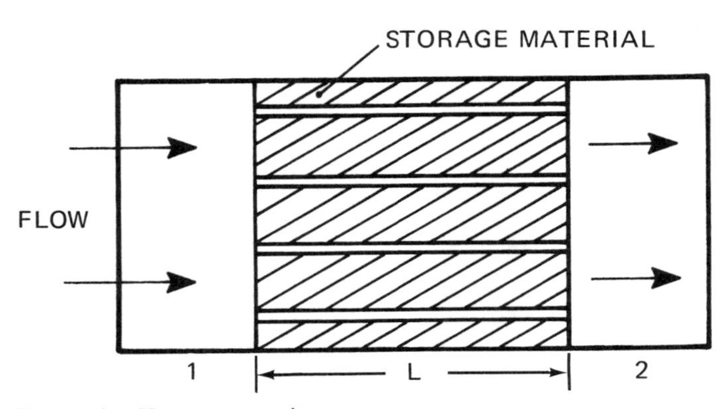

Figure 14.1 Heat storage unit.

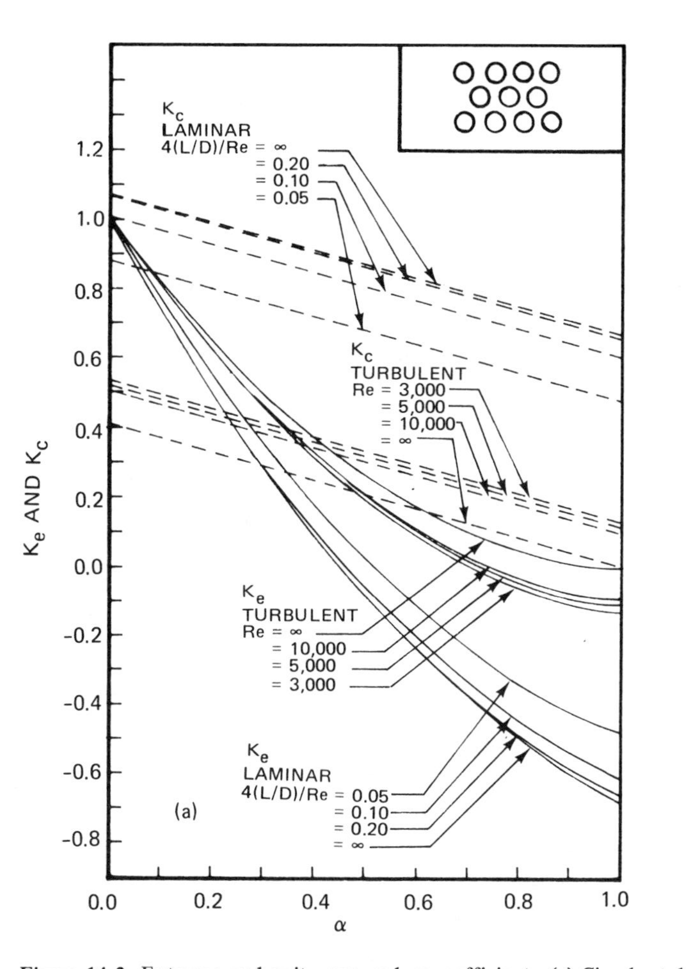

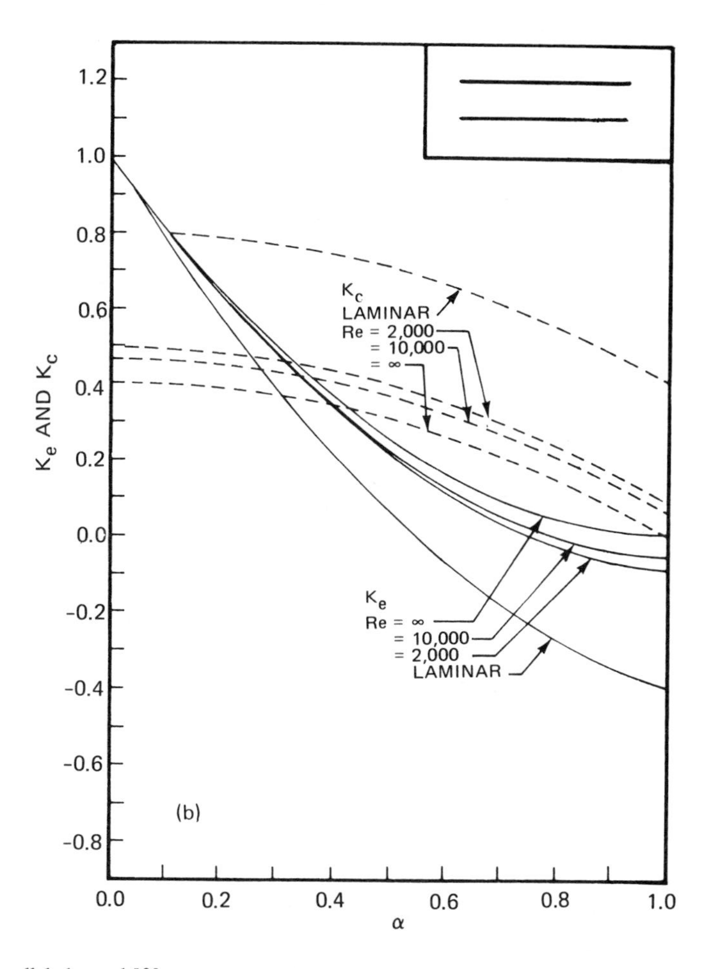

Figure 14.2 Entrance and exit pressure loss coefficients. (*a*) Circular tubes. (*b*) Parallel channel [2].

The Fanning friction f can be obtained for fully developed turbulent flow in a smooth channel from the following equation:

$$4f = [1.82 \log_{10}(\text{Re}) - 1.64]^{-2} \tag{14.2}$$

For rough channels the friction factor can be estimated by using the Moody diagram presented in Fig. 14.3. Values of the surface roughness e are given in Table 14.1 for several different materials.

Although all of the above relationships are limited to flow within a pipe, they can be used for flow in channels with other cross sections if the characteristic length used in the nondimensional numbers is the hydraulic diameter.

The effects of heating on the friction factor may be taken into account by using the following relationship:

$$\text{Liquids:} \qquad f = f_{\text{iso}}\left(\frac{\mu_w}{\mu_m}\right)^{0.05} \tag{14.3}$$

$$\text{Gas:} \qquad f = f_{\text{iso}}\left(\frac{T_w}{T_m}\right)^{0.1} \tag{14.4}$$

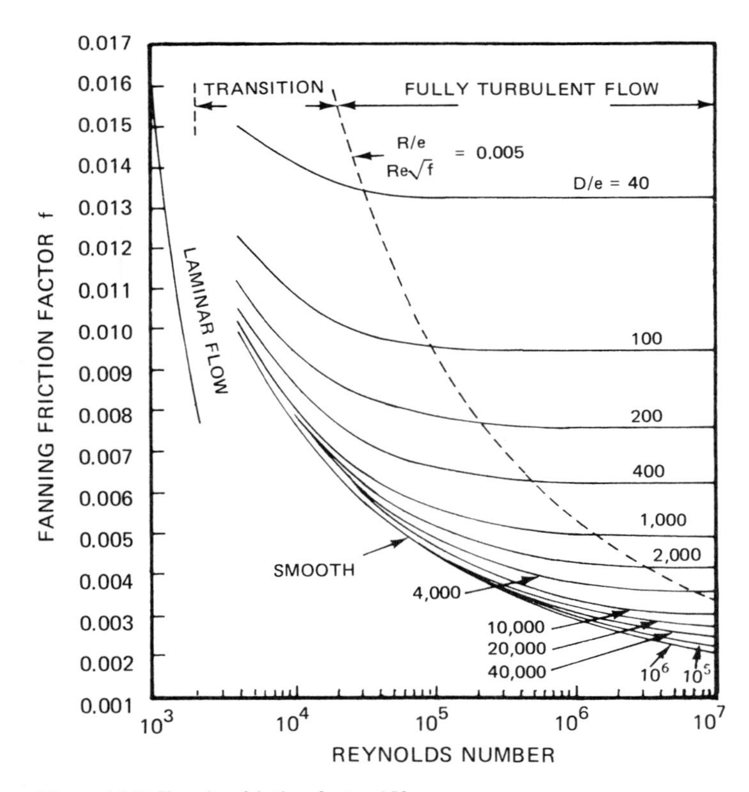

Figure 14.3 Fanning friction factor [5].

Table 14.1 Surface roughness, *e*

Material	Surface roughness $\times 10^3$	
	m	(ft)
Commercial steel	0.046	(0.15)
Cast iron	0.259	(0.85)
Concrete	0.305–3.05	(1.0–10.0)

where T is the absolute temperature and the subscripts w and m refer to the wall and the bulk fluid, respectively.

Example 14.1 Air at 80.0°C (176.0°F) flows through a heat storage unit composed of rectangular cross-sectional flow channels, Fig. 14.1. The channels are 1.0 m (3.28 ft) wide, 1.5 cm (0.59 in) high, and separated by storage material 8.0 cm (3.15 in) thick. The unit is 5.0 m (16.4 ft) long. The initial temperature of the unit is 10.0°C (50.0°F). Estimate the pressure drop across the unit if the mass flow rate per channel is 0.25 kg/s (0.55 lb_m/s).

SOLUTION The properties of the air needed in these calculations are as follows:

Air @ 10.0°C	Air @ 45.0°C	Air @ 80.0°C
$\rho_f = 1.257$ kg/m^3	$\rho_f = 1.106$ kg/m^3	$\rho_f = 0.998$ kg/m^3
(0.0785 lb_m/ft^3)	(0.069 lb_m/ft^3)	(0.0623 lb_m/ft^3)
$\mu_f = 1.815 \times 10^{-5}$ kg/m s	$\mu_f = 2.016 \times 10^{-5}$ kg/m s	$\mu_f = 2.088 \times 10^{-5}$ kg/m s
(1.22×10^{-5} lb_m/ft s)	(1.355×10^{-5} lb_m/ft s)	(1.403×10^{-5} lb_m/ft s)

The hydraulic diameter is

$$D_h = \frac{4A}{P} = \frac{(4.0)(0.015)(1.0)}{2.0} = 0.030 \text{ m} \qquad (0.098 \text{ ft})$$

The Reynolds number is

$$\text{Re} = \frac{\rho v D_h}{\mu_f} = \frac{\dot{m}}{A}\frac{D_h}{\mu_f}$$

At 10°C,

$$\text{Re} = 27{,}548$$

At 45°C,

$$\text{Re} = 24{,}802$$

At 80°C,

$$\text{Re} = 23{,}946$$

The value of α is $0.015/0.095 = 0.158$, and the mass velocity G is 16.67 kg/m^2 s. The values of the entrance and exit correction factors and the Fanning friction factor are as follows:

	At 10.0°C	At 45.0°C	At 80.0°C
K_e	0.7	0.7	0.7
K_c	0.44	0.44	0.44
f	0.006	0.00615	0.0062

The calculations have been carried out with the properties evaluated at the three temperatures to illustrate the effect of the change in physical properties on the values of K_e, K_c, and f.

The pressure drop at the beginning of storage when the outlet fluid temperature is still 10°C is

$$\Delta P = \frac{G^2}{2g_c\rho_{f1}}\left[(K_c + 1 - \alpha^2) + 2\left(\frac{\rho_{f1}}{\rho_{f2}} - 1\right) + \frac{4fL}{D_h}\frac{\rho_{f1}}{\rho_{fm}} - (1 - \alpha^2 - K_e)\frac{\rho_{f1}}{\rho_{f2}}\right]$$

$$= \frac{(16.66)^2}{(2.0)(1.0)(0.998)}\left\{[0.44 + 1.0 - (0.158)^2] + 2\left(\frac{0.998}{1.257} - 1.0\right) + \frac{(4.0)(0.00615)(5.0)(0.998)}{(0.030)(1.106)} - [1.0 - (0.158)^2 - 0.70]\frac{0.998}{1.257}\right\}$$

$$= 139.06(1.415 - 0.412 + 3.70 - 0.218)$$

$$= 623.68 \text{ N/m}^2 \quad (0.09 \text{ psi})$$

The pressure drop at the end of the storage period with the outlet fluid temperature at 80°C is

$$\Delta P = 139.06(1.415 + 0.0 + 4.1 - 0.275)$$

$$= 728.7 \text{ N/m}^2 \quad (0.106 \text{ psi})$$

14.2.2 Heat Transfer Coefficient Correlation

In most thermal storage units, gases are the energy transporting fluid and the flow is normally in the turbulent flow region. When liquids are used, however, the high heat capacity of the fluid greatly reduces the temperature drop in the fluid as it passes through the storage unit. In order to increase the temperature drop and store a larger fraction of the available energy, the velocity of the fluid is decreased. The flow may have a very low Reynolds number well within the laminar flow region.

In laminar flow the Nusselt number, $\mathrm{Nu} = hD_h/k_f$, is greatly influenced by the geometry of the section, the thermal boundary conditions, and the velocity profile. A description of the thermal flow characteristics at a given location can be classified as:

1. Developing—both the thermal and velocity profiles are developing and thus the convective film coefficient is a function of axial location
2. Thermal developing—the velocity profile is invariant with axial distance, whereas the thermal profile and the convective film coefficient are functions of axial location
3. Fully developed—the velocity and thermal profiles are invariant with axial location, and the convective heat transfer coefficient is a constant.

The influence of geometry and thermal boundary conditions on the Nusselt number is illustrated by the tabulated results for fully developed laminar flow:

	Constant wall temperature	Constant heat flux
Circular tube	3.66	4.364
Parallel-plate channel	7.54	8.235

It is recommended that Shah and London [4] be consulted to obtain information necessary to estimate accurately the convective film coefficient when laminar flow is involved.

In turbulent flows the convective heat transfer coefficient is less sensitive to the flow geometry and boundary condition. The characteristic length used is usually the hydraulic diameter of the cross section, which is defined as

$$D_h = \frac{4(\text{area})}{\text{perimeter}} \tag{14.5}$$

The following expressions have been recommended by Legkiy and Makarov [6] for predicting the heat transfer coefficient of a gas in the thermal entry region of tubes and rectangular cross-sectioned channels:

$$\mathrm{Re}_L < 25 \times 10^3; \frac{L}{D_h} < 60: \qquad h = 0.22 \frac{k_f}{L} \mathrm{Re}_L^{0.6} \left(\frac{L}{D_h}\right)^{0.08} \tag{14.6}$$

$$\mathrm{Re}_L > 25 \times 10^3; \frac{L}{D_h} < 60: \qquad h = 0.029 \frac{k_f}{L} \mathrm{Re}_L^{0.8} \left(\frac{L}{D_h}\right)^{0.08} \tag{14.7}$$

An alternative expression for the mean heat transfer coefficient in the entrance region was given by McAdams [7]:

$$\text{For } 2 < \frac{L}{D_h} < 20: \qquad h = h_\infty \left[1 + \left(\frac{D_h}{L}\right)^{0.07}\right] \tag{14.8}$$

$$\text{For } 20 < \frac{L}{D_h}: \qquad h = h_\infty \left(1 + \frac{CD_h}{L}\right) \tag{14.9}$$

where h_∞ is the fully developed flow coefficient. The values of C for two flow configurations of a gas are

Fully developed velocity profile: 1.4
Abrupt contraction entrance: 6.0

There are a number of correlations used for the determination of the heat transfer coefficient in fully developed flow. The most popular are those of Dittus and Boelter [8], Colburn with Seider-Tate modifications [9], and Petukhov [10].

Dittus-Boelter correlation [8]

$$\text{Nu} = 0.023\ \text{Re}^{0.8}\ \text{Pr}^{0.4} \tag{14.10}$$

for $0.7 < \text{Pr} < 120$ and $10{,}000 < \text{Re} < 120{,}000$;

Colburn correlation with Seider-Tate modifications [9]. For gases,

$$\text{Nu} = C\ \text{Re}^{0.8}\ \text{Pr}^{1/3} \left(\frac{T_b}{T_w}\right)^N \tag{14.11}$$

and $N = 0.55$ heating

$= 0.0$ cooling

For liquids,

$$\text{Nu} = C\ \text{Re}^{0.8}\ \text{Pr}^{1/3} \left(\frac{\mu_b}{\mu_w}\right)^N \tag{14.12}$$

and $N = 0.36$ heating

$= 0.20$ cooling

The value of C is

$C = 0.020$ constant wall temperature

$= 0.021$ constant heat flux

Petukhov correlation [10]

$$\text{Nu} = \frac{f\ \text{Pe Pr}}{2[1.07 + 12.7\sqrt{f/2}(\text{Pr}^{2/3} - 1)]} \left(\frac{\mu_b}{\mu_w}\right)^N \tag{14.13}$$

where $N = 0.11$ for heating and 0.25 for cooling. The friction factor is obtained from Eq. (14.2).

The effect of surface roughness can be considered by using the correlation proposed by Nunner [11] for air,

$$\mathrm{Nu} = \frac{(f/2)\ \mathrm{Re\ Pr}}{1 + 1.5\ \mathrm{Re}^{-1/8}\ \mathrm{Pr}^{-1/6}[\mathrm{Pr}(f/f_s) - 1]} \tag{14.14}$$

where f_s is the smooth tube friction factor.

It must once again be emphasized that the correlations presented in this section are valid for most flow situations. The reader is encouraged to consult the technical literature if the flow configuration is such that the validity of the correlation presented is questionable.

Example 14.2 Determine the convective heat transfer coefficient for the heat storage unit described in Example 14.1.

SOLUTION The physical properties of the air will be evaluated at an average temperature of 45°C.

$$\rho_f = 1.106\ \mathrm{kg/m^3} \quad (0.069\ \mathrm{lb_m/ft^3})$$

$$\mu_f = 2.016 \times 10^{-5}\ \mathrm{kg/m\ s} \quad (1.355 \times 10^{-5}\ \mathrm{lb_m\ ft\ s})$$

$$k_f = 0.0276\ \mathrm{W/m\ {}^\circ C} \quad (0.0159\ \mathrm{Btu/ft\ hr\ {}^\circ F})$$

$$\mathrm{Pr} = 0.704$$

The hydraulic diameter was found to be 0.030 m and the L/D_h ratio is equal to 167. The flow may be considered to be fully developed and the Nusselt number can be calculated from Eq. (14.10). The Reynolds number is

$$\mathrm{Re} = \frac{GD_h}{\mu} = \frac{(16.66)(0.030)}{2.016 \times 10^{-5}} = 24{,}802$$

and the Nusselt number is

$$\begin{aligned}\mathrm{Nu} &= 0.023(\mathrm{Re})^{0.8}(\mathrm{Pr})^{0.4} \\ &= 0.023(24{,}802)^{0.8}(0.704)^{0.4} \\ &= 65.49\end{aligned}$$

The convective heat transfer film coefficient is

$$h = \frac{\mathrm{Nu}k_f}{D_h} = \frac{(65.49)(0.0276)}{0.030}$$

$$= 60.25\ \mathrm{W/m\ {}^\circ C} \quad (10.61\ \mathrm{Btu/ft^2\ h\ {}^\circ F})$$

14.3 HEAT TRANSFER AND PRESSURE DROP CORRELATIONS FOR FLOW IN PACKED BEDS

As the fluid flows through the packed bed, it follows a very tortuous path, continuously turning to pass through the flow channels formed by the packing material. The size and distribution of the channels will be determined by the shape of the packing material and the porosity of the packed bed. The pressure drop calculations must take these factors into consideration in order to obtain a reasonable degree of accuracy.

The heat transfer coefficient depends on those items previously mentioned as affecting the flow, as well as the thermal properties of the storage material. The resistance to the transfer of heat is encountered at the surface-fluid interface, inversely proportional to the convective film coefficient, and internally within the material in the bed. Heat is also transferred directly between particles in the bed, since adjacent particles are in physical contact with each other. This phenomenon is called "interparticle" effects. A detailed discussion on the correlations to be used in evaluating the heat transfer coefficient will now be presented.

14.3.1 Pressure Drop Calculations

The primary references for the calculation of pressure drop in packed beds are the papers by Ranz [12] and Ergun [13]. The method proposed by Ergun will be used in this discussion. After completing a careful analysis of the pressure losses associated with the flow of a fluid through a packed bed, Ergun concluded that it was composed of simultaneous kinetic and viscous energy losses. The following expression for the pressure drop was proposed:

$$\frac{\Delta P}{L} g_c = 1.75 \left[\frac{(1-\epsilon)}{\epsilon^3} \frac{Gv}{D_p}\right] + 150.0 \frac{(1-\epsilon)^2}{\epsilon^3} \frac{\mu v}{D_p^2} \tag{14.15}$$

where the first term on the right-hand side is associated with the kinetic energy losses and the second term is associated with the viscous energy losses.

This equation can be rearranged by multiplying by

$$\frac{\epsilon^3 D_p^2}{(1-\epsilon)^2 \mu v}$$

and introducing the Reynolds number, $\mathrm{Re} = d\rho v/\mu$, to yield

$$\frac{\Delta P}{L} g_c \frac{\epsilon^3}{(1.0-\epsilon)^2} \frac{D_p^2}{\mu v} = 1.75 \frac{\mathrm{Re}}{1-\epsilon} + 150.0$$

This equation can be expressed in a general form as

$$\frac{\Delta P}{L} g_c \frac{D_p \epsilon^3}{(1.0-\epsilon)^2} \frac{\mathrm{Re}}{\rho v^2} = C_1 \frac{\mathrm{Re}}{1-\epsilon} + C_2 \tag{14.16}$$

where C_1 is associated with the kinetic energy losses and C_2 is associated with the viscous losses. Handley and Heggs [14] have correlated their experimental results by using Eq. 14.16. The appropriate constants are given in Table 14.2. The constants are valid for the region $1.0 < \text{Re} < 13.0 \times 10^3$. If the Reynolds number is less than 1, the Kozeny-Carman equation is valid. The constant $C_1 = 0.0$ and $C_2 = 150.0$, and the expression becomes

$$\frac{\Delta P}{L} g_c \frac{\epsilon^3}{(1.0-\epsilon)^2} \frac{D_p^2}{\mu v} = 150.0 \tag{14.17}$$

If the Reynolds number is greater than 13.0×10^3, the appropriate expression for the pressure drop is given by the Burke-Plummer relationship,

$$C_1 = 1.75 \qquad C_2 = 0$$

or

$$\frac{\Delta P}{L} g_c \frac{\epsilon^3}{(1.0-\epsilon)^2} \frac{D_p^2}{\mu v} = 1.75 \frac{\text{Re}}{1-\epsilon} \tag{14.18}$$

Example 14.3 Air at an average temperature of 50°C passes through a random packed bed of spheres 1 cm (0.3937 in) in diameter. The frontal area of the bed is 0.1 m² (1.076 ft²). The mass flow rate of the air is 0.1 kg/s (0.22 lb_m/s), and the porosity of the bed is 0.4. If the length of the bed is 2 m (6.562 ft), determine the pressure drop in the air flowing through the bed.

SOLUTION The physical properties of the air are

Air @ 50°C (122°F):

$$\rho_f = 1.106 \text{ kg/m}^3 \quad (0.069 \text{ lb}_\text{m}/\text{ft}^3)$$

$$c_f = 1.007 \text{ kJ/kg °C} \quad (0.24 \text{ Btu/lb}_\text{m} \text{ °F})$$

$$k_f = 0.0278 \text{ W/m °C} \quad (0.016 \text{ Btu/ft h °F})$$

$$\mu_f = 1.94 \times 10^{-5} \text{ kg/m s} \quad (1.304 \times 10^{-5} \text{ lb}_\text{m}/\text{ft s})$$

$$\text{Pr} = 0.70$$

The mass velocity is

$$G = \frac{\dot{m}_f}{S_{fr}} = \frac{0.1}{0.1} = 1 \text{ kg/m}^2 \text{ s}$$

Table 14.2 Pressure drop correlation [14], $1 < \text{Re} < 13 \times 10^3$

Packing	Size, in	C_1	C_2
Cylinder	$\frac{1}{4} \times \frac{1}{4}$	1.25	598.0
Cylinder	$\frac{3}{16} \times \frac{3}{16}$	1.28	458.0
Cylinder	$\frac{1}{4} \times \frac{1}{2}$	1.54	1083.0
Rings, steel	$\frac{1}{4} \times \frac{1}{4}$	3.15	410.0
Rings, porcelain	$\frac{1}{4} \times \frac{1}{4}$	2.37	356.0
Rings, porcelain	$\frac{3}{8} \times \frac{3}{8}$	1.72	452.0
Miscellaneous [13]		1.75	150.0

and the Reynolds number is

$$\mathrm{Re} = \frac{GD_p}{\mu} = \frac{(1)(0.01)}{1.94 \times 10^{-5}} = 515$$

The superficial velocity is 0.90 m/s. The pressure drop is given by

$$\Delta P = \left(C_1 \frac{\mathrm{Re}}{1.0 - \epsilon} + C_2\right) \frac{(1.0 - \epsilon)^2 \rho v^2 L}{D_p \epsilon^3 \,\mathrm{Re}\, g_c}$$

From Table 14.2, the value of $C_1 = 1.75$ and $C_2 = 150$. Thus,

$$\Delta P = \left[\frac{1.75(515.0)}{1.0 - 0.4} + 150.0\right] \frac{(1.0 - 0.4)^2(1.106)(0.90)^2(2.0)}{(0.01)(0.4)^3(515.0)(1.0)}$$

$$= (1652.08)(1.957)$$

$$= 3.23 \times 10^3 \ \mathrm{N/m^2} \quad (0.469 \text{ psi})$$

The drop in pressure associated with the kinetic loss is an order of magnitude greater than that associated with the viscous energy loss.

14.3.2 Heat Transfer Correlations

Several different techniques have been used to determine the heat transfer coefficients in packed beds. They may be classified as either steady state or transient. The heat transfer coefficients are measured directly or are obtained by using an analogy between heat and mass transfer. An extensive survey of all heat transfer coefficient correlations available in 1965 was presented by Barker [15].

Different bed configurations were studied, and considerable spread in the data was found when the Colburn j factor was plotted versus the Reynolds number. This was an expected result, since data were obtained from different test procedures; the beds were composed of particles with different configurations, sizes, materials, types of packing, and void fractions; and several different test fluids were employed. It would be unreasonable to assume that a correlation for j could be obtained as a function of Reynolds number only for all of the different cases reported.

Steady-state methods for the determination of the convective film coefficient have generally used the mass-heat transfer analogy or involved the experimental measurement of the film coefficient directly. In this latter method, the particles are heated either by imbedding electrical resistance heating elements in them or by using induction heating of the particles. Constant-drying techniques as described by Sen Gupta and Thodes [16] are used when the mass-heat transfer analogy method is utilized. Correlations of the heat transfer coefficient in beds of steel cubes and cylinders and commercial ceramic packing may be obtained by using the information presented in Fig. 14.4. The commercial packings tested are shown in Fig. 14.5. The porosity of the beds with cylinders was in the range of 0.37 to 0.48, and the cubes had a porosity range from 0.41 to 0.47. In Table 14.3 characteristics of random packed beds composed of commercial packings are given. The surface area per unit

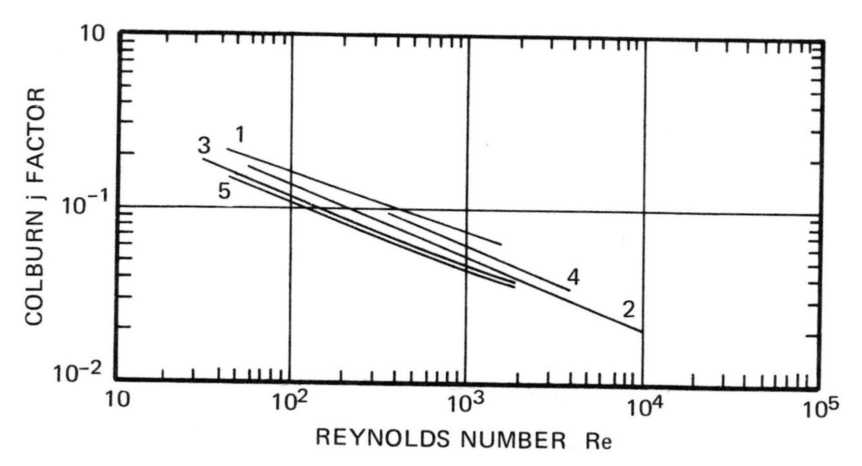

Figure 14.4 Heat transfer correlations–cubes, cylinders, and commercial packings [15]. 1, Berl saddles; 2, Raschig and Pall rings; 3, cylinder (metal); 4, cylinder (celite); 5, cubes (metal). Reprinted with permission from J. J. Barker, "Heat Transfer in Packed Beds," *Ind. Eng. Chem.*, vol. 57, 1965, p. 43. Copyright 1965 American Chemical Society.

volume is denoted by a. The Colburn j factor is defined as

$$j = St\, \mathrm{Pr}^{2/3} = \frac{h}{Gc_f} (\mathrm{Pr})^{2/3} \tag{14.19}$$

therefore the expression for the convective film coefficient is

$$h = \frac{jGc_f}{(\mathrm{Pr})^{2/3}} \tag{14.20}$$

The characteristic length dimension used in these correlations is the diameter of a sphere that would have the same surface area as the actual packing material. The mass velocity is based on the superficial velocity, $G = \dot{m}_f / S_{fr}$.

The correlation for the heat transfer coefficient in a random packed bed of spheres is given in Fig. 14.6. The porosity of the beds was in the range 0.35 to 0.48. The single crosshatched area represents the ranges in which data have been reported. Most of these data were taken with metallic particles. The boxed crosshatched area is for packed beds composed of gravel, where the porosity varied from 0.43 to 0.45.

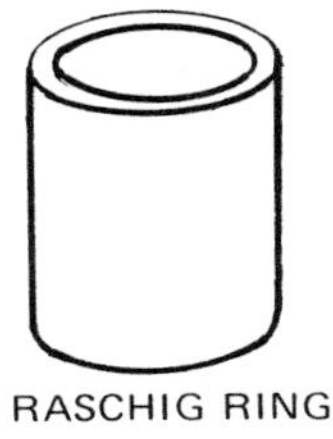

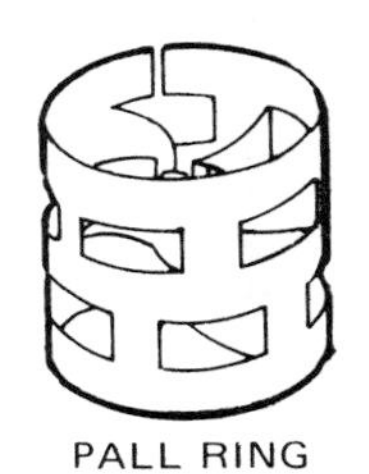

Figure 14.5 Commercial packings.

Table 14.3 Characteristics of random packing[a]

Packing	Nominal size, in										
	$\frac{1}{4}$	$\frac{3}{8}$	$\frac{1}{2}$	$\frac{5}{8}$	$\frac{3}{4}$	1	$1\frac{1}{4}$	$1\frac{1}{2}$	2	3	$3\frac{1}{2}$
Raschig rings:											
Ceramic:											
ϵ	0.73	0.68	0.63	0.68	0.73	0.73	0.74	0.71	0.74	0.78	
a	240	155	111	100	80	58	45	38	28	19	
Metal:											
$\frac{1}{32}$-in wall:											
ϵ	0.69		0.84		0.88	0.92					
a	236		128		83.5	62.7					
$\frac{1}{16}$-in wall:											
ϵ			0.73		0.78	0.85	0.87	0.90	0.92	0.95	
a			118		71.8	56.7	49.3	41.2	31.4	20.6	
Pall rings:											
Plastic:											
ϵ				0.88		0.90		0.905	0.91		
a				110		63.0		39	31		23.4
Metal:											
ϵ				0.902		0.938		0.953	0.964		
a				131.2		66.3		48.1	36.6		
Berl saddles:											
Ceramic:											
ϵ	0.60		0.63		0.66	0.69		0.75	0.72		
a	274		142		82	76		44	32		

[a] Table from the United States Stoneware Company.

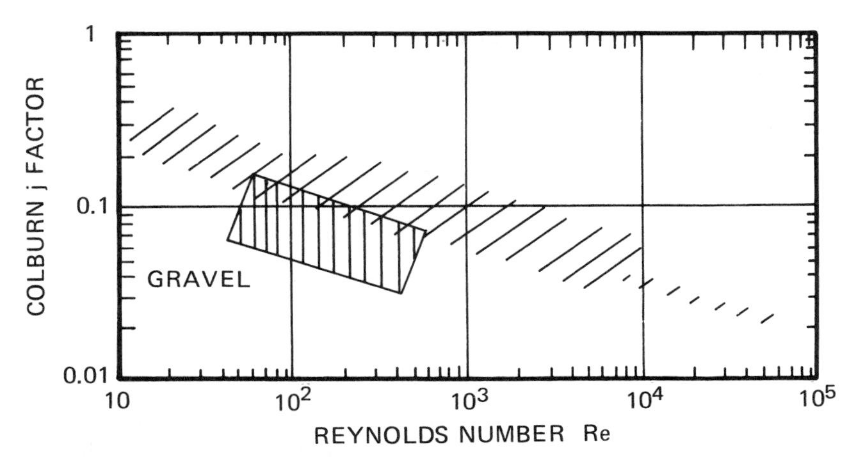

Figure 14.6 Heat transfer correlations–spheres, random packing.

The above results can be used for the evaluation of the convective film coefficient, but it can easily be seen that the values obtained may be off by ±25%. If more accuracy is required, it is suggested that Barker's survey article be read carefully and the appropriate references checked to obtain more accurate correlations.

It will also be necessary to review the literature between 1965 and the present time in order to determine if additional results have become available for the particular packing configuration of interest to the reader. In 1968, Handley and Heggs [14] reported on a study of the momentum and heat transfer mechanisms in regular-shaped packed beds. The heat transfer coefficients were determined experimentally by using a transient method based on the iterative solution of Schumann's mathematical model of a packed bed with axial dispersion and intraparticle conduction neglected. The results were correlated as

$$j = \frac{C_1(\text{Re}_m)^{C_2}}{\epsilon} \tag{14.21}$$

These correlations are based on an equivalent sphere with a volume equal to that of the packed particles. The hydraulic diameter is defined as the void volume of the bed divided by the surface area of the packing. For the equivalent sphere with a diameter d_e, the expression for the hydraulic diameter is

$$D_h = \frac{2D_p\epsilon}{3(1-\epsilon)} \tag{14.22}$$

The modified Reynolds number is

$$\text{Re}_m = \frac{D_h \rho v_a}{\mu} = \frac{2}{3}\frac{D_p}{(1-\epsilon)}\frac{\rho v}{\mu}$$

$$= \frac{2}{3}\frac{D_p \rho v}{(1-\epsilon)\mu} \tag{14.23}$$

Table 14.4 Heat transfer correlations for cylinders, rings, and spheres [14]

Packing	Size, in	C_1	C_2	ϵ
Cylinder	$\frac{1}{4} \times \frac{1}{4}$	0.13	−0.27	31
Cylinder	$\frac{3}{16} \times \frac{3}{16}$	0.09	−0.20	30
Cylinder	$\frac{1}{4} \times \frac{1}{2}$	0.29	−0.36	34
Rings, steel	$\frac{1}{4} \times \frac{1}{4}$	0.24	−0.31	77
Rings, porcelain	$\frac{1}{4} \times \frac{1}{4}$	0.29	−0.35	62
Rings, porcelain	$\frac{3}{8} \times \frac{3}{8}$	0.29	−0.37	56
Sphere (metalic)		0.26	−0.33	37

The superficial mean gas velocity is $v = \dot{m}_f/\rho S_{fr} = \epsilon v_a$, and the superficial mass velocity is $G = v\rho$. The values of C_1 and C_2 are given in Table 14.4. When the thermal conductivity of the packing material is low, intraparticle conduction effects become important and the experimental results will deviate from those predicted by the above-noted correlations.

Example 14.4 Estimate the convective heat transfer coefficient for the packed bed described in Example 14.3.

SOLUTION The mass velocity is 1 kg/m² s, and the Reynolds number is 515.0. From Fig. 14.6, $j = 0.1$. The convective film heat transfer coefficient is

$$h = \frac{jGc_f}{(\text{Pr})^{2/3}} = \frac{(0.1)(1.0)(1007.0)}{(0.70)^{2/3}} = 127.7 \text{ W/m}^2\,{}^\circ\text{C}$$

The relationship given by Handley and Heggs could also be used to determine the convective film heat transfer coefficient. The modified Reynolds number is

$$\text{Re}_m = \frac{2}{3}\frac{D_p \rho v}{(1-\epsilon)\mu} = \frac{2}{3}\frac{(0.01)(1.0)}{(1.0-0.4)(1.94 \times 10^{-5})} = 572.7$$

Therefore,

$$j = \frac{0.26(\text{Re}_m)^{-0.333}}{\epsilon} = \frac{0.26(572.7)^{-0.333}}{0.4} = 0.0785$$

The convective film coefficient is $h = 100.24$ W/m² °C. The correlation recommended by Handley and Heggs falls toward the bottom of the crosshatch area shown in Fig. 14.6.

REFERENCES

1. H. Hausen, *Wärmeübertragung im Gegenstrom, Gleichstrom und Kreuzstrom,* 2d ed., Springer-Verlag, Berlin, 1978.

2. W. M. Kays and A. L. London, *Compact Heat Exchangers*, 2d ed., McGraw-Hill, New York, 1964.
3. W. M. Rohsenow and J. P. Hartnett (eds.), *Handbook of Heat Transfer*, McGraw-Hill, New York, 1973.
4. R. K. Shah and A. L. London, *Laminar Flow Forced Convection in Ducts*, Academic Press, New York, 1978.
5. L. F. Moody, "Friction Factors for Pipe Flow," *Trans. ASME*, vol. 66, 1944, p. 671.
6. V. M. Legkiy and A. S. Makarov, "Heat Transfer in the Thermal Entry Region of Circular Pipes and Rectangular Ducts with Stabilized Turbulent Flow of Air," *Heat Transfer–Soviet Res.*, vol. 3, 1971, p. 1.
7. W. H. McAdams, *Heat Transmission*, 3d ed., McGraw-Hill, New York, 1954, pp. 225–226.
8. F. W. Dittus and L. M. K. Boelter, University of California-Berkeley, Pub. Eng. 2, 1930, p. 443.
9. E. M. Seider and C. E. Tate, "Heat Transfer and Pressure Drop in Liquids in Tubes," *Ind. Eng. Chem.*, vol. 28, 1936, p. 1429.
10. B. S. Petukhov, "Heat Transfer and Friction in Turbulent Pipe Flow with Variable Properties," *Adv. Heat Transfer*, vol. 6, 1970, p. 503.
11. W. Nunner, *VDI Forschungsheft 455*, vol. 22, no. 5 (AERE Lib/Trans. 786), 1956.
12. W. E. Ranz, "Friction and Heat Transfer Coefficients for Single Particles and Packed Beds," *Chem. Eng. Proc.*, vol. 48, 1952, p. 247.
13. S. Ergun, "Fluid Flows Through Packed Columns," *Chem. Eng. Proc.*, vol. 48, 1952, p. 89.
14. D. Handley and P. J. Heggs, "Momentum and Heat Transfer Mechanism in Regular Shaped Packing," *Trans. Inst. Chem. Eng.*, vol. 46, 1968, p. T251.
15. J. J. Barker, "Heat Transfer in Packed Beds," *Ind. Eng. Chem.*, vol. 57, 1965, p. 43.
16. A. Sen Gupta and G. Thodos, "Direct Analogy Between Mass and Heat Transfer to Beds of Spheres," *Amer. Inst. Chem. Eng. J.*, vol. 9, 1963, p. 751.

INDEX

Acceleration of calculations (regenerators), 186–191
Allen, D. N. de G., 174

Babcock, R. E., 292
Bahnke, G. D., 137
Balanced regenerators, 119, 144, 228
Barker, J. J., 344
Beet, J., 220
Benjamin, M. K., 163
Biancardi, F. R., 220
Boelter, L. M. K., 340
Box, M. J., 303
Burns, A., 220, 228, 233, 237, 240
Butterfield, P., 150

Cast iron, 311–312, 337
Chao, R., 292
Chato, W. W., 242
Cima, R. M., 220, 244, 247
Complex optimization method, 303
Concrete, 311–313, 337
Conduction parameter, 137
Constraints:
- explicit, 308, 324
- implicit, 309

Cowper stoves, 106–110, 117, 154, 160, 162

Delpero, P., 269
Denton, J. C., 6
Dispersion effect, 280, 290–294
Dittus, F. W., 340
Dwell time, 251

Edwards, J. F., 292
Effectiveness:
- heat exchanger, 210
- regenerator (thermal ratio), 113, 137, 208–218

Egbert, R. B., 165
Elshout, J., 220
Energy storage applications:
- commercial, 3
- industrial, 4
- residential, 3
- utilities, 2

Energy storage material;
- operating characteristics, 5
- phase change, 6
- sensible heat, 7
- types, 6

Energy storage types, 4
Energy storage units:
- economics, 2
- operating characteristics, 2

Ergum, S., 325, 342

Feolite, 8, 311, 316–319
Field, A. A., 3
Finite difference solution, 125, 140, 155–159, 166, 169–174, 186–191
Fixed bed regenerator, 106–110, 113, 221
Furnas, C. C., 284

Gibbs, M. G., 3
Geometrical configuration;
flat slab, 11, 13, 31–51, 74–83
hollow cylinder, 11, 15, 51–75
Glenn, D. R., 5
Goldstern, W., 6
Green, D. R., 222, 232
Green, D. W., 292

Hanley, D., 82, 292, 325, 342, 347
Harmonic means:
length, 119, 144
period, 119, 144, 208
Hausen, H., 18, 28, 114, 123, 129, 144, 149, 174, 194, 196, 203, 284, 333
Hawley, R. W., 326
Heat pole, 196–202
Heat storage exchanger:
heat flux–single fluid, 266–277
steady-state, 249, 250, 269
two-fluid, 243–266
Heat storage models,:
finite conductivity, 31–84, 128–153
infinite heat capacity, 12–17, 75–83
simplified, 17–29, 75–83
Heat transfer coefficient:
channel flow, 338–341
modified, 27–29
overall (bulk), 129–132, 149–153
packed bed, 344–348
volumetric, 284
Heggs, P., 82, 292, 325, 342, 347
Hoelscher, H. E., 292
Hottel, H. C., 165
Howard, C. P., 137

Illife, C. E., 114, 123, 125, 179, 184, 189
Initial temperature distribution, arbitrary, 93–97
Interparticle effect, 280
Intraparticle effect, 280, 290–294

Jeffreson, C. P., 292

Kardas, A., 194
Kauffman, K. W., 6
Kays, W. M., 310, 334
Klinkenberg, A., 20, 284
Kulakowski, B., 186
Kumar, M., 195, 204–214

Laananen, D. H., 311
Lampsell, P. F., 220, 244
Larsen, F. W., 18, 85, 284
Legkiy, V. M., 309, 339
Leung, P. K., 292
Ljungstrom (rotary) regenerator, 106–110
Lof, G. O. G., 326
London, A. L., 220, 222, 244, 334, 339
Longitudinal conduction, 82, 136
Lorsch, H. G., 6

Makarov, A. S., 309, 339
McGowan, J. G., 220, 244
Mitchell, J. W., 220

Nahavandi, A. N., 114, 123, 125, 175, 181–186, 202
National Research Council Report, 2
NTU, 210, 251–262
Nunner, W., 341
Nusselt, W., 18, 124, 169, 175

Packed beds, 279–302, 322–331, 342–348
Parszewski, M., 326
Perry, R. H., 292
Petukhov, B. S., 340
Pollard, R., 3
Porosity, 279
Pressure drop, 310, 333–338, 342–344

Radiation heat transfer, 163–167
Raiz, M., 281
Ranz, W. E., 342
Razelos, P., 163
Reay, D. A., 6
Reduced length, 112
Reduced period, 112
Regenerator equilibrium criteria, 224
Reversal effect, 112
Richardson, J. F., 292
Rosen, J. B., 292
Rotary regenerator (Ljungstrom), 106–110, 113, 117, 221

Quasi-linearization (regenerator), 158
Quon, D., 292

Schofield, J. S., 150
Schumann, T. E. W., 18, 281, 284
Seider, E. M., 340

Sen Gupta, A., 344
Shah, R. K., 334, 339
Smith, S. A. H., 28
Solution method–regenerator:
  closed, 123–127, 180–186
  open, 123–127, 169–180
Somers, R. R., 305, 308, 311, 320
Strausz, I., 220
Superposition–arbitrary varying:
  fluid mass flow rate, 97–104, 300
  heat flux, 275–277
  inlet fluid temperature, 86–93, 207, 263, 294
  initial temperature distribution, 93–97, 296–299
Symmetric regenerator, 113, 144, 148, 181, 195, 203, 208–214, 222–228, 232–236
Szego, J., 33, 37, 75, 85, 244, 248, 311

Tate, C. E., 340
Thermal ratio (effectiveness), 113, 116, 117
Thodos, G., 344
Thomas, R. J., 125, 186
Three-dimensional model, 143–148, 151–153
Timewise variations:
  fluid mass flow rate, 97–104
  inlet fluid temperature, 86–93
Two-dimensional model, 143–148, 151–153

Unbalanced regenerator, 119–123, 214, 230, 235–239
Unsymmetric regenerator, 119, 182, 228, 234

Variable flow (regenerator), 159–163

Weinstein, A. S., 114, 123, 125, 175, 187–189, 202

Young, P. A., 150

Zuidema, P., 220